System Modeling in Cell

System Modeling in Cell Biology
From Concepts to Nuts and Bolts

edited by
Zoltan Szallasi
Jörg Stelling
Vipul Periwal

A Bradford Book
The MIT Press
Cambridge, Massachusetts
London, England

MIT Press may be purchased at special quantity discounts for business or sales promotional use. For information, please email special_sales@mitpress.mit.edu or write to Special Sales Department, The MIT Press, 55 Hayward Street, Cambridge, MA 02142.

This book was set in LaTeX by the authors and was printed and bound in the United States of America

Front cover painting: "*Vit (∞) all*" by Astrid Colomar.

Library of Congress Cataloging-in-Publication Data

System modeling in cellular biology: from concepts to nuts and bolts / edited by Zoltan Szallasi, Jörg Stelling, Vipul Periwal.
    p.   cm.
  ISBN 0-262-19548-8 (alk. paper)
  1. Cytology—Data processing.
  2. Cytology—Mathematical models.
  3. Cytology—Computer simulation.
  4. Biological systems.
  I. Szallasi, Zoltan.
  II. Stelling, Jörg.
  III. Periwal, Vipul.
QH585.5.D38S97 2006
571.601'1—dc22   2005054539

Contents

III MODELS AND REALITY   199

IV COMPUTATIONAL MODELING   313

Preface

System Modeling: Why and Why Now?

Vipul Periwal, Zoltan Szallasi and Jörg Stelling

Introduction

Biology is the study of self-replicating chemical processes. Biology is the study of systems accurately transmitting a genetic blueprint. Biology is the study of complex adaptive reproducing systems.

What is systems biology if all definitions of biology implicitly or explicitly refer to the study of a whole object, whether it is a virus, a cell, a bacterium, a protozoan or a metazoan? We treat systems biology as the *quantitative* study of biological systems, aided (or hindered) by technological advances that both permit molecular observations on far more inclusive scales than possible even 15 years ago, and permit computational analysis of such observations. Thus, for the purposes of this book, systems biology is the promise of biology on a larger and quantitatively rigorous scale, a marriage of molecular biology and physiology. Concretely, this defines the focus of the book: data-centric quantitative modeling of biological processes and systems.

Biology is an experimentally driven science simply because evolutionary processes are not understood well enough to allow theoretical advances to rest on terra firma. Systems biology is experimentally driven, computationally driven, *and* knowledge driven. It is experimentally driven because the complexity of biological systems is difficult to penetrate without large-scale coverage of the molecular underpinnings; it is computationally driven because the data obtained from experimental investigations of complex systems need extensive quantitative analysis to be informative; and it is knowledge driven because it is not computationally feasible to analyze the data without incorporating all that is already known about the biology in question. Furthermore, the use of data, computation and knowledge must be concurrent. Available knowledge guides experiment design, novel knowledge is generated by the computational analysis of new data in light of available knowledge, and the cycle repeats.

The difference between knowledge and data is central to understanding the underpinnings of systems biology. The sequencing of whole genomes is a good example. Any given genome is data. Without extensive analysis, it is just as uninformative about biological processes as a photograph of the night sky. First steps in transforming a genome into knowledge include identifying genes, identifying transcription factor binding sites, finding the transcription factor complexes that control the expression of the genes, and finding the chromatin structure in the cell being studied, to determine which genes are accessible for transcription. While this is wildly optimistic in terms of the knowledge that can be extracted from the genome data, it is still nowhere close to the level of understanding required to make predictions about the response of an organism to a specific stimulus. A reductionist approach to biology is bootless because complex adaptive systems are inherently *nonlinear*, so their behavior is well summarized by the statement: the whole is more than the sum of the components.

Handicapping the Bout

From a quantitative perspective, there are striking features of biological dynamics that make analysis challenging:

1. Large range of spatial scales

2. Large range of temporal scales

3. A lack of separation between responses to external stimuli versus internal programs

4. Multiple functionalities of constituents

5. Multiple levels of signal processing

6. Incomplete evolutionary record

7. Wide range of sensitivities to perturbations

8. Genotypic variation

None of these challenges is an absolute barrier to progress. Nevertheless, these challenges must be addressed to make real progress.

From an experimental perspective, the challenges of biology are better understood:

1. Coverage in terms of components and interactions

2. Reproducibility

3. Spatial resolution

4. Temporal resolution

5. Cross-validation

6. Combinatorial perturbations

7. Accuracy

From a knowledge perspective, there are four central problems:

1. Find an appropriate level of abstraction for a given analytic problem.

2. Find a common basis to relate knowledge gained using different experimental techniques on the same system.

3. Find a common basis to relate knowledge gained from the same experiment on different model systems.

4. Incorporate knowledge incrementally as new data is analyzed.

Taking all these difficulties together, it is not surprising that researchers traditionally have considered the study of biological systems rather resistant to quantitative approaches. It is, therefore, worth pointing out to skeptics that in some cases thorough quantitative analysis has produced insights into or explanations of biological phenomena that would have been impossible without the application of advanced mathematical tools. Various chapters in this book will discuss a great variety of, often counterintuitive, examples. For instance, the advantages of a more extensive mathematical analysis over simpler approaches are emphasized in chapter 8 (pp. 170–173). When circadian oscillators are analyzed by formal logic, the traditional analytical tool of molecular biology, or by macroscopic descriptors such as differential equations, the experimentally observed behavior cannot be reconstructed from the molecular machinery. Stochastic analysis, however, demonstrates how, by random fluctuations, the system escapes the macroscopic point-attractor and thus oscillatory behavior is maintained. Examples such as this will probably contribute to the long-awaited common ground for discussions between biologists and quantitative scientists. The mutual suspicion on both sides, which has been difficult to overcome by intellectual curiosity alone, will probably be eliminated by the mutual need for each other's expertise.

"My Complications Had Complications"

The goal of systems biology is a predictive understanding of the whole. If the whole is more than the sum of its parts, it follows that acquiring a catalog of all the parts is *not* necessarily the first order of business. In a caricature, there are two avenues of attack possible: either one focuses on subsystems governing a specific function in arbitrary conditions and gains a predictive understanding of the system, one subsystem at a time, or one focuses on the system in a restricted set of conditions and gains an understanding by gradually increasing the set of conditions and, as required, the level of detail in the model of the system. The analogy is with molecular biology in the former approach and with physiology in the latter.

The modeling associated with each approach is distinct. In the molecular biology type approach, the aim is to go beyond traditional pathway-centric points of view and deal with the challenges of feedback loops formed either directly or indirectly due to interactions with other pathways. In the physiology type approach, interactions between the components in the model are added as needed to maintain contact with the experimental data. The components in this approach are not necessarily directly related to biochemical species. Eventually, these bottom-up and top-down approaches should meet. However, each has its own strengths and weaknesses and they complement each other.

Why Read This?

The importance of feedback loops and crosstalk in almost all facets of biological systems has been apparent for several decades. The cell cycle control circuitry or the developmental programs in bilaterians are prime examples of this. The ability of cancerous cells to evade targeted therapies results largely from biological systems having evolved in ways that place a premium on robustness and adaptability. Such properties, as yet only nebulously defined, are not localizable to a small set of interactions. They reside in the network as a whole, as has been clearly demonstrated in predictions on metabolic networks.

Modeling biological systems faces the challenge of appropriate abstractions—levels on which to focus, and details to be left out. For instance, molecular biology abounds with mechanistic analogies, but on a more detailed level often the underlying interactions are driven by chemistry. This makes modeling subtle since statistical biases are often the driving force in what superficially appears to be a mechanical process, for example, chemotaxis. At what level does such detail become relevant, and at what level can one ignore it? This is not *a priori* obvious, and one needs rigorous approaches to model parsimony to answer such questions. Indeed, the answer to the model selection question depends to a great extent on the predictions required. This is an important point in all biological modeling: The model, its purpose, and the experimental data are intimately related. A model that predicts hepatic glucose uptake precisely but insulin levels with greater uncertainty is not a useful selection if the only measurement available is insulin levels.

There are two main approaches to computational analysis of biological data. The *causal* approach makes concrete deterministic or stochastic models (differential equations, stochastic differential equations, Boolean networks, et cetera) of biological processes. The *probabilistic* view is associated with probabilistic inference approaches, using pattern recognition or learning algorithms (such as neural networks and graphical models) for analysis of data from large-scale experimental methods. These two approaches rest on a large part of applied mathematics (including numerical integration, optimization, interpolation, and control theory) and computer science (search theory, coding theory, and database design). This breadth necessitates collaborations between people with diverse backgrounds, but an inadequate understanding of the limitations and applicability of techniques and concepts from different fields hinders such collaborations. The background information required makes biological modeling a difficult task, but the real challenge remains that of making computational models *effective* and *efficient* representations of biological systems.

What's Included and What's Not

This book starts with generalities and progresses towards practicalities. Thus, the first section is conceptual, with attempts to define the role of modeling in biology, as well as attempts to cut through the miasma that surrounds the use of the terms *robust, complex, adaptive,* and *module* in the systems biology literature. As will be evident, these are important notions that need much further work to crystallize to the point where they can be assigned the honorific *concept*. Nevertheless, these terms may ultimately be quantitatively used as concrete guiding principles in modeling.

The next section provides introductions to general approaches to making models of biology: qualitative models, constraint-based models, dynamical systems based on differential equations, and stochastic models, as well as models with spatial structure. The other side of the modeling coin, probabilistic inference aimed at inference from large-scale data sets, is also introduced. The section proceeds from relatively simple towards mathematically more demanding approaches. Although each chapter tries to convey its central messages in an intuitive as well as in a mathematically rigorous way, readers arriving from biology will have to realize that each method has a certain minimum difficulty level associated with it. While ordinary differential equation–based or qualitative models can be quite readily introduced in an intuitive manner, stochastic or spatial modeling cannot be described in simple terms and require an appropriate level of background in quantitative sciences. Key applications of the various modeling approaches are also widely covered. Taken together, this section will provide the reader with an overall impression of the relationship between the potential utility of quantitative approaches and their associated analytical cost.

Reality bites. And models model biological reality. The section that follows next contains introductions to the data that is available for systems biology and the caveats that go with the data. It also contains introductions to inferring model architecture from data, using control theory in models, and studying synthetic gene networks. The antidote to these computational limitations is multi-level modeling, and this is also introduced in this section. Limitations in observability, accuracy, and coverage of biological data are widely recognized. One of the goals of this section is to guide the readers through various data interpretation methods while emphasizing what the data will or will not allow in terms of quantitative analysis.

The last section of the book contains the computational issues and techniques for practical application of the preceding approaches: numerical methods for simulating biochemical systems, and the software infrastructure for representing models in a reusable and exchangeable manner. Biological data quality is not the only obstacle systems biology is facing. The various numerical methods also have their well known strengths and limitations and these should be considered when designing experiments and their associated models. For instance, computational limitations form barriers to increasing model size arbitrarily.

The book ends with an eclectic list of the software tools that the contributing authors of this book find useful.

While this book contains a plethora of approaches to biological modeling, we are keenly aware that there are many that we have not covered. For instance, we have eschewed much discussion of pattern recognition because this is only really useful when combined with domain specific biological knowledge—for which no general technique exists. Likewise, we do not cover approaches such as neural networks or Petri nets that have either limited application in systems biology so far, or are problematic regarding model interpretation. Our attempt has been to provide broad basic coverage of fundamental approaches and techniques. In our view, picking some of the techniques introduced in this book and combining them artfully leads to almost complete coverage of modeling in systems biology.

Enjoy

Systems biology is an approach to quantitatively understand biological systems that attempts to embrace the complexity of life as a fact of life. There is no hope of understanding biological systems at the predictive level required for disease detection, prevention, or cure other than by this means. Nevertheless, it would serve us well to temper Burnham's maxim of grand thinking, "Make no little plans . . ." with the story of the emperor's new clothes.

I GENERAL CONCEPTS

1   The Role of Modeling in Systems Biology

Douglas B. Kell and Joshua D. Knowles

The use of models in biology is at once both familiar and arcane. It is familiar because, as we shall argue, biologists presently and regularly use models as abstractions of reality: diagrams, laws, graphs, plots, relationships, chemical formulae and so on are all essentially models of some external reality that we are trying to describe and understand (fig. 1.1). In the same way we use and speak of "model organisms" such as baker's yeast or *Arabidopsis thaliana*, whose role lies in being similar to many organisms without being the same as any other one. Indeed, our theories and hypotheses about biological objects and systems are in one sense also just models (Vayttaden et al., 2004). Yet the use of models is for most biologists arcane because familiarity with a *subset* of model types, especially *quantitative mathematical models*, has lain outside the mainstream during the last 50 years of the purposely reductionist and qualitative era of molecular biology. It is largely these types of model that are an integral part of the "new" (and not-so-new) systems biology and on which much of the rest of this book concentrates. Since all such models are developed for some kind of a purpose, our role in part is to explain why this type of mathematical model is both useful and important, and will likely become part of the standard armory of successful biologists.

1.1   Philosophical Overview

When one admits that nothing is certain one must, I think, also admit that some things are much more nearly certain than others.

Bertrand Russell, *Am I an Atheist or an Agnostic?*

It is conventional to discriminate (as in fig. 1.2) (a) the world of ideas, thoughts, or other mental constructs and (b) the world of observations or data, and most scientists would recognize that they are linked in an iterative cycle, as drawn: we improve our mental picture of the world by carrying out experiments that produce data, and such data are used to inform the cogitations that feed into the next part

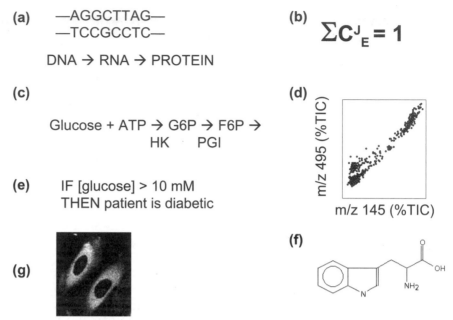

Figure 1.1 Models in biology. Although we shall be concentrating here on a subset of mathematical models, we would stress that the use of all sorts of models is entirely commonplace in biology—examples include (a) diagrams (here a sequence of DNA bases and the "central dogma"), (b) laws (the flux-control summation theorem of metabolic control analysis), (c) graphs—in the mathematical sense of elements with nodes and edges (a biochemical pathway), (d) plots (covariation of 2 metabolites in a series of experiments), (e) relationships (a rule describing the use of the concentration of a metabolite in disease diagnosis), (f) chemical formulae (tryptophan), and (g) images (of mammalian cells).

of the right-hand arc, that designs and performs the next set of experiments as part of an experimental program. Such a cycle may be seen as a "chicken and egg" cycle, but for any *individual* turn of the cycle there is a clear distinction between the two essential starting points (ideas or data). This also occurs in scientific funding circles—is the activity in question *ideas*- (that is, *hypothesis*-)driven or is it *data*-driven? (Until recently, the latter, hypothesis-*generating* approach was usually treated rather scornfully.)

From a philosophical point of view, then, the hypothetico-deductive analysis, in which an idea is the starting point (however muddled or wrongheaded that idea may be), has been seen as much more secure, since deductive reasoning is sound in the sense that if an axiom is true (as it is supposed to be by definition) and the observation is true, we can conclude that the facts are at least consistent with the idea. If the hypothesis is "all swans are white" then the prediction is that a measurement of the whiteness of known swans will give a positive response. By contrast, it has been known since the time of Hume that inductive reasoning, by which we seek to generalize from examples ("swan A is white, swan B is white, swan C is white ... so I predict that all swans are white") is insecure—and a

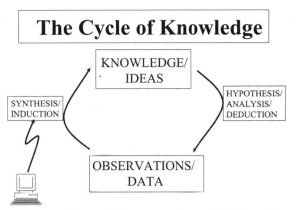

The Cycle of Knowledge

Figure 1.2 The iterative relationship between the world of ideas/hypotheses/thoughts and the world of data/observations. Note that these are linked in a cycle, in which one arc is not simply the reverse of the other (Kell, 2002, 2005; Kell and Welch, 1991).

single black swan shows it. Nothing will ever change that, and the "problem of induction" probably lies at the heart of Popper's insistence (see Popper (1992) and more readable commentators such as Medawar (1982)) that theories can only be disproved. Note of course that it is equally true for the hypothetico-deductive mode of reasoning that a single black swan will disprove the hypothesis. This said, the ability of scientists to ignore any number of ugly facts that would otherwise slay a beautiful hypothesis is well known (Gilbert and Mulkay, 1984), and in this sense—given that there are no genuinely secure axioms (Hofstadter, 1979; Nagel and Newman, 2002)—the deductive mode of reasoning is not truly much more secure than is induction.

Happily, there is emerging a more balanced view of the world. This recognizes that for working scientists the reductionist and ostensibly solely hypothesis-driven agenda has not been as fruitful as had been expected. In large measure in biology this realization has been driven by the recognition, following the systematic genome sequencing programs, that the existence, let alone the function, of many or most genes—even in well-worked model organisms—had not been recorded. This could be seen in part as a failure of the reductionist agenda. In addition there are many areas of scientific activity that have nothing to do with testing hypotheses but which are exceptionally important (Kell and Oliver, 2004); perhaps chief among these is the development of novel methods. In particular there are fields—functional genomics not least among them (Kell and King, 2000), although this is very true for many areas of medicine as well—that are data-rich but hypothesis-poor, and are best attacked using methods that are data-driven and thus essentially inductive (Kell and King, 2000).

A second feature that has emerged from a Popperian view of the world (or at least from his attempt to find a means that would allow one to discriminate "science'" from "pseudo-science" (Medawar, 1982; Popper, 1992)) is the intellectual significance of prediction: if your hypothesis makes an experimentally testable (and thus falsifiable) prediction it counts as "science," and if the experimental prediction is consistent with the prediction then (confidence in) the "correctness" of your hypothesis or worldview is bolstered (see also Lipton (2005)).

1.2 Historical Context

The history of science demonstrates that both inductive and deductive reasoning occur at different stages in the development of ideas. In some cases, such as in the history of chemistry, a period of almost purely inductive reasoning (stamp-collecting and classification) is followed by the development of more powerful theories that seek to explain and predict many phenomena from more general principles. Often these theories are reductionist, that is to say, complicated phenomena that seem to elude coherent explanation are understood by some form of breaking down into constituent parts, the consideration of which yields the required explanation of the more complicated system. A prime example of the reductionist mode is the explanation of the macroscopic properties of solids, liquids, and gases—such as their temperature, pressure, and heat— by considering the average effect of a large number of microscopic interactions between particles, governed by Newtonian mechanics. For the first time, accurate, quantitative predictions with accompanying, plausible explanations were possible, and unified much of our basic understanding of the physical properties of matter.

The success of early reductionist models in physics, and later those in chemistry, led in 1847 to a program to analyze (biological) processes, such as urine secretion or nerve conduction, in physico-chemical terms proposed by Ludwig, Helmholtz, Brucke, and du Bois-Reymond (Bynum et al., 1981). However, although reductionism has been successful in large part in the development of physics and chemistry, and to a great extent in acquiring the parts list for modern biology—consider the gene—the properties of many systems resist a reductionist explanation (Solé and Goodwin, 2000). This ultimate failure of reductionism in biology, as in other disciplines, is due to a number of factors, principal among them being the fact that biological systems are inherently complex.

Although *complexity* is a phenomenon about which little agreement has been reached, and certainly for which no all-encompassing measure has been established, the concept is understood to pertain to systems of interacting parts. Having many parts is not necessary: it is sufficient that they are coupled in some way, so that the state of one of them affects the state of one or more others. Often the interactions are *nonlinear* so, unlike systems which can be modeled by considering averaged effects, it is not possible to reduce the system's behavior to the sum of its parts (Davey and Kell, 1996). Common interactions in these systems are *feedback loops*, in which, as the name suggests, information from the output of a system transformation is sent back to the input of the system. If the new input facilitates and accelerates the transformation in the same direction as the preceding output, they are positive feedback —their effects are cumulative. If the new data produce an output in the opposite direction to previous outputs, they are negative feedback—their effects stabilize the system. In the first case there is exponential growth or decline; in the second there is maintenance of the equilibrium. These loops have been studied in a variety of fields, including control engineering, cybernetics, and economics. An understanding of them and their effects is central to building and understanding models of complex systems (Kell, 2004, 2005; Milo et al., 2002).

Negative feedback loops are typically responsible for regulation, and they are obviously central to homeostasis in biological systems. In control engineering, such systems are conveniently described using Laplace transforms—a means of simplifying the combination and manipulation of ordinary differential equations (ODEs), and closely related to the Fourier transform (Ogata, 2001); Laplace transforms for a large variety of different standard feedback loops are known and well-understood, though analysis and understanding of non-linear feedback remains difficult (see chapter 12 for details). Classical negative feedback loops are considered to provide stability (as indeed they do when in simple systems in which the feedback is fast and effective), though we note that negative feedback systems incorporating delays can generate oscillations (for example (Nelson et al, 2004)).

Positive feedback is a rather less appreciated concept for most people and, until recently, it could be all but passed over in even a control engineer's education. This is perhaps because it is often equated with undesired instability in a system, so it is just seen as a nuisance; something which should be reduced as much as possible. However, positive feedback should not really be viewed in this way, particularly from a modeling perspective, because it is an important factor in the dynamics of many complex systems and does lead to very familiar behavior. One very simple model system of positive feedback is the *Polya urn* (Arthur, 1963; Barabási and Albert, 1999; Johnson and Kotz, 1977). In this, one begins with a large urn containing two balls, one red and one black. One of these is removed. It is then replaced in the urn, together with another ball of the same color. This process is repeated until the urn is filled up. The system exhibits a number of important characteristics with respect to the distribution of the two colors of balls in the full urn: early, essentially random events can have a very large effect on the outcome; there is a lock-in effect where later in the process, it becomes increasingly unlikely that the path of choices will shift from one to another (notice that this is in contrast to the "positive feedback causes instability" view); and accidental events early on do not cancel each other out. The Polya urn is a model for such things as genetic drift in evolution, preferential attachment in explaining the growth of scale-free networks (Barabási and Albert, 1999), and the phenomenon whereby one of a variety of competing technologies (all but) takes over in a market where there is a tendency for purchasers to prefer the leading technology, despite equal, or even inferior, quality compared with the others (for example QWERTY keyboards and Betamax versus VHS video). (See also Goldberg (2002) and Kauffman et al. (2000) for the adoption of technologies as an evolutionary process.)

Positive feedback in a resource-limited environment also leads to familiar behavior. The fluctuations seen in stock prices, the variety of sizes of sandpiles, and cycles of population growth and collapse in food-chains all result from this kind of feedback. There is a tendency to reinforce the growth of a variable until it reaches a value that cannot be sustained. This leads to a crash which "corrects" the value again, making way for another rise. Such cyclic behavior can be predictably periodic but in many cases the period of the cycle is *chaotic*—that is, deterministic but essentially unpredictable. All chaotic systems involve nonlinearity, and this is most frequently the result of some form of positive feedback, usually mixed with negative feedback (Glendinning, 1994; Tufillaro et al., 1992; Strogatz, 2000).

Behavior involving oscillatory patterns may also be important in biological signaling (Lahav et al., 2004; Nelson et al., 2004), where the downstream detection may be in the frequency rather than the amplitude (that is, simply concentration) domain (Kell, 2005). All of this said, despite encouraging progress (for example (Tyson et al., 2003; Wolf and Arkin, 2003; Yeger-Lotem et al., 2004)), we are far from having a full understanding of the behavior of concatenations of these simple motifs and loops. Thus, the Elowitz and Leibler oscillator (Elowitz and Leibler, 2000) is based solely on negative feedback loops but is unstable. However, this system could be made comparatively stable and robust by incorporating positive feedback loops, which led to some interesting work by Ferrell on the cell cycle (Angeli et al., 2004; Pomerening et al., 2003).

It is now believed that most systems involving interacting elements have both chaotic and stable regions or phases, with islands of chaos existing within stable regions, and vice versa (for a biological example, see (Davey and Kell, 1996)). Chaotic behavior has now been observed even in the archetypal, clockwork system of planetary motion, whereas the eye at the heart of a storm is an example of stability occurring within a wildly unpredictable whole.

Closely related to the vocabulary of complexity and of chaos theory is the slippery new (or not so new?) concept of *emergence* (Davies, 2004; Holland, 1998; Johnson, 2001; Kauffman, 2000; Morowitz, 2002). Emergence is generally taken to mean simply that the whole is more than (and maybe qualitatively different from) the sum of its parts, or that system-level characteristics are not easily derivable from the "local" properties of their constituents. The label of *emergent phenomenon* is being applied more and more in biological processes at many different levels, from how proteins can fold to how whole ecosystems evolve over time. A central question that the use of the term *emergence* forces us to consider is whether it is only a convenient way of saying that the behavior of the whole system is difficult to understand in terms of basic laws and the initial conditions of the system elements (weak emergence), or whether, in contrast, the whole *cannot* be understood by the analysis of the parts, and current laws of physics, *even in principle* (strong emergence). The latter view would imply that high level phenomena are not reducible to physical laws (but may be consistent with them) (Davies, 2004). If this were true, then the modeling of (at least) some biological processes should not follow solely a bottom-up approach, hoping to go from simple laws to the desired phenomenon, but might eventually need us to posit high-level organizing principles and even downward causality. Such a worldview is completely antithetical to materialism and remains as yet on the fringes of scientific thought.

In summary, reductionism has been highly successful in explaining some macroscopic phenomena, purely in terms of the behavior of constituent parts. However, this was predicated (implicitly) on the assumption that there were few parts (for example, the planets) and that their interactions were simple, or that there were many parts but their interactions could be neglected (for example, molecules in a gas). However, the scope of a reductionist approach is limited because these assumptions are not true in many systems of interest (Kell and Welch, 1991; Solé and Goodwin, 2000). The advent of computers and computer simulations led to the insight that even relatively small systems of interacting parts (such as the Lorenz model) could exhibit very complex (even chaotic) behavior. Although the behavior may be deterministic, complex systems are hard to analyze using traditional mathematical and analytical methods. Prediction, control, and understanding arise mainly from modeling these systems using iterated computer simulations. Biological systems, which are inherently complex, *must* be modeled and studied in this way if we are to continue to make strides in our understanding of these phenomena.

1.3   The Purposes and Implications of Modeling

We take it as essentially axiomatic that the purposes of academic biological research are to allow us to understand more than we presently do about the behavior and workings of biological systems (see also Klipp et al. (2005)) (and in due time to exploit that knowledge for agricultural, medical, commercial, or other purposes). We consider that there are several main reasons why one would wish to make models of biological systems and processes, and we consider each in turn. In summary, they can all be characterized as variations of simulation and prediction. By simulation we mean the production of a mathematical or computational model of a system or subsystem that seeks to represent or reproduce some properties that that system displays. Although often portrayed as substantially different (though we consider that it is not), prediction involves the production of a similar type of mathematical model that simulates (and then predicts) the behavior of a system related to the starting system described above. Clearly simulation and prediction are thus related to each other, and the important concept of *generalization* describes the ability of a model derived for one purpose to predict the properties of a related system under a separate set of conditions. Thus some of the broad reasons—indeed probably the main reasons—why one would wish to model a (biological) system include:

- Testing whether the model is accurate, in the sense that it reflects—or can be made to reflect—known experimental facts

- Analyzing the model to understand which parts of the system contribute most to some desired properties of interest

- Hypothesis generation and testing, allowing one rapidly to analyze the effects of manipulating experimental conditions *in the model* without having to perform complex and costly experiments (or to restrict the number that are performed)

- Testing what changes in the model would improve the consistency of its behavior with experimental observations

Our view of the basic bottom-up systems biology agenda is given in fig. 1.3.

1.3.1   Testing Whether the Model Is Accurate

A significant milestone in a modeling program is the successful representation of the behavior of the "real" system by a model. This does not, of course, mean that the model is accurate, but it does mean that it might be. Thus the dynamical behavior of variables such as concentrations and fluxes is governed by the parameters of the systems such as the equations describing the local properties and the parameters of those equations. This of itself is not sufficient, since generalized equations (for example, power laws, polynomials, perceptrons with nonlinear properties) with no mechanistic or biological meaning can sometimes reproduce well the kinetic behavior of complex systems without giving the desired insight into the true constitution of the system.

Such models may also be used when one has no experimental data, with a view to establishing whether a particular design is sensible or whether a particular experiment is worth doing. In the former case, of engineering design, it is nowadays commonplace to design complex structures such as electronic circuits and chips, buildings, cars, or aeroplanes entirely inside a computer before committing them to reality. Famously, the Boeing 777 was designed entirely *in silico* before being tested first in a wind tunnel and then with a human pilot. It is especially this kind of attitude and experience in the various fields of engineering that differs from the current status of work in biology that is leading many to wish to bring numerical modeling into the biological mainstream. Another example is the development of "virtual" screening, in which the ability of drugs to bind to proteins is tested *in silico* using structural models and appropriate force fields to calculate the free energy of binding to the target protein of interest of ligands in different conformations (Böhm and Schneider, 2000; Klebe, 2000; Langer and Hoffmann, 2001; Shen et al., 2003; Zanders et al., 2002), the most promising of which may then be synthesized and tested. The attraction, of course, is the enormous speed and favorable economics (and scalability) of the virtual over the actual "wet" screen.

1.3.2   Analyzing Subsystem Contributions

Having a model allows one to analyze it in a variety of ways, but a chief one is to establish those parts of the model that are most important for determining the behavior in which one is particularly interested. This is because simple inspection of models with complex (or even simple) feedback loops just does not allow one

(a)

Basic Bottom-up-Driven Systems Biology Pipeline

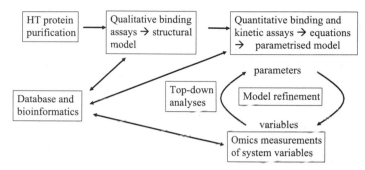

(b)

Bottom-up Systems Biology Pipeline (Dry)

1. Qualitative (structural) model—who talks to whom as substrate, product, or effector →
2. Quantitative model including "real" or approximate equations describing individual steps →
3. Parametrisation of those equations →
4. Run the model and assess its most important parameters
5. Iteratively, with wet data, GOTO 1....

(c)

Systems Biology Experiments (Including the Wet Side)

- Set up a well-defined system
- Effect systematic perturbations (genetic, environmental, chemical)
- Measure a time series of as many concentrations of variables, especially RNAs, proteins, metabolites (the 'omes) as possible
- Model the system and compare the experimental time series to those generated by the model
- Repeat iteratively

Figure 1.3 The role of modeling in the basic systems biology agenda, (a) stressing the bottom-up element while showing the iterative and complementary top-down analyses. (b) The development of a model from qualitative (structural) to quantitative, and (c) its integration with ("wet") experimentation.

to understand them (Westerhoff and Kell, 1987). Techniques such as sensitivity analysis (see below) are designed for this, and thus indicate to the experimenter which parameters must be known with the highest precision and should be the focus of experimental endeavor. This is often the focus of so-called top-down analyses in which we seek to analyze systems in comparatively general or high-level terms, lumping together subsystems in order to make the systems easier to understand. The equivalent in pharmacophore screening is the QSAR (quantitative structure-activity relationship) type of analysis, from which one seeks to analyze those features of a candidate binding molecule that best account for successful binding, with a view to developing yet more selective binding agents.

1.3.3   Hypothesis Generation and Testing

Related to the above is the ability to vary, for example, parameters of the model, and thereby establish combinations or areas of the model's space that show particular properties in which one might be interested (Pritchard and Kell, 2002), and then to perform that small subset of possible experiments that it is predicted will show such interesting behavior. An example here might be the analysis of which multiple modulations of enzymatic properties are best performed for the purposes of metabolic engineering (Cascante et al., 2002; Cornish-Bowden, 1999; Fell, 1998). We note also that when modeling can be applied effectively it is far cheaper than wet biology and, as well as its use in metabolic engineering, can reduce the reliance on *in vivo* animal/human experimentation (a factor of significant importance in the pharmaceutical industry).

1.3.4   Improving Model Consistency

In a similar vein, we may have existing experimental data with which the model is inconsistent, and it is desirable to explore different models to see which changes to them might best reproduce the experimental data. In biology this might, for example, allow the experimenter to test for the presence of an interaction or kinetic property that might be proposed. In a more general or high-level sense, we may use such models to seek evidence that existing hypotheses are wrong, that the model is inadequate, that hidden variables need to be invoked (as in the Higgs Boson in particle physics, or the invocation of the existence of Pluto following the registration of anomalies in the orbit of Neptune), that existing data are inadequate, or that new theories are needed (such as the invention of the quantum theory to explain or at least get round the so-called "ultraviolet catastrophe"). In kinetic modeling this is often the case with "inverse problems" in which one is seeking to find a ("forward") model that best explains a time series of experimental data (see below).

1.4   Different Kinds of Models

Most of the kinds of systems that are likely to be of interest to readers of this book involve entities (metabolites, signaling molecules, etc.) that can be cast as "nodes" interacting with each other via "edges" representing reactions that may be catalyzed via other substances such as enzymes. These will also typically involve feedback loops in which some of the nodes interact directly with the edges. We refer to the basic constitution of this kind of representation as a *structural model* (not, of course, to be confused with a similar term used in the bioinformatic modeling of protein molecular structures). A typical example of a structural model is shown in fig. 1.4.

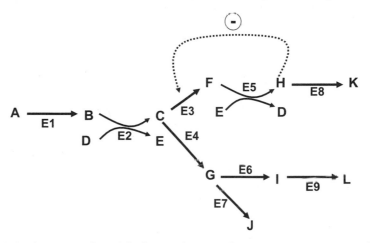

The elements of a model always include the structural relationships (such as shown), the "local" equations describing the behaviour of each step (not shown) and the values of their parameters (not shown)

Figure 1.4 A structural model of a simple network involving nine enzymes (E1 to E9), four external metabolites (A,J,K,L—whose concentration must be assumed to be fixed if a steady state is to be attained), and eight internal metabolites (B,C,D,E,F,G,H,I). D and E are effectively cofactors and are part of a 'moiety-conserved cycle' (Hofmeyr et al., 1986) in that their sum is fixed and they cannot vary their concentrations independently of each other.

The classical modeling strategy in biology (and in engineering), the ordinary differential equation (ODE) approach (discussed in chapter 6) contains three initial phases, and starts with this kind of structural model, in which the reactions and effectors are known. The next level refers to the kinetic rate equations describing the "local" properties of each edge (enzyme), for instance that relate the rate of the reaction catalyzed by, say, E1 to the concentrations of its substrates; a typical such equation (which assumes that the reaction is irreversible) is the Henri-Michaelis-Menten equation $v = V_{max} \cdot [S]/([S] + K_m)$. The third level involves the parameterization of the model, in terms of providing values for the parameters (in this case V_{max} and K_m. Armed with such knowledge, any number of software packages can predict the time evolution of the variables (the concentrations and fluxes of the metabolites) until they may reach a steady state. This is done (internally) by recasting the system as a series of coupled ordinary differential equations which are then solved numerically. We refer to this type of operation as *forward modeling*, and provided that the structural model, equations, and values of the parameters are known, it is comparatively easy to produce such models and compare them with an experimental reality. We have been involved with the simulator Gepasi, written by Pedro Mendes (Mendes, 1997; Mendes and Kell, 1998, 2001), which allows one to do all of the above, and that in addition permits automated variation of the parameters with which to satisfy an objective function such as the attainment of a particular flux in the steady state (Mendes and Kell, 1998).

In such cases, however, the experimental data that are most readily available do not include the parameters at all, and are simply measurements of the (time-dependent) *variables*, of which fluxes and concentrations are the most common (see chapter 10). Comparison of the data with the forward model is much more difficult, as we have to solve an *inverse modeling*, *reverse engineering* or *system identification* (Ljung, 1999b) problem (discussed in chapter 11). Direct solution of such problems is essentially impossible, as they are normally hugely underdetermined and do not have an analytical solution. The normal approach is thus an iterative one in which a candidate set of parameters is proposed, the system run in the forward direction, and on the basis of some metric of closeness to the desired output a new set of parameters is tested. Eventually (assuming that the structural model and the equations are adequate), a satisfactory set of parameters, and hence solutions, will be found (see table 1.1). These methods are much more computer-intensive than those required for simple forward modeling, as potentially many thousands or even millions of candidate models must be tested. Modern approaches to inverse modeling use approaches from heuristic optimization (Corne et al., 1999) to search the model space efficiently. Recent advances in multiobjective optimization (Fonseca and Fleming, 1996) are particularly promising in this regard, since the quality of a model can usually be evaluated only by considering several, often conflicting criteria. Evolutionary computation approaches (Deb, 2001) allow exploration of the Pareto front, that is the different trade-offs (for example, between model simplicity and accuracy) that can be achieved, enabling the modeler to make more informed choices about preferred solutions.

We note, however, that there are a number of other modeling strategies and issues that may lead one to wish to choose different types of model from that described. First, the ODE model assumes that compartments are well stirred and that the concentrations of the participants are sufficiently great as to permit fluctuations to be ignored. If this is not the case then stochastic simulations (SS) are required (Andrews and Bray, 2004) (which are topics of chapter 8 and chapter 16). If flow of substances between many contiguous compartments is involved, and knowledge of the spatial dynamics is required (as is common in computational fluid dynamics), partial differential equations (PDEs) are necessary. SS and PDE models are again much more computationally intensive, although in the latter case the designation of a smaller subset of representative compartments may be effective (Mendes and Kell, 2001).

If the equations and parameters are absent, it may prove fruitful to use qualitative models (Hunt et al., 1993), in which only the direction of change (and maybe rate

Table 1.1   10 Steps in (Inverse) Modeling.

1.	Get acquainted with the target system to be modeled
2.	Identify important variable(s) that changes over time
3.	Identify other key variables and their interconnections
4.	Decide what to measure and collect data
5.	Decide on the form of model and its architecture
6.	Construct a model by specifying all parameters. Run the model forward and measure behavior.
7.	Compare model with measurements. If model is improving return to 6. If model is not improving and not satisfactory, return to 3, 4, and 5.
8.	Perform sensitivity analysis. Return to 6 and 7 if necessary.
9.	Test the impact of control policies, initial conditions, etc.
10.	Use multicriteria decision-making (MCDM) to analyze policy trade-offs.

of change) is recorded, in an attempt to constrain the otherwise huge search space of possible structural models (see chapter 7). Similarly, models may invoke discrete or continuous time, they may be macro or micro, and they may be at a single level (such as metabolism, signaling) or at multiple levels (in which the concentrations of metabolites affect gene expression and vice versa (ter Kuile and Westerhoff, 2001). Models may be top-down (involving large "blocks") or bottom-up (based on elementary reactions), and analyses beneficially use both strategies (fig. 1.3). Thus a "middle-out" strategy is preferred by some authors (Noble, 2003a) (see chapter 14). Table 1.2 sets out some of the issues in terms of choices which the modeler may face in deciding which type of model may be best for particular purposes and on the basis of the available amount of knowledge of the system.

Table 1.2: Different types of model, presented as choices facing the experimenter when deciding which strategy or strategies may be most appropriate for a given problem.

Dimension or Feature	Possible choices	Comments
Stochastic or deterministic	Stochastic: Monte Carlo methods or statistical distributions Deterministic: equations such as ODEs	Phenomena are not of themselves either stochastic or deterministic; large-scale, linear systems can be modeled deterministically, while a stochastic model is often more appropriate when nonlinearity is present.
Discrete versus continuous (in time)	Discrete: Discrete event simulation, for example, Markov chains, cellular automata, Boolean networks. Continuous: Rate equations.	Discrete time is favored when variables only change when specific events occur (modeling queues). Continuous time is favored when variables are in constant flux.

Table 1.2: Different types of model, presented as choices facing the experimenter when deciding which strategy or strategies may be most appropriate for a given problem.

Dimension or Feature	Possible choices	Comments
Macroscopic versus microscopic	Microscopic: Model individual particles in a system and compute averaged effects as necessary. Macroscopic: Model averaged effects themselves, for example, concentrations, temperatures, etc.	Are the individual particles or subsystems important to the evolution of the system, or is it enough to approximate them by statistical moments or ensemble averages?
Hierarchical versus multi-level	Hierarchical: Fully modular networks. Multi-level: Loosely connected components.	Can some processes/variables in the system be hidden inside modules or objects that interact with other modules, or do all the variables interact, potentially? This relates to reductionism versus holism.
Fully quantitative versus partially quantitative versus qualitative	Qualitative: Direction of change modeled only, or on/off states (Boolean network). Partially quantitative: Fuzzy models. Fully quantitative: ODEs, PDEs, microscopic particle models.	Reducing the quantitative accuracy of the model can reduce complexity greatly and many phenomena may still be modeled adequately.
Predictive versus exploratory/explanatory	Predictive: Specify every variable that could affect outcome. Exploratory: Only consider some variables of interest.	If a model is being used for precise prediction or forecasting of a future event, all variables need to be considered. The exploratory approach can be less precise but should be more flexible, for example, allowing different control policies to be tested.
Estimating rare events versus typical behavior	Rare events: Use importance sampling. Typical behavior: Importance sampling not needed.	Estimation of rare events, such as apoptosis times in cells is time-consuming if standard Monte Carlo simulation is used. Importance sampling can be used to speed up the simulation.
Lumped or spatially segregated	Lumped: Treat cells or other components/compartments as spatially homogeneous. Spatially segregated: Treat the components as differentiated or spatially heterogeneous.	If heterogeneous it may be necessary to use the computationally intensive partial differential equation, though other solutions are possible (Mendes and Kell, 2001)

1.5   Sensitivity Analysis

-Sensitivity analysis for modelers?
-Would you go to an orthopaedist who didn't use X-ray?

Jean-Marie Furbringer

Sensitivity analysis (Saltelli et al., 2000) represents a cornerstone in our analysis of complex systems. It asks the generalized question "what is the effect of changing something (a parameter P) in the model on the behavior of some variable element M of the model?" To avoid the magnitude of the answer depending on the units used we use fractional changes ΔP and observe their effects via fractional changes (ΔM) in M. Thus the generalized sensitivity is $(\Delta M/M)/(\Delta P/P)$ and in the limit of small changes (where the sensitivity is then independent of the size of ΔP) the sensitivity is $(dM/M)/(dP/P) = d(\ln M)/d(\ln P)$. The sensitivities are thus conceptually and numerically the same as the control coefficients of metabolic control analysis (MCA) (see Fell (1996); Heinrich and Schuster (1996); and Kell and Westerhoff (1986)).

Reasons for doing sensitivity analysis include the ability to determine:

1. If a model resembles the system or process under study

2. Factors that may contribute to output variability and so need the most consideration

3. The model parameters that can be eliminated if one wishes to simplify the model without altering its behavior grossly

4. The region in the space of input variables for which model variation is maximum

5. The optimal region for use in a calibration study

6. If and which groups of factors interact with each other.

A basic prescription for performing sensitivity analysis (adapted from (Saltelli et al., 2000)) is:

1. Identify the purpose of the model and determine which variables should concern the analysis.

2. Assign ranges of variation to each input variable.

3. Generate an input vector matrix through an appropriate design (DoE).

4. Evaluate the model, thus creating an output distribution or response.

5. Assess the influence of each variable or group of variables using correlation/regression, Bayesian inference (chapter 4), machine learning, or other methods.

Two examples from our recent work illustrate some of these issues. In the first, (Nelson et al., 2004; Ihekwaba et al., 2004), we studied a refined version of a model (Hoffmann et al., 2002) of the NF-κB pathway. This contained 64 reactions with their attendant parameters, but sensitivity analysis showed that only 8–9 of them exerted significant influence on the dynamics of the nuclear concentration of NF-κB in this system, and that each of these reactions involved free IκBα and free IKK. An entirely different study (White and Kell, 2004) asked whether comparative genomics and experimental data could be used to rank candidate gene products in terms of their utility as antimicrobial drug targets. The contribution of each of the submetrics (such as essentiality, or existence only in pathogens and not hosts or commensals) to the overall metric was analyzed by sensitivity analysis using 3 different weighting functions, with the top 3 targets— which were quite different from those of traditional antibiotics—being similar in all cases. This gave much confidence in the robustness of the conclusions drawn.

1.6   Concluding Remarks

The purpose of this chapter was to give an overview of some of the reasons for seeking to model complex cellular biological systems, and this we trust that we have done. We have also given a very brief overview of some of the methods, but we have not dwelt in detail on: their differences, the question of which modeling strategies to exploit in particular cases, the problems of overdetermination (where many models can fit the same data) and of model choice (which model one might then prefer and why), nor on available models (for example, at http://www.biomodels.net/) and model exchange using, for example, the systems biology markup language (SBML) (http://www.sbml.org) (Finney and Hucka, 2003; Hucka et al., 2003; Shapiro et al., 2004) or others (Lloyd et al., 2004). These issues are all covered well in the other chapters of this book.

Finally, we note here that despite the many positive advantages of the modeling approach, biologists are generally less comfortable with, and confident in, models (and even theories) than are practitioners in some other fields where this is more of a core activity, such as physics or engineering. Indeed, when Einstein was once informed that an experimental result disagreed with his theory of relativity, he famously and correctly remarked "Well, then, the experiment is wrong!" It is our hope that trust will grow, not only from a growing number of successful modeling endeavors, but also from a greater and clearer communication of models enabled by new technologies such as Web services and the SBML.

Acknowledgments

We thank the BBSRC and EPSRC for financial support, and Dr. Neil Benson, Professor Igor Goryanin, Dr. Edda Klipp and Dr. Jörg Stelling for useful discussions.

2   Complexity and Robustness of Cellular Systems

Jörg Stelling, Uwe Sauer, Francis J. Doyle III, and John Doyle

The daunting complexity of cellular systems appears as a major hurdle for large-scale modeling efforts. This complexity resides not only in the sheer number of components and interactions, but also in the operations on multiple levels and time-scales. Guidelines for meaningful modeling such as underlying organizational and design principles are thus required. A key to derive guidelines could be the high internal organization and the selection for function that distinguish cellular systems from complex physical systems; both factors considerably shrink the space of possible designs. One prominent aspect of cellular functions is their robustness, that is, their insensitivity to a wide range of perturbations. Here, we focus on connections between cellular complexity and robustness—with robustness requirements being the driving forces for complexity. Since only a rather limited set of mechanisms establishes robustness in biological circuits, understanding robustness can provide a key for understanding cellular organization. Practical implications for the modeling task are, for instance, the emphasis on network structures over exact values of kinetic parameters. Thus, we advocate that qualitative or structural modeling approaches may already yield deep insights by identifying important versus less important parts of a system for the purpose of more detailed modeling.

2.1   Introduction

Complexity is a hallmark of cellular systems, with great challenges for the development and analysis of cellular networks at the system level. Without appropriate conceptional frameworks for dealing with that complexity, the vision of ultimately going from the description of entire cells to organs and organisms will not be achievable. Hence, it is important to think about rather high-level abstractions of cellular properties that could help in system modeling and analysis. In general, complex systems may either show a behavior or a design that is difficult to understand (Weng et al., 1999). While the behavior of biological systems is, in most cases, relatively simple, the numbers of metabolic and regulatory genes shows that complexity in biology arises mainly from abundant control circuits, that is, from the system's design.

For maintaining simple behavior under real-life conditions, biological systems have to cope with a constantly varying environment, be it changing physico-chemical conditions or noisy external signals that have to be processed. Moreover, their internal properties are also subject to uncertainty, since they can, for instance, be changed by mutations, and because stochastic noise is an important source of cellular variability. Therefore, evolution must have strongly favored robustness, that is, a system's ability to maintain (key) functional characteristics despite potentially harmful external or internal perturbations. A now widely accepted notion is that many (or most) cellular sub-systems are robust (Kitano, 2002a; Stelling et al., 2004b; Kitano, 2004b). Examples for this capacity can already be found in simple organisms such as the bacterium *Escherichia coli*, which displays robust perfect adaptation in its search for nutrients (see chapter 12) and also a high resistance to gene deletions (see section 2.4).

Robustness has long been recognized as an important property of biological systems, for instance described as "canalization" (towards a specific outcome despite uncertain starting conditions) in developmental biology. However, the understanding of how robustness is accomplished at the cellular or molecular level is still limited (Hartman et al., 2001), mainly because robustness is intimately linked to the apparent complexity of cellular systems. For instance, the main purpose of cellular control systems seems to be to guarantee reliable performance of vital functions under conditions of uncertainty (Lauffenburger, 2000; Csete and Doyle, 2002). Hence, elucidating high-level cellular design principles that could be exploited in systems modeling will require the simultaneous consideration of complexity and robustness in cellular networks—which is the topic of the present chapter.

We will start with describing the sources and types of cellular complexity in more biological detail, before attempting to distinguish the type of complexity that is present in biological and physical systems by focusing on functional and organizational principles that underly this complexity at a more abstract level (section 2.2). Robustness as a concept for understanding biological function and behavior will require a more in-depth exposition of the theoretical concept (section 2.3), before we discuss two biological example systems, namely central metabolism and circadian clocks (section 2.4). These examples are intended to explain how and why robustness can help in modeling cellular complexity (section 2.5).

2.2 Complexity of Cellular Networks

2.2.1 Sources of Complexity

Biological complexity arises at several levels. At the molecular level, heterogeneous regulation networks control individual cell responses to environmental changes. The basic biological information flow from DNA to biochemical activities—with interconnected control mechanisms—is illustrated here for metabolic networks (figs. 2.1 and 2.2). About a quarter of the around 4,000 genes in a typical microbe encode the enzymes that catalyze approximately 1,000 biochemical reactions. While all cells share essentially the same DNA, the rate of transcription (synthesis of mRNA from DNA) varies greatly for each gene. Dynamically controlled by overlapping networks of repressors and activators, transcription is further affected by the hard-wired location of the gene in an operon (or on the genome), promoter or initiation site quality, or more general mechanisms like DNA topology and epigenesis. Typically, regulatory proteins themselves are subject to negative and positive feedback regulation through interaction with other proteins or metabolites. Next, mRNA is translated into protein, which again is regulated at multiple levels by different mechanisms that include mRNA stability, active degradation, attenuation (premature termination as a function of the initial rate of translation), rare tRNAs, anti-sense RNA, quality of the ribosome binding site, etc.

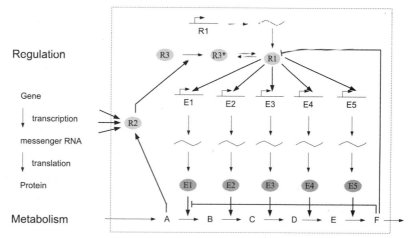

Figure 2.1 Complexity in cellular networks. Flow of information (left) and example interaction network (right). Cellular components are, for example, regulatory proteins (ellipses, R), enzymes (ellipses, E), and metabolites (capital letters). Bold arrows indicate regulatory influences (activation or inhibition), while normal arrows denote chemical reactions.

Essentially each step of protein synthesis is affected by multiple and overlapping regulation loops that operate both at the global cellular and a pathway/reaction

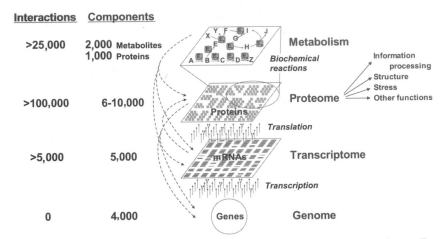

Figure 2.2 Complexity in cellular networks for a typical microbe such as *E. coli*. Regulatory interactions are indicated by dashed lines. Transcript interactions are based on operon structures and ribosomal RNA interactions. Proteome interactions include an average of 6–7 protein-protein interactions as well as protein-DNA, protein-RNA, and protein-membrane interactions (see chapter 10 for details). Metabolic interactions include biochemical transformations and regulatory interactions between metabolites, RNA, and protein. Protein numbers encompass differences in folding, size, and covalent modifications. Note that not all proteins are necessarily present at the same time.

specific level. Activity and stability of the synthesized proteins may then be modulated by posttranslational modification (for example, phosphorylation), aggregation to multimers, or complex formation with other proteins. Beyond such genetically determined regulation, enzyme activity is often regulated by feedback inhibition. This is a common regulatory principle in biosynthetic pathways, where endproducts inhibit the first enzyme in the pathway. In the multipurpose central metabolic pathways, several key enzymes are subject to feedback and feedforward inhibition and activation through multiple metabolites. Temporal coordination of control is achieved by combining rapid and sensitive regulation through feedback loops (seconds) with somewhat slower protein modification (seconds to minutes) and transcriptional/translational regulation (minutes). Almost no individual mechanism achieves on/off effects but rather modulates processing rates in a 2–20 fold range. Thus, much of the complexity is based on multi-level combination of heterogeneous control systems that tune strength and speed of cellular responses to stimuli.

Unlike most technical systems, *individual* biological processes are extremely sensitive to the exact physico-chemical conditions because slight changes in, for example, temperature, pH, or the concentration and nature of the surrounding protein/membrane matrix influence the availability of substrates, products, and the kinetic properties of the enzymes themselves. Rarely are all physico-chemical parameters identical in independent experiments, but enzymes are also exposed to different micro-environments within a single cell that cannot be determined exactly. An extreme, but not exclusive case is spatial separation into several distinct intracellular compartments—a distinguishing feature between eukaryotes and simpler prokaryotes.

An additional level of complexity is the organization of different cell types into tissues and organs and finally of multiple tissues and organs into higher organisms (for instance, humans, plants). Not even in steady state cultures of single-celled microbes, however, are all cells necessarily in identical states. Driven by a not overly stringent control design, often subpopulations enter a resting state or simply exhibit different phenotypes, which increases chances to propagate the genetic offspring in an ever-changing environment. On longer timescales (days to years), the enormous potential of biological systems for evolutionary adaptation adds yet a different level of complexity. Random imprecisions in copying the genetic source code during cell duplication continuously increase the genetic diversity within a population. While the overall precision of the duplication process is extraordinary high—about 0.003 point mutations occur per microbial genome (2–8 million base pairs) and round of replication—short generation times (minutes to hours) rapidly lead to recognizable genetic differences (Sauer, 2001). While most random differences have no apparent effect or are harmful, some variants bear the potential for improved survival upon drastic environmental changes. In contrast to most technical systems, biological systems thus continuously adapt by "redesigning" their makeup through the evolutionary process of mutation and selection.

2.2.2 "Organized" versus "Emergent" Complexity

The staggering complexity of cellular networks makes appropriate abstractions mandatory for meaningful mathematical modeling. An obvious pragmatic approach consists of decomposing the networks into smaller units that allow for the development of models of limited complexity. Likewise, models for cellular networks are not built at atomic resolution of individual biochemical species. More generally, however, with an ultimate goal of modeling entire cells and organs, we will need a deeper understanding of the specific type of complexity prevalent in biology to develop rigorous analysis methods. Here, we aim at outlining such a characterization by contrasting biological (and engineered) systems with complex physical systems.

Complexity has become a field of intensive research in physics through the notion that systems with many components and interactions can show complicated collective ("emergent") behavior. For instance, when adding sand to apparently stable sand piles, we cannot predict at which point the system reaches its "margin of stability" and avalanches are generated. This does not mean that the behavior is not deterministic; we simply do not have complete knowledge of the initial conditions when starting such an experiment. As the system is extremely sensitive to changes in those conditions, the apparent behavior is chaotic. Similarly, simple sets of interacting particles can generate complicated spatial structures. Rationalizing these emergent properties often abstracts from the real systems by assuming homogeneous components that interact randomly; analysis methods for characterizing the collective behavior are often rooted in statistics (Goldenfeld and Kadanoff, 1999). Such approaches were, for instance, used in revealing rich and complex dynamic behaviors that could be generated by simplified models of cellular signaling networks (Amaral et al., 2004).

A different issue is whether this type of abstraction is useful for a deeper understanding of biological complexity. At the first glance, biological systems differ in several aspects from the type of physical systems mentioned above. One hallmark is their *heterogeneity* of components and interactions. They are *highly structured*, which encompasses, among other things, sophisticated spatial organization and layering of different types of control mechanisms. Finally, their complexity resides in these two features as well as in the sheer numbers of components and interactions. From a dynamic point of view, real biological systems are rather boring in that homeostasis and simple switching of states prevail, while complex behavior such as chaos mainly occurs under conditions when the systems are not working properly. Hence, today's biological systems could perhaps best be understood as *rare*, extremely improbable outcomes of emergent processes leading to primitive forms of life, and their subsequent shaping through evolution.

Functional requirements constitute the main differences between complex physics and biology/engineering. In physics, they do not exist. Biological and engineered systems, in contrast, are evolved or designed to fulfill functions, and are constantly evaluated with respect to how well they perform. In both cases, insufficient performance will lead to extinction of a specific species, irrespective of whether this occurs through evolutionary or human design processes. The immediate consequence of a purpose is a considerably smaller design space, in which network structures that could be effective and reliable implementations are likely to be rare. Hence, we will face a more structured (instead of randomly connected) system. A hope for understanding complexity in biology then is to uncover operational principles through a "calculus of purpose" (Lander, 2004)—by asking teleological questions such as *why* cellular networks are organized as observed, given their known or assumed function.

2.2.3   Function and Organization Principles

The purpose or function as one hallmark of cellular networks itself is a rather complicated concept. Attributing a particular function to a subnetwork may not be easy because it is in many cases *context dependent*. For instance, a particular signaling pathway may have roles in counter-acting biological processes such as the regulation of cell proliferation and apoptosis. Owing to the *multiscale* organization in biology, we need a precise notion of function at the different scales. For the example above, at the organismic level the pathway may coherently serve to achieve homeostasis of cells in an organ. Hence, we will need a *hierarchical description* of functions and corresponding organization principles at different levels, from coarse-grained overall architectures to detailed insight into individual network motifs (Shen-Orr et al., 2002). This corresponds to the modularization of explanations as a final aim of dealing with biological complexity. Here, we consider the global

architecture of metabolism as an example. In metabolism, analyses based on the
networks' stoichiometry alone (neglecting unknown kinetics and regulation) have
revealed a close relation between network structure, function, and regulation at
least for bacteria (see chapter 5 for details), which makes it suitable for high-level
abstractions of organization principles. One possible principle has been proposed
recently, focusing on "bow tie" structures as shown in fig. 2.3 (Csete and Doyle,
2004).

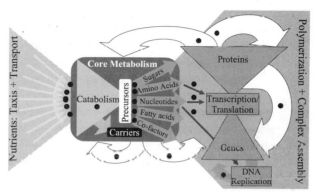

Figure 2.3 Bow tie abstraction of cellular organization. Open arrows denote cellular
regulation and control. Involvement of carriers such as ATP and NAD(P)H in individual
processes is indicated by •.

In the bow tie view, the basic network organization is a combination of fans
of possible inputs (such as nutrients that can be processed) and possible outputs
(for example, the variety of biomass components) that are linked through the core
of central metabolism. Fans and core have rather different structural properties:
while the former show many specialized, mostly linear pathways for catabolism and
anabolism, the highly interconnected network of central metabolism generates and
distributes only 12 metabolites as building blocks and a few carrier molecules (such
as ATP and NADH) that are precursors for all biosynthetic processes. The carriers,
in addition, serve as common currencies for all (energy- or redox-dependent) cellular
processes. Hence, standard interfaces (such as the currency metabolites) and shared
protocols (for instance, using (A)TP for energy-dependent reactions) establish
coherence of the network. Cellular regulation relies on a similar structure, with
a core of general transcription/translation/degradation processes mediating the
information flow from genetic diversity to the large numbers of proteins and their
variants. Nesting of the two bow ties is achieved through material flux and—more
importantly—by abundant feedback regulation. Functional advantages of such an
organization become especially clear when comparing it to a "flat" architecture with
individual pathways leading from every substrate to every product. Such a solution
would be very inefficient due to the number or complexity of enzymes required.
Coordination of pathways and buffering of fluctuations in the environment could
be achieved only with a massive overhead of regulation connecting all the individual

entities. In the long run, such a design would severely impede evolution because it would have to operate on entire pathways and their associated control systems.

The bow tie organization, in contrast, can accommodate the divergent demands on metabolic systems. The core facilitates high-throughput of metabolites with only a few specialized enzymes. At all timescales, ranging from the fast regulation of the high-flux backbone by allosteric control to the slower expression control for individual pathways in the periphery, the structure facilitates systems integration and regulation. It appears not merely that biology uses the available control mechanisms but that the stoichiometry itself is highly structured and organized to facilitate the effectiveness of these control mechanisms to create coherent and global responses to variations, while allowing implementation in the local mechanisms. The shared interfaces and protocols, finally, create "plug-and-play" features, where less central reactions and pathways can easily be exchanged or added. Apparently, bow tie architectures are associated with risks not present in the simpler type of networks, such as high fragility when failures in the core affect the entire system (Csete and Doyle, 2004). It also means low variability of the core, as documented by the universality of the tricarboxylic acid (TCA) cycle at the heart of catabolism in all living organisms (Smith and Morowitz, 2004). Hence, the structures may primarily allow for *optimal trade-offs* between a variety of requirements such as efficiency, robustness, and evolvability (Csete and Doyle, 2004).

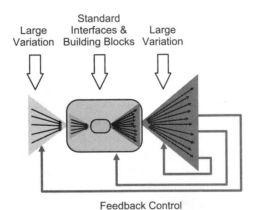

Figure 2.4   General features of bow tie structures.

At a more abstract level, we see highly organized and structured networks that facilitate global and coordinated responses to variations in the environment on all time scales, using local and decentralized mechanisms. Fig. 2.4 illustrates the key features of the organization. The basic framework is employed in many advanced technological systems. The power grid, for instance, coordinates many producers and consumers with highly variable production and demand, respectively, by employing a common exchange protocol, namely 220 V AC. TCP/IP would be its equivalent for the internet. Clearly, from an engineering point of view, biology is a marvel of technological "design." We argue that analogies with engineered systems, in particular regarding how to generate appropriate responses to variations, are one major requirement on all highly integrated systems that can help us grasp biological complexity.

2.3   Robustness in Cellular Networks

The notion of robustness has recently received considerable interest in diverse fields for which the existence of complex networks is characteristic. Examples include the internet, social networks, and biology (Strogatz, 2001; Stelling et al., 2004b; Kitano, 2004a). Not surprisingly, the term *robustness* has been associated with different, sometimes conflicting interpretations. Here, starting from a broad definition we aim at an operational concept that proves suitable for analyzing the properties of cellular networks.

2.3.1   The Concept of Robustness

In general, robustness means the persistence of a system's characteristic behavior under perturbation or conditions of uncertainty. Robustness is, hence, defined for a specific system, which, however, may have arbitrary structural and behavorial features. The concept is closely related to stability in dynamical systems theory, but usually employed with respect to a broader class of phenomena (Kitano, 2002b; Carlson and Doyle, 2002). In engineering, the task of determining a system's robustness is often accomplished by transformation into a suitable stability problem. However, compared to stability theory in systems dynamics, no elaborate theory of robustness exists yet.

It has to be noted that robustness (such as stability) encompasses a relative, not an absolute, property of a system. No system can maintain stability for all its functions when encountering any kind of perturbation. Any operational definition of robustness, and systems analysis thereof, thus, requires two additional specifications. Namely, it has to be explicitly clarified, (i) which characteristic behavior or function remains unchanged, and (ii) for which type of disturbances or uncertainties this invariance property holds. For relatively simple systems, the characteristic behavior can often be captured by definition of a dynamical regime. Investigations of oscillators may thus focus on the persistence of a regular periodic solution (see section 2.4.2 for an example). Moreover, robustness is a qualitative property, and does not preclude quantitative changes (in period or amplitude of the oscillations) to occur (Barkai and Leibler, 1997). For engineered or biological systems, one often understands by characteristic behavior the "desired system characteristics" (Carlson and Doyle, 2002) to be maintained. Here, robustness directly connects to functionality. In technical as well as in living systems, it makes sense to protect key functions by design, or as a result of evolution. Especially in biology, however, function can, in many cases, not easily be assigned to a particular subsystem of a cell or organism (Morohashi et al., 2002). In bacterial chemotaxis, for instance, maintaining the ability to adapt to changing nutrient concentrations, whereas adaptation times are allowed to fluctuate, is intuitively understandable. As a counter-example, signal transduction relies upon sensitive detection, amplification, and decoding of input signals. It would not be sensible to react identically irrespective of the signals received. Identification of key inputs and outputs for specific sub-systems, however, may not be evident from the complex overall network structure, and cellular signaling requires both robustness and precision (Freeman, 2000). The claim of higher-order behavior or entire modules to be robust and, hence, imply functional advantage, therefore needs careful justification.

Similar considerations apply for the specification of perturbations. Cellular systems face three broad classes of uncertainties, namely (i) externally induced perturbations owing to variable environments, (ii) internal perturbations arising from changes in the structure of the system (such as mutations affecting kinetic properties of proteins, or leading to the lack of components), and (iii) intrinsic noise as a consequence of the low copy number of many cellular components. The first two classes of disturbance can be dealt with in a deterministic framework. External perturbations may directly influence the solutions of a dynamical system; resistance to these influences equals the notion of stability in dynamic systems theory. Perturbations affecting the structure of the systems itself, but which do not result in qualitatively different dynamics, reveal structural stability of a system (see chapter 6). These two types of perturbations can, hence, be mapped on changes in inputs and system parameters, respectively. Stochastic effects resulting from the random character of biochemical reactions (see chapter 8) in principle require an explicit inclusion of noise in robustness analysis (Rao et al., 2002). In gene expression, for instance, intrinsic noise considerably contributes to overall variation, with potential amplification and propagation by regulatory dynamics (Thattai and van Oudenaarden, 2001; Elowitz et al., 2002). Hence, also the theoretical methods for analyzing robustness have to be tailored to the specific aspects of a system under investigation.

2.3.2   Mechanisms for Robustness

Mainly four ingredients are currently discussed as cellular design elements for the protection against deleterious disturbances. These encompass (i) back-up systems (redundancy), (ii) disturbance rejection through feedback control, (iii) structuring of complex systems into semi-autonomous functional units (modularity), and (iv) their reliable coordination via establishment of hierarchies and protocols (Csete and Doyle, 2002; Kitano, 2002a; Stelling et al., 2004b). We will discuss their potential contributions for conferring robustness to cellular networks—and for the analysis thereof—in this section.

The simplest strategy to protect against failure of a specific component is to provide for alternative ways to carry out the function the component performs. However, genes that do not diverge in functionality or regulation would not survive during evolution (Krakauer and Plotkin, 2002). In particular the genomics revolution with comprehensive gene knockout libraries of entire organisms has initiated the quest to identify mechanisms that underlie the seemingly surprising number of phenotypically silent deletion mutations; that is, only about 1,100 knockouts of the 5,700 genes are lethal in haploid *S. cerevisiae*. In this context, the term *genetic robustness* was coined to describe the condition in which a gene may be deleted without qualitatively compromising cell growth (Gu et al., 2003). The explanation is trivial for at least the approximately 1,000 genes with metabolic functions: about 45% of the known metabolic genes are simply not active under the investigated condition (Papp et al., 2004; Blank et al., 2005). For the remaining 207 viable *S. cerevisae* mutants of active reactions during glucose catabolism, network redundancy through duplicate genes was the major (3/4) and alternative pathways the minor (1/4) molecular mechanism of genetic network robustness (Blank et al., 2005). Although duplicate genes clearly contribute to the robustness of metabolic networks to gene deletions, the argument cannot be turned around that this is indeed their function because this would imply that it is a distinct mechanism. Quantitative analyses of the 105 duplicate gene families in *S. cerevisiae* clearly demonstrated that no particular dominant function maintains duplicate genes in the genome (Kuepfer et al., 2005). In particular the putative back-up function is not favored by evolutionary selection because duplicates do not occur more frequently in essential reactions than singleton genes. Hence, redundancy plays some role in biological robustness, but it may be largely overrated and misunderstood.

More importantly, feedback loops can account for robustness in cellular network function. By using the output of a function to be controlled in order to determine appropriate input signals, feedback enables a system to adjust the output by monitoring it. In general, negative feedback is employed in reducing the difference between actual output and a given set-point, thereby dampening noise and rejecting perturbations. For instance, a simple, engineered feedback loop relying on negative autoregulation of a transcription factor stabilized steady-state gene expression levels despite the inherent noise in gene expression. This autoregulation proved advantageous over unregulated transcription for a range of biologically plausible parameters (Becskei and Serrano, 2000). The role of positive feedback (or autocatalysis) in conferring robustness is less obvious, since it may cause instabilities. However, decisions for example in development need to be derived from noisy and graded input signals and have to be maintained (see chapter 1). In one example from engineered gene networks (see chapter 13), two genes mutually repressing each other's expression (double-negative feedback) proved sufficient to construct a reliable irreversible switch (Gardner et al., 2000). Enhanced sensitivity through positive feedback also speeds up stress responses. Depending on which cellular functionalities require protection from perturbations, both forms of feedback and combinations thereof can contribute to robustly achieve a desired behavior (Freeman, 2000). Therefore, in many cases where highly precise and reliable behavior is indispensable for overall cellular functionality, multiple intertwined feedback loops operate (Ferrell Jr., 2002) (see also section 2.4.2). True redundancy is most useful when it is part of feedback control systems that can sense variations and failures, and coordinate the use of multiple resources. Trivially, there are lots of copies of enzymes at the protein level, even when there is one gene, and the number is controlled.

Focusing on the internal structure of cellular systems, one central, increasingly discussed notion is that these systems are composed of "functional units" or "modules". Modules can be understood as semi-autonomous entities that show dense internal functional connections, but looser connections with their environment (Kremling et al., 2000; Girvan and Newman, 2002). With respect to robustness, modularity can lead to a benefit for overall functionality of complex systems. Encapsulation of simpler functions can reduce the risk of catastrophic failure by preventing the spread of damage in one module throughout the network (Hartwell et al., 1999; Albert et al., 2000). However, two critical issues have yet to be clarified, namely to prove the existence (or absence) of modularity in cellular systems, and to establish methods for the unanimous identification of modules (Lauffenburger, 2000). As discussed in detail in chapter 3, both problems are intimately linked.

Protocols encompass the set of rules aiming at an efficient management of relationships between the parts (for example, modules) that constitute a system. They include, for instance, the organizational structures for embedding modules and the interfaces between modules that allow for system function (Csete and Doyle, 2002). A common protocol in biology, for instance, is "protein phosphorylation relies on ATP." Using only 12 basic building blocks in metabolism is a similar convention. Protocols, hence, are of primary importance for an understanding of how information in complex systems is integrated (Hartwell et al., 1999). One efficient means for coordination in complex systems is to organize a system hierarchically, namely to establish different layers of integration (Mesarovic et al., 1970). This architecture, for instance, helps to reduce the costs of information transmission (Guimerà et al., 2001). Several lines of evidence suggest that hierarchical structures confer robustness to cellular systems. One major proposition is that separation of functions, and their integration at higher levels, reduces the average damage owing to arbitrary perturbations of the network. Analysis of dynamical networks with overall structures similar to those of cellular networks demonstrated a superior systems performance and controllability when feedback control specifically operates on higher levels of integration (Wang and Chen, 2002). Moreover, as we argued already in section 2.2.3, well-designed hierarchies and protocols can contribute to robustness, for instance, by constraining the effects of local deregulation or by providing common standards for coordination of cellular functions.

2.3.3 Robustness, Fragility, and Complexity

With the variety of mechanisms for incorporating robustness into cellular systems available, it appears surprising that cells are sensitive to quantitatively minor, but extremely powerful perturbations such as oncogenic mutations that enable profound changes at a genomic scale. Two possible explanations would be that either evolution has yet to attain optimal robustness, or that *principal* limitations exist regarding how robust the systems can be made. The overwhelming evidence speaks for the latter hypothesis, which we will discuss now.

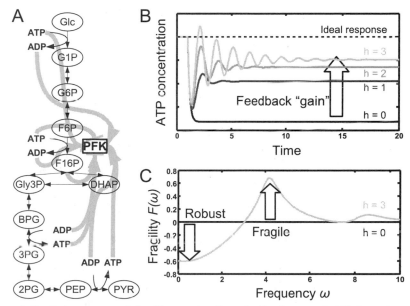

Figure 2.5   Robustness and fragility trade offs in feedback control. (A) Reaction scheme of glycolysis with activating (by ADP and F16P, grey arrow heads) and inhibiting (ATP, gray bar heads) influences of co-factors and metabolites on Phosphofructokinase (PFK) as a key glycolytic enzyme. (B) Response of the system to a sudden up-shift in ATP demand at different feedback gains h. (C) Relative fragility F as a function of the frequency ω.

Consider the following example from control of glycolysis (fig. 2.5A): phospho-fructokinase (PFK) at the center of the pathway is a highly regulated enzyme, with activation by the products of the reaction it catalyzes (ADP and fructose-1,6-bisphosphate (F16P)), and inhibition through its co-substrate ATP. Among others, this feedback structure allows the cell to adapt to varying ATP demands, while keeping the cellular ATP concentration tightly regulated. As shown in fig. 2.5B, the effect of a step increase in ATP demand—and thereby a sudden decrease in the concentration of PFK-inhibiting ATP—eventually leads to an increased flux through glycolysis and corresponding ATP production. The degree of recovery in ATP concentration apparently depends on the strength (or "gain") of the feedback h. Higher feedback gain eventually reduces the steady-state deviation between ideal and predicted response. However, increased precision in the long run is accompanied by more pronounced transient responses to the perturbation.

For a more quantitative analysis of this connection, let us employ the absolute sensitivity for a frequency ω, $|S(\omega)|$, as a measure of the deviation from perfect control. By defining a fragility $F(\omega) = \log|S(\omega)|$, the sign of $F(\omega)$ indicates if perturbations will be attenuated ($F(\omega) < 0$) or amplified ($F(\omega) > 0$). Analysis in the frequency domain shows that the effect of feedback in glycolysis in fact is two-fold: it increases robustness at low frequencies (for example, steady-state), but introduces fragility at higher frequencies (figure 2.5C). This is indicative of a certain "conservation of robustness"—increased robustness somewhere will be compensated by increased fragility elsewhere. For certain types of (linear) systems, the co-called Bode sensitivity integral (Bode, 1945) even describes this trade-off quantitatively as a conservation law:

$$\int_0^\infty F(\omega)d\omega = 0 \ . \tag{2.1}$$

Note, however, that a formal proof currently is only possible for a very limited class of dynamical systems.

The concept of "highly optimized tolerance" (HOT) relies on the very idea that robustness has to be regarded as a limited and conserved resource. This quantity (tolerance) requires careful distribution, adapted to the function a system is intended to perform, and the associated uncertainties. High optimization refers to a strategy of simultaneously achieving high performance and error-tolerance by a high degree of internal organization. The management and overall conservation of robustness lead to a "robust yet fragile" behavior of such systems, namely a high robustness ("barriers to cascading failures") in the face of anticipated or usually encountered disturbances, but hypersensitivity to unexpected perturbations, design flaws, or hijacking (Carlson and Doyle, 1999, 2000). In addition, HOT emphasizes a necessary connection between complexity and robustness. Making certain functions of a system more insensitive to disturbances, for instance, may require additional control loops. This, in turn, leads to higher complexity and to new potential sources of fragility. The effect is a "spiraling complexity" in which new features expose new fragilities to be "fixed" by further additions to the system (Carlson and Doyle, 2002). Hence, the distribution of robustness/fragility may be key for understanding system design in cell biology.

2.3.4   From Robustness to Evolvability

An often noted reservation against the type of analogies between biological and engineered systems we brought forward states that these two types of complex systems arise in fundamentally different ways, namely through evolution versus purpose-driven, top-down design (see, for example, Bosl and Li (2005)). Clearly, evolvability is of paramount importance for living systems (Kirschner and Gerhart, 1998). Here, we think of evolvability simply (maybe naively) in the sense of controlled and structured change in lineages, rather than cells, on long time scales in response to perhaps large variations in the environment. At the population level (of all engineered systems of one type), evidently progress in engineering fulfills similar criteria. More importantly, the generic mechanisms and structures responsible for robustness do not operate at the expense of evolvability—in fact they facilitate both.

Genetic redundancy allows duplicate genes to acquire new functions without perturbing the cells under most conditions. Feedback control, for instance, supports the normal operation even during evolutionary changes. The exchange of modules such as biosynthetic pathways through lateral gene transfer (for instance, via plasmids carrying the corresponding operons) lets organisms easily gain completely new functions. Finally, protocols are of paramount importance for facilitating plug-and-play mechanisms. Protein kinases, for instance, gain new functions by changing substrate specificity and control of their activity, but the common currencies of ATP and phosphate groups as effectors remain functional. In the realm of technology, Lego is one of the best examples of an evolvable system on many time scales (Csete and Doyle, 2002); the common carrier for the power grid, which facilitates control in response to short term fluctuations in supply and demand, also facilitates long term evolvability by providing a simple protocol for suppliers and consumers. Hence, we could think of evolvability as robustness on longer time scales, which is also subject to selection during evolution (Earl and Deem, 2004).

2.4   Biological Examples

2.4.1   Robustness in Central Metabolism

Although complex in their operation, metabolic networks are structurally organized such that a large variety of biochemical products and complex macromolecules are synthesized from myriads of nutrients by conversion through relatively few common intermediate metabolites. This so-called bow-tie architecture (section 2.2.3) results in an ubiquitous and interconnected core set of central reactions that constitute the backbone of high metabolic fluxes. The central carbon metabolism, in particular, provides a plethora of alternative routes for generating essential precursor molecules and the carrier molecules for energy (ATP) and reduction equivalents (NAD(P)H). Here, we describe two complementary—theoretical and experimental—approaches for analyzing the robustness of central metabolism.

At the theoretical level, elementary flux mode (EFM) analysis decomposes the metabolic network into meaningful smaller units or pathways. These EFMs can be defined as the smallest sub-networks enabling the metabolic system to operate in steady state (Schuster et al., 1999) (see chapter 5). The high number of EFMs in *E. coli* central metabolism on different substrates (see figure 2.6A,B) directly reflects the flexibility of this network (Stelling et al., 2002). Although all substrate regimes comprise the same number of reactions and metabolites, the EFMs differ by two orders of magnitude (figure 2.6B). When considering only single-substrate regimes, glucose can be utilized in approximately 45 times more different ways than acetate, which corresponds to biological intuition. Simultaneous utilization of all substrates enhances the number of alternative pathways by a factor of ten.

A plausible hypothesis concerning the connection between network flexibility and robustness is that the degrees of freedom of a network could be used to predict

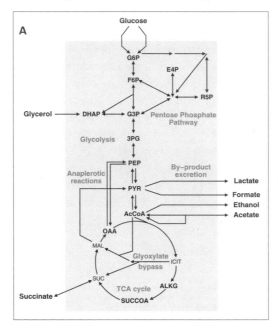

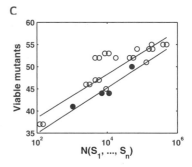

Figure 2.6 Flexibility and robustness in central bacterial metabolism. A stoichiometric model for *E. coli* with 89 substances and 110 reactions was decomposed into elementary flux modes (EFMs) (Stelling et al., 2002). (A) Schematic network representation. Shaded areas indicate the intracellular space. Only major nodes and twelve precursor metabolites (bold face) (Neidhardt et al., 1990) are labeled; reactions were partially combined. (B) Number and distribution of EFMs for different substrates. (C) Effect of arbitrary gene deletions on viability for single (•) and for multiple (○) substrates as a function of the total number of EFMs in wild type $N(S_1, \ldots, S_n)$.

its sensitivity to disturbances. For different single-substrate uptake regimes, the organism's resistance to arbitrary gene deletions correlates well with the number of EFMs N for the corresponding wild type (figure 2.6C). Similar results are obtained when more then one substrate can be utilized. Here, in general, the number of viable mutants is higher than for single-substrate regimes showing a comparable number of elementary modes. Most likely, this represents the effect of higher degrees of independence of metabolic pathways for the multi-substrate case. The ability to utilize different carbon sources simultaneously could, thus, be advantageous for the organism's robustness.

Mechanistically, robustness in central metabolism can be assessed by ^{13}C-tracer-based flux experiments (Sauer, 2004). The particular strength of quantitative flux data is their high degree of integrative information on regulatory and biochemical interactions within the network. A recent systematic *in vivo* flux analysis investigated flexibility and optimal performance in central metabolism of the model prokaryote *Bacillus subtilis* by selecting a near random choice of 137 knockout mutants that roughly reflect the proportion of all major functional gene categories (Fischer and Sauer, 2005). The data revealed a remarkably robust distribution of intracellular carbon fluxes, as shown exemplarily for three key fluxes (figure 2.7). The flux partitioning between alternative pathways was generally very robust against genetic perturbations and, for several pathways, completely independent of the absolute flux through the branch point. The only detected branch point that featured any significant flexibility in flux partitioning to different pathways was acetyl-CoA, the entry substrate into the TCA cycle.

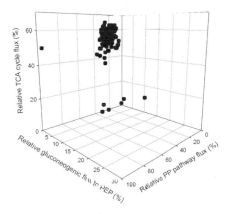

Figure 2.7   Relative fluxes through the pentose phosphate (PP) pathway, the TCA cycle, and gluconeogenesis from oxaloacetate to PEP in 137 *B. subtilis* knockout mutants during growth on glucose (Fischer and Sauer, 2005).

This control architecture of metabolism that maintains an unexpectedly stable metabolic state under a given environmental condition appears to be designed to provide a rigid flux distribution. While this state was robust against random genetic perturbations, it was sensitive to regulatory mutations because several regulator knockouts specifically affected flux partitioning at the acetyl-CoA branch point; that is, reduced the TCA cycle flux. The combination of high robustness and suboptimal efficiency also illustrates the need for trade-offs between different functional requirements.

2.4.2 Control Architectures in the Circadian Clock

Circadian clocks provide endogenous cellular rhythms of approximately 24 hours that directly or indirectly control many physiological processes and have been observed in species across four kingdoms. At the molecular level, however, they show an apparently complex regulatory architecture with multiple intertwined positive and negative feedback loops. For the fly and the mouse, the cellular genetic networks contain delayed transcriptional feedback mechanisms (Hastings, 2000). The core of the heavily studied *Drosophila* transcriptional feedback network is shown in figure 2.8 (Hastings, 2000; Reppert and Weaver, 2000; Young and Kay, 2001).

The transcription rates of the genes *per* (period) and *tim* (timeless) are accelerated when protein *dCLK* binds to their promoter regions. The transcribed *per*

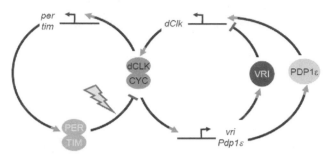

Figure 2.8 Core genetic network of the *Drosophila* circadian clock, adapted from (Hastings, 2000; Reppert and Weaver, 2000; Young and Kay, 2001).

and *tim* mRNAs are exported from the nucleus and translated into proteins *PER* and *TIM*, respectively. In the cytoplasm the protein *DBT* (doubletime) binds to *PER*. *DBT* either phosphorylates *PER*, causing it to be degraded, or allows *PER* to bind to *TIM* after a delay, thereby protecting it from degradation. After the *DBT-PER-TIM* complex is formed, it is imported into the nucleus where it represses the transcription of *per* and *tim* and activates the transcription of *dClk* (clock). The *dClk* mRNA is exported from the nucleus and translated into protein *dCLK*. Protein *dCLK* is imported into the nucleus where it represses the transcription of *Clk* and activates the transcription of *per* and *tim*. This system can be characterized by a two loop transcriptional feedback network, where *DBT-PER-TIM* negatively feeds back on *per* and *tim* transcription and activates *dClk* transcription, and *dCLK* negatively feeds back on *Clk* transcription and activates *per* and *tim* transcription. In addition to the main (double) negative feedback loop, there are loops involving the genes *vri* and *Pdp1e*. This multi-loop architecture is shared by mammals, although some homologous proteins play different roles (Reppert and Weaver, 2000; Leloup and Goldbeter, 2003; Forger and Peskin, 2003).

Model-based analyses of these networks have pointed out their remarkable robustness in the presence of molecular noise (Barkai and Leibler, 2000; Ueda et al., 2001; Gonze et al., 2002) and with respect to parametric perturbations (Smolen et al., 2001; Leloup and Goldbeter, 1999; Stelling et al., 2004a). Different models display model-specific robustness and fragility properties (Zak et al., 2001; Stelling et al., 2004a). Employing tools from systems engineering, Stelling et al. (2004) performed a comparative analysis of the global robustness and fragility properties of two published mathematical models for the *Drosophila* circadian clock. Both deterministic models relied upon negative autoregulatory feedback for generating sustained oscillations. A less complex 5-state model with only one branch (Goldbeter, 1995) and a 10-state model including two distinct branches of the control system for *per* and *tim* (Leloup and Goldbeter, 1999) were considered. To gain insight into the structure-function relationship, they studied robustness towards parametric perturbations by numerical computation of the parameter sensitivities (see chapter 1).

For the detailed analyses of both fly clock models, model parameters were organized in functional categories (for example, transcription, translation, phosphorylation, etc.), as well as into hierarchical categories. For the latter, *global* parameters reflected characteristics of well regulated core cellular machineries (such as the maximal capacity of the general transcriptional apparatus embodied in maximal transcription rates for all genes), while *local* parameters were primarily confined to the circadian oscillator. The analysis revealed clearly that global parameters were more fragile, in comparison to the more robust local parameters. Furthermore the separation between the two was sharpened by the complex hierarchical organization underlying the fly dual feedback clock model (as opposed to the single feedback engineering model). In agreement with the bow tie proposition for cellular organization (section 2.2.3), these results suggest a design principle of cellular regulation, in which robustness of specific (local) functions is achieved by delegation of fragilities to global control circuits (Stelling et al., 2004a) (figure 2.9). The same trade-offs are observed in the mammalian clock architecture (F. Doyle, unpublished results).

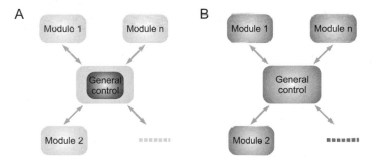

Figure 2.9 Scenarios for distribution of robustness and fragility. (A) Concentration of fragilities in a central core, whereas functional modules are error tolerant; gray levels correspond to levels of fragility. (B) Equal distribution of fragilities.

One important consideration in the analysis of robustness, particularly with regard to the circadian rhythm circuit, is the selection of a performance attribute for evaluation of its robustness characteristics. For example, Stelling et al. (2004a) compared the rank ordering of sensitivities (robustness) for both period of the proteins and transcripts, and their amplitude. Not surprising, the order is changed significantly, with transcriptional/translational regulation having a larger impact on amplitude, while phosphorylation/dephosphorylation have a larger impact on the clock's period. Additional attributes may be considered for robustness, including entrainment, phase response sensitivity, and relative phase timing of key proteins. In general, the conclusions drawn about robustness may vary as different attributes are evaluated. Moreover, the scale of the network analysis may influence the conclusions: single-cell attributes are likely to be quite different from whole organism robustness.

2.5   Consequences for Systems Modeling

In this chapter, we focused on rather abstract concepts that deal with cellular complexity, function and structure. As the biological examples in the previous section showed, such concepts provide an organizational framework for modeling—and finally understanding—cellular systems. In particular, insight into the robustness of cellular networks can guide us in *what* and *how* to model for such systems. In general, given the small repertoire of mechanisms providing robustness, modeling could specifically probe these mechanisms; the analysis of bacterial central metabolism suggests that network properties rather than redundancy of individual components should be in the focus of such efforts. The circadian clock examples revealed structuring of sensitivities in agreement with the predictions made by the bow tie hypothesis. For model development, such features help in identifying the important/less important parts of the system. For instance, they would suggest devoting more efforts to detailed modeling of the core processes because these are likely to be highly sensitive to design flaws in the models. In addition, characteristic distributions of robustness and fragility can be exploited to decompose larger networks into manageable subunits. Such an approach proved successful, for instance, for the analysis of signal transduction processes in apoptosis (Bentele et al., 2004).

More generally, robustness may facilitate model development because exact parameter values are not required in many instances, and sensitive parameters could possibly be predicted from network architectures. In other words, robustness implies an importance of accurately describing the structure of a system as opposed to identification of the associated parameters. A classical study on the segment polarity network in *Drosophila* revealed that without the appropriate (feedback) structures, despite large freedom in choosing parameter values, even the qualitative behavior of this developmental control circuit could not be reproduced (von Dassow et al., 2000). For practical purposes, the importance of network structures may imply that the transferring quantitative models between similar systems, such as different cell lines from one organism, might reduce to adjusting a few key parameters. These are some justifications for why qualitative and structural modeling methods—although devoid of parameters—may yield deep insight into the relations between network structure and behavior (see chapter 7 and chapter 5). Given our incomplete knowledge on cellular circuits, moreover, we often face the challenge of evaluating sets of different hypotheses on cellular network structures. The robustness property allows for relatively easy discrimination between hypotheses because exact parameter values are not important in many cases. For instance, the models' ability to perform robust control tasks could be used in elucidating network structures underlying morphogen gradients in embryonic patterning (Eldar et al., 2002, 2003). In the model world, thus, robustness could be employed as one criterion for assessing the plausibility of a particular model (Morohashi et al., 2002). However, we always have to be aware that prior knowledge is essential for determining the exact nature of robustness for a system.

Finally, representation of a cellular network through a mathematical model is always only the first step towards understanding—subsequently we have to ask *why* the models perform as they do, and which are the underlying design principles (Lander, 2004). The analysis of robustness properties can lead to abstractions in this direction, for instance, by revealing a common operating principle in bacterial chemotaxis despite different molecular implementations in different organisms (Rao et al., 2004) (see also chapter 12). From engineering, it is known that feedback control (plus feedforward control) enabled by fast and if possible remote advanced-warning sensing is the most powerful mechanism for providing robustness to fluctuations in the environment and the component parts. The heat-shock response in *E. coli* appears to employ exactly the same principles as shown by detailed modeling and subsequent model reduction to the core elements (El-Samad et al., 2005). Future studies can make use of such principles by searching for this type of mechanism. Hence, abstract concepts on complexity and robustness have broad implications both for systems modeling and systems analysis.

2.6 Concluding Remarks

In this chapter, we adopted a high-level view of cellular systems by combining biology and engineering approaches. This perspective does not want to disguise large differences between the two types of systems; in fact, biology often shows a more remarkable "design" than technology. However, it appears as though there are universal principles in biology and technology that facilitate robustness, efficiency, and evolvability. We do not yet have a clear and concise characterization of them all, but we can say some things: (i) feedback control is the most powerful mechanism for providing robustness to fluctuations in the environment and the component parts; (ii) redundancy plays some role in robustness to component variations and failures but is most useful when it is part of feedback control systems that can sense variations and failures, and coordinate the use of multiple resources; (iii) protocols that enable carrier and building block–based metabolism facilitate both decentralized control and supply chain management for short term fluctuations as well as plug-and-play modularity for long term evolution. Taking such abstractions into account for systems analysis in biology—as several examples showed—can provide the necessary guidelines for modeling and analyzing biological complexity. In our view, the close relations between complexity and robustness requirements may imply that living cells are complicated, yet comprehensible systems.

3 On Modules and Modularity

Zoltan Szallasi, Vipul Periwal, and Jörg Stelling

The enormous complexity of biological systems begs for unifying, simplifying concepts that might allow a predictive understanding of their functioning. Suggestions for such concepts include "modularity" along with robustness, discussed in the previous chapter. No comprehensive survey of system modeling can ignore these concepts, even if it means pointing out the lack of consistent and clear definitions in the field. Modularity is without doubt an enticing concept that may hold promise for helping to overcome some of the computational limitations of dynamic modeling of biological systems. The list of what cannot be achieved far exceeds the utility of this concept as demonstrated thus far. As this chapter outlines, the long and arduous task of laying its rigorous quantitative foundations is in its infancy.

3.1 Introduction

Biological systems are often said to be "complex." Is this a precise logical concept, in the sense that given a set of systems we can unambiguously separate the complex systems from the simple ones, or is this merely an adjective assigned on the basis of the user's inability to comprehend the relation between inputs and outputs of the system? In the quantitative science literature, a more or less standard definition exists: A complex system is a system whose properties are not fully explained by an understanding of its component parts. Complex systems consist of a large number of mutually interacting and interwoven parts, entities, or agents. As is evident from this standard definition, there is considerable ambiguity implicit in its negative character. A variant of this definition, interesting for biology, posits that *both* understanding and verification of design and/or function is difficult in complex systems.

As a counterpoint, modeling a biological system is an exercise in understanding how the outputs arise from the inputs. In light of the preceding definition, we might then suppose that biological modeling is the process of moving systems from the complex systems set to the simple systems set. A natural notion in modeling complex systems is to replace some of the parts being modeled with an abstraction, while maintaining the fidelity of the model with the given experimental data. This requires, usually, the maintenance of the functional interface of the replaced parts with the rest of the system. For example, the engine in an automobile is an abstraction for a large number of components. The functional interface with the rest of the automobile is provided by the drive shaft and various hoses and wiring harnesses. Such functional abstractions are often called modules.

Almost all artifacts of evolved human engineering are modular through and through: their entire architecture is composed of parts packaged within bigger parts with clear functional interfaces. This is true of electronics, it is true of houses, and it is true of software. In these systems, all parts are members of some module, and the entire architecture is modular. It would seem, based on this experience, that the way forward in simplifying the complex biochemistry of life is to encapsulate complexity in similar modules. Certainly, the computational limitations of dynamic gene network modeling are much easier to evade and an understanding of complex networks in terms of (higher-level) functional interactions is easier to achieve if a modular architecture underlies the network. The question at hand is: To what extent does modularity provide realistic and useful abstractions for systems shaped by biological evolution?

There are two separate issues here – the existence of modules in biology, and the utility of this concept, although it will most likely be an approximation of reality (as in any abstract model). Regarding the existence of modules, even in the engineering example, autonomy is never absolute. We therefore have to consider subsystems of limited (quasi-)autonomy. One subissue is whether we can identify modules in the biochemical interaction network that work quasi-autonomously, like the engine in our analogy. The other subissue is the existence of an overall modular architecture for the entire biochemical interaction network. The modules evident in the morphological structures present in each eukaryotic cell, such as the nucleus, the mitochondria, and other organelles, as well as the presence of specialized organs in metazoans are certainly evidence of a modular architecture at a high level. In this sense, the success of organ transplants can be considered as taking advantage of the existence of modules in the human body for medical intervention. Close to fundamental biochemistry, biological concepts of a "gene," a "protein," or a "protein domain" are widely employed abstractions from the underlying chemistry. In the context of interaction networks, for instance, protein functions are usually not discussed in terms of the protein's atomic coordinates. A given protein is thus considered to be a module of all of its constituent atoms. Hence, at these two very separate levels of cell biology we find evidence for modularity. But does this hold for modularity at all levels in biochemistry?

3.2   The Concept of Modules in Other Biological Disciplines

Evolutionary biologists have long considered a type of modularity in which the animal body is composed of units, which integrate functionally related characters (with characters in genetics defined as structures, functions, or attributes determined by a gene or group of genes) into units of evolutionary transformation (Wagner, 1996). They have also investigated extensively the origins of this modularity, either from evolution or from a priori principles of organization for reproducing systems. In terms of the evolutionary origins of modularity, modularity could arise from specialization resulting in the elimination of some pleiotropic effects from a more integrated phylogenctically primitive state, that is, a bigger module splitting into two more independent submodules or from the opposite process when a linked functional role leads to differentially greater integration of evolutionary characters, preventing independent variation. These two processes are acting in independent, and potentially opposing, directions. Thus an evolving system never exhibits perfect nonoverlapping modularity, just as a matter of simple irreversible statistical mechanical relaxation.

The presence of modules may well enhance the rate of evolution due to noninterference between functional roles, though this mechanism is unlikely to be of interest in multilocus systems because it is hard to maintain the necessary level of linkage disequilibrium in multilocus systems (Wagner, 1996). Stabilizing selection, likely the mode of selection experienced most of the time, is blind to modular organization in systems with multiple characters, neither enhancing nor washing out distinctions. Directional selection forces, on the other hand, may result in the adaptation of a small number of linked characters, preserving other characters under the influence of stabilizing selection. Pleiotropic effects may interfere with adaptation, perhaps leading to mutations that decrease pleiotropic effects linking the genes associated with the adapting characters to other characters, and thereby leading to the appearance of modules. Thus evolutionary biology favors the appearance of modules, but not necessarily modularity in the overall organization. It is also apparent in the argument for the appearance of modules that the environmental circumstances which favored the decrease in pleiotropic effects are integral to the definition of these modules. Therefore, the response of a biological system may reflect the existence of certain modules only in specific contexts. This is not necessarily the case for modules in human engineering constructs.

Modularity has also been investigated extensively in neurobiology. In fact, the notion has been considered independently in the three fields of psychology, neuroscience, and artificial intelligence, which can be regarded as the neurobiology analogues of physiology, molecular biology, and biological modeling in systems biology, respectively. It is instructive to note that workers in each of these fields have their own definitions of modularity (Bryson, 2005). Reconciling these definitions is an important part of understanding the actual behavior of organisms, and it is just as likely that modules found in cellular physiology are intricately related to modules found in molecular biology. From a physiological point of view, modularity might be considered in the form of the hypothesis that the cell contains independent input systems that are restricted in the range of environmental and cell-state information that they can access.

3.3   The Concept of Modularity in Systems Biology

The interest in modules in the systems biology context was expressed clearly in (Hartwell et al., 1999), albeit mainly invoking a hypothetical parallel between human and evolutionary design and providing little in the way of evidence. As is evident from other disciplines interested in biological modules, there is a lack of well-defined, quantitatively applicable definitions. One reason for this lack is that the concept of "modular design" is borrowed from human engineering and therefore has an essentially forward looking, goal-oriented nature. Complex engines and networks are constructed from modules while the final overall behavior of the system is kept in mind. It is much more difficult to identify a "modular architecture" in an already existing complex network, such as a cell, especially in an unsupervised fashion. Overlapping modules and multiple "hidden" or ill-defined functions of subsystems pose additional, potentially insurmountable, difficulties (see also Figure 3.1 and text below). We would encounter similar difficulties in human designed systems if we were only presented with the results without an appropriate understanding of the functions. Advanced engineered systems are rather frequently modular in their overall design, but for evolved systems we do not even have the appropriate analytical tools to address the issue of modular decomposition.

As a consequence, most studies on modularity in systems biology rely on operational definitions that reflect to a large extent the biological system or data set from which the modules were extracted, as well as data quality and the available computational tools. The abstraction of a "module" will always be an approximation to reality. This already holds for the concept of a gene and – as discussed above – this will be true to an even larger extent for complex cellular networks. Hence, these operational definitions should be judged by their value in facilitating the development of dynamic models and by the extent they enhance our understanding of these systems. Two extremes in definitions and analysis of modularity can be found in "bottom-up" and "top-down" approaches (see chapter 1) that we will discuss in the following.

3.3.1   Bottom-up approaches

Bottom–up approaches to a large extent build on existing biological knowledge. The function of the proposed module is well defined (at least under a limited set of conditions) and the individual members of the module are determined by detailed biochemical or molecular biological analysis, such as testing the effect of individual gene knock-outs on the function in question. In an ideal case the dynamic interactions between the various components in a module are also known. This allows the validation of the proposed module in a dynamic context. If the quantitative behavior of this module, when studied in relative isolation from the rest of the entire intracellular regulatory network, provides an accurate and comprehensive description of the specific function in question, then the proposed module can be considered validated. Hence, the approach essentially constitutes a direct test of the "quasi-autonomy" that is characteristic of most definitions of a module.

An excellent example of this approach is given by von Dassow and colleagues (von Dassow et al., 2000). In their paper on the quantitative analysis of the segment polarity network of *Drosophila*, they first defined a module as a quasi-autonomous subsystem of a complex genetic circuit with a specific function. Their proposed module was built on a large body of knowledge of *Drosophila* differentiation from which they created a dynamic mathematical representation. With a few subsequent corrections and modifications, this modular representation robustly reproduced the qualitative *in vivo* dynamics of the specific differentiation process in question. Moreover, its predictions were consistent with a wide array of experimental observations. Note that the components of the module participate in other cellular processes as well, so the modular character of the subsystem is specific to the process being modeled.

It is evident from the description above, that bottom-up approaches require considerable effort in terms of assembling and validating modules. High throughput, large dataset–based, computationally aided efforts for module identification, therefore, hold considerable appeal. The rationale of such approaches may be motivated by the following analogy: In the classical age of genetics, genes were traditionally identified by individually sequencing DNA fragments of limited size that were isolated based on the fact that the nucleotide sequence in question had some functional relevance in biological experiments. More recently, however, entire genomes have been sequenced in a wholesale fashion, and genes have to be extracted from a deluge of sequence information, often resulting in erroneous gene identification. Furthermore, in most cases the function of the putative genes has to be determined *a posteriori*. In other words, in the first case a nucleotide sequence is matched to a function of interest, and in the second case functions have to be found for existing sequences.

3.3.2 Top-down Approaches

Fueled by the overall accessibility of genome scale data sets, several top-down approaches have been proposed for the high throughput identification of putative modules. These methods usually rely on the concept that intramodular "connections" (whatever they may be) are more frequent than intermodular ones. The underlying assumption is that the number of interactions provides an indicator for how well embedded a certain component is into a subsystem; most approaches do not consider the "strength" of these interactions. Graph theoretical approaches,

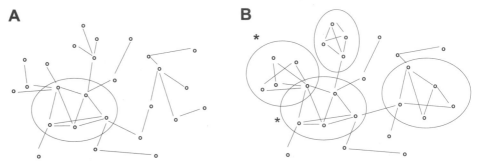

Figure 3.1 Identifying one or more modules in a genetic regulatory network versus modular organization of the entire genetic network. The figures show a static graph representation of two genetic regulatory networks. The circles represent various proteins or genes whereas the edges represent regulatory interactions. In the genetic network of panel A, a module (circled) can be identified by the following criteria: each member of the module is connected to at least two other members of the same module. Note that no other module can be identified in this network. In the genetic network of panel B, a far larger number of modules are present. Note the overlapping modules marked by stars. Since the majority of protein and genes can be included into one or more modules, a certain level of modular organization is apparent.

for instance, statically represent components as nodes and interactions as edges between these nodes (see chapter 7) (Figure 3.1).

Graph representations of large-scale biological data sets are especially attractive targets for analysis of modularity because these simple representations can be analyzed for very large networks. In one study, von Mering and coworkers performed whole genome bioinformatics analysis of protein interaction networks (von Mering et al., 2003). In their work a functional module is a tight cluster of proteins in the protein interaction network. A similar approach was followed by Spirin and colleagues (Spirin and Mirny, 2003) while studying protein complexes in molecular networks: molecular modules were defined as sets of proteins that have more interactions amongst members of the set than with the rest of the protein interaction network. The logical end point of these static approaches to the identification of modules is in the definition given by Guimera and colleagues (Guimera et al., 2004), with modules assigned by a partitioning of the nodes in an interaction graph that maximizes a modularity cost function defined entirely in graph-theoretical terms (intramodule versus intermodule links in the graph, the sum of the node degrees within a module).

In addition to direct physical interactions, modular connections may reflect regulatory relationships, such as shared regulatory inputs. A regulatory module, therefore, can be defined as a set of genes that are regulated in concert by a shared regulatory program that governs their behavior (Segal et al., 2003). Both the behavior and the modules assigned through analysis of the behavior may be dynamic and overlapping. Similarly, a transcriptional module may be defined as a self-consistent regulatory unit consisting of a set of coregulated genes as well as the experimental conditions that induce their coregulation, with modules decomposing into higher-resolution modules when a resolution parameter is varied (Ihmels et al., 2002, 2004a).

Several points should be emphasized in connection with the above-described top down approaches:

1. They produce only putative modules; their relevance has to be validated in a detailed manner as done for the bottom up approaches. This is especially true for methods relying on static graph representations, such as representations of protein-protein interaction networks (Spirin and Mirny, 2003). This is also true for modules extracted from time-course data, although, depending on the manner in which they were defined, such modules may carry over to dynamic modules more readily.

2. Finding a large number of putative modules in a high-throughput analysis does not automatically translate into the modular organization of an intracellular network. For one reason, most high-throughput studies only consider one level of biological regulation (for example, transcriptional control, protein-protein interactions).

3. Top-down approaches may often produce putative modules without a well-defined associated function, without which a reliable validation is significantly more difficult.

4. The limitations of identifying modules by the above-described top down approaches are evident. For example, a more or less linear signal transduction pathway will not show dense intramodular connectivity in protein interaction networks, and will therefore be missed by these methods.

3.4   Definition of Modules for Dynamic Networks

The bottom up method for module identification described above produces modules that could be, at least in principle, readily incorporated into dynamic network models – with the caveat that a dynamic model is needed as a prerequisite for the identification of a module. This would involve replacing the detailed dynamic model of the given module with a simpler system that would still correctly characterize the dynamic behavior of the associated function and also provide a sufficiently accurate description of the dynamic interactions between the module and its functional environment. Ideally, it would also allow deducing higher order cellular functions by combinations of modules. However, this module identification method, in addition to its essentially low throughput nature, comes with several caveats. It relies on the existence of a biologically interpretable "function," and it barely takes into consideration that the module is sitting in the middle of, and has to be extracted from, a complex dynamic network. Modules for cellular level dynamic network modeling, however, are expected to satisfy other criteria. The main goals are:

1. The modules should provide a significant level of abstraction, aiding in the simplification of an otherwise barely tractable dynamic network.

2. The various functions of the *entire* biological network are expected to be described even if the individual "dynamic network modules" cannot be associated with an easily interpretable or observable function, such as a given differentiation pattern.

The biological function is thus approximated and replaced by an appropriate "mathematical function."

At a level of low complexity – that is, for small modules comprising few components and interactions – biochemical "building blocks" that perform (a small number of) characteristic dynamic functions can be identified. For instance, the graph theoretical analysis of transcriptional regulation networks in *E. coli* (Milo et al., 2002; Shen-Orr et al., 2002) and budding yeast (Lee et al., 2002) identified small over-represented "motifs" that can be attributed distinct functions, such as filtering noise or speeding up transcriptional responses in the case of the "feedforward motif" (Mangan and Alon, 2003) (see also chapter 7). From a theoretical perspective, one can determine small "building blocks" that are required for obtaining classes of dynamic behavior such as adaptation, homeostasis, switching, and oscillations (see chapter 6 for details).

Both approaches, however, are limited by the network size that can be assigned a distinct function. For instance, cellular circuits rarely employ one prototypic device to establish a biological function because robustness and efficiency of the function usually need additional complexity, for example, in the form of interwoven feedback circuits (see chapter 2 for examples). Hence, while the abstraction of "motifs" can provide insight into constituents of a "functional" module (in terms of biology), in general it is sufficient for neither the dynamic analysis nor the definition of a module.

An interesting, "middle-out" approach has been proposed recently by El-Samad et al. (El-Samad et al., 2005). They studied the heat shock response system in *E. coli*, which as a first step involved developing a detailed mechanistic model for the entire system as defined by the traditional biological definition of the module, taking into account individual proteins and their interactions. In a top-down manner within this module, the authors have performed a systematic model reduction, and they have proposed the existence of certain functional submodules based on characteristics of the overall behavior of the entire system, such as robustness (see chapter 2) or optimal performance. This approach closely follows the method of modular decomposition routinely used in system engineering, namely the identification of submodules or devices based on their dynamic functions. Although the computational analysis suggests intriguing insight and circumstantial evidence for the proposed overall design and modular structure of the heat shock response system, experimental assignment of the various proteins to the various submodules and their functional validation remains to be performed.

In a similar spirit, Kholodenko and coauthors proposed methods for the modular analysis of complex (signaling) networks, in particular with respect to the quantitative identification of network topologies (Bruggeman et al., 2002; Kholodenko et al., 2002). The system to be analyzed is a given network, the boundaries of which are determined, for example, by a biological function or according to the notion of a traditional signaling pathway. At a lower level, however, details of this network need not be resolved. Instead, operational modules are the subject of analysis. Responses of the modular system in steady state to perturbations are then described through interactions between the modules alone in order to quantitatively analyze signal transfer through the entire network (Bruggeman et al., 2002). Similarly, the abstraction of modules can be employed for identifying networks of (partly) unknown structure through perturbations affecting one module at a time (Kholodenko et al., 2002). Apparently, the definition of modules proceeds first irrespective of whether these modules correspond to a biological function. However, this abstraction potentially enables one to develop the dynamic models that are required for a more unambiguous definition of modules. Note that this necessarily involves an iterative process – a hallmark of systems biology in general (see chapter 1).

In addition, the middle-out approaches discussed above also provide some guidance in determining whether a given dynamic intracellular regulatory network is modularly organized: if a wide variety of higher level functions of the entire dynamic network can be comprehensively and accurately characterized by replacing the majority of individual genes and proteins by a significantly smaller number of dynamic modules, then a modular organization is likely to exist.

The reader will notice gaps and tensions in this section on the identification of dynamic modules in biology. Current approaches cannot cover the huge gap between the levels of a few interacting components and of the cell as a whole. Tensions are evident because a proper definition of biological modules would require dynamic models, for the development of which focusing on a small part of the cellular networks is necessary – a classic catch-22. There are algorithmically implementable definitions available – see the top-down approaches using graph theory discussed above or the mathematically well-defined concept of metabolic pathways discussed in chapter 5 – but it is largely unclear if these definitions have any relevance for biological functions (which by themselves often require a more rigorous definition). Hence, for this field as for systems modeling in biology in general, only iterative processes may ultimately lead to a framework of methods by which parts of large dynamic networks could be collapsed into and replaced by relatively simple modules. A point to ponder, illustrating the challenges ahead, is that in the modeling domain in general there is no universal recipe for the task of model reduction.

3.5   Conclusions

Abstractions, such as modules, are required for analyzing complex systems, but they have obvious limitations, usually more often recognized by failure than foresight. Studying individual modules, especially those identified by bottom–up approaches, is appealing. Through such studies, one can make predictions and design and test desired changes in biological functions. Various approaches along these lines are documented throughout this book. The behavior of biologically existing modules is studied, for example, in chapter 6, and human designed modules, with potential biotechnological consequences, are described in chapter 13. As biologists have identified an increasing number of genes associated with functions, the study of individual modules has started in earnest, reaching the level of quantitative dynamic approaches during the past couple of years. Models of the Ran nucleocytoplasmic transport (Smith et al., 2002), and the EGF receptor pathway (Schoeberl et al., 2002) provided an accurate, predictive description of their respective modules. However, these modules have not been coupled to others in order to attempt a higher–level integration of cellular functions. Therefore, the integration of dynamic functional modules as well as their rigorous definition and identification remain to be investigated.

II MODELING APPROACHES

4 Bayesian Inference of Biological Systems: The Logic of Biology

Vipul Periwal

Systematic model selection and inference in modeling biological systems must deal with the specific problems of incomplete prior knowledge, limited heterogeneous data, and similar but not identical model systems. In addition, the model selection process must allow incremental updates as new data becomes available. Probability theory as embodied in Bayes' theorem is the unique logically consistent framework for such reasoning. The foundations of Bayesian inference are summarized with some excursions into information theory and search theory. Some recent examples taken from the recent literature are reviewed.

4.1 Introduction

Reasoning in biology imposes three general desiderata on the reasoning process:

1. We must reason with incomplete prior knowledge of and limited data on the biological system under study. For example, we may have microarray or proteomics data with little knowledge of cellular localization.

2. We must be able to update our inferences taking into account new data, without having to revisit the entire reasoning process. For example, we should be able to add mass-spectroscopy data to our inference based on expression data.

3. We must be able to combine observations in multiple model systems, with no sense in which the different systems are merely repetitions. For example, we should be able to use knowledge of expression levels in a pathway in different bacteria to make more trustworthy inferences of aspects of the regulation of the pathway, even though there is no sense in which the observations are repetitions.

The mathematical rules of probability theory (Jaynes, 2003; D'Agostini, 2003) are the *unique consistent rules* for conducting plausible reasoning in such a setting. It must be emphasized here that, given partial observability and incomplete knowledge characteristic of biological systems, all probabilities may be considered conditional probabilities, *especially* conditioned on the state of knowledge of the biologist. The mathematical rules of probability theory, applied consistently, will lead to a consistent and optimal revision of the probabilities in light of new evidence. Given adequate data and possibly quite different initial assignments of probabilities, two different experimenters will usually arrive at convergent inferences, provided that the rules of probability theory are consistently applied.

The aim of this chapter is to explain the basics of plausible reasoning relevant for systems biology. As with the rest of this book, the foundational material presented here is intended to facilitate understanding between scientists with different backgrounds and to allow workers access to more specialized tracts with a basic understanding of the issues. The application of plausible reasoning to biological systems is not a novel idea and is well-documented in the medical literature (Lusted, 1968). The same desiderata given above in the systems biology context also apply to medicine, so this should not come as a surprise. Probabilistic reasoning has also been applied to systems biology in many papers, under the terminology Bayesian networks or graphical models (Jordan, 1998; Pearl, 2000). This chapter is intended to provide a foundational perspective on the logic that underlies such applications.

Notation: A *proposition* is a statement that may be true or false, for example $A =$ "The upregulation of Erbb1 leads to increased expression of β-catenin." The term *probability* is used in this chapter in the sense of a quantitative assignment of a degree of plausibility to a proposition. Clearly such a probability has nothing logically to do with the number of times a proposition is observed to hold in a repetition of an experiment. $p(A|B)$ is the probability that proposition A is true, given that proposition B is true. The negation of a propostion is denoted $\bar{A}$. The proposition "A and B" is denoted AB, and the proposition "A or B" is denoted $A + B$.

A *sampling distribution* or *likelihood function* is a rule for assigning probabilities to data, given that a hypothesis is true, $p(\text{data}|\text{hypothesis})$. Thus, sampling theory is concerned prototypically with problems of the form: given the contents of a cell, determine the probabilities of drawing a certain set of messages. Scientific inference, on the other hand, is concerned with problems of the form: knowing the observed expression data, determine the contents of the cell. Much of this chapter is devoted to this inverse problem: how do we calculate $p(\text{hypothesis}|\text{data})$?

In the most general context, the biological system under investigation may be characterized by a set of unobserved and unknown variables, X, for example the phase of the cell cycle, and localization and concentration information for a large set of proteins, but the available experimental data, D, may be only a few protein and mRNA measurements. We expect a functional relationship of the form $D = F(X)$, and we hope to extract X from D by inverting $F : X = F^{-1}(D)$, but typically the data is not sufficient to allow this inversion. Biological systems are never completely observed experimentally. Thus experiments exhibit variability that is often termed "noise", as a short-hand for uncontrollable effects. Biological systems often exhibit fundamental stochasticity too in their mechanisms of action, but this stochasticity and the former noise are completely unrelated to the probabilities that we are concerned with here. Finally, there is the randomness associated with experimental protocols, for example small differences in aliquots of mRNA extract, which for the purposes of inference is in the same category as the other unknown variables in X.

We may have useful information relating D and X in the form of likelihood functions, which give us the probability of observing D given a certain set of values X. For example, supppose we observe a certain level of the phosphorylated form P_f of a protein P. Using the Heaviside $\Theta(x)$ function, which vanishes for $x < 0$ and is equal to 1 for $x > 0$, our prior probability for P_f given P is

$$p(P_f|P) = \Theta(P - P_f)/P, \tag{4.1}$$

reflecting the probability that, given the total level of P, the observed level of P_f must be less than P. Suppose we know that only the phosphorylated form P_f is stable, and that the concentration of the message corresponding to P is M. The set of reactions is, to the best of our knowledge,

$$
\begin{array}{ccc}
M & \to \quad P & \leftrightarrow \quad P_f \\
\downarrow & \downarrow & \\
D_M & D &
\end{array}
\tag{4.2}
$$

where D_M and D are the products of other reactions/decays of M and P, respectively. What would be a plausible prior probability for $p(P|P_f, M)$? Knowing nothing else, we could start with

$$p(P|P_f, M) = \Theta(P - P_f)\Theta(30\,M - P)/(30\,M - P_f), \tag{4.3}$$

which quantitatively expresses our expectation that P must be higher than P_f and lower than M at equilibrium, and incorporates our knowledge that P is ubiquitinated in the unphosphorylated state. The factor 30 might reflect our ignorance of the precise translational control of M, based on a review of the literature. These different functional forms for the likelihood function are reflections of our differing biological knowledge in the two cases: Message does not necessarily get translated into protein, but protein does not get made without message being expressed. These examples may be too simple to be of practical value, but the key point is central: any quantitative model or hypothesis linking unobserved quantities and observed quantities can be translated into a likelihood function. In a sense, this is implicit in the whole idea of a quantitative model and gives the data a meaning in the context of the system under study.

There are two basic rules for the evaluation of probabilities:

1. Product Rule: $p(AB|C) = p(B|C)p(A|BC)$.

2. Sum Rule: $p(A|B) + p(\bar{A}|B) = 1$.

From these two rules, it is possible to derive relations between probabilities, for example

$$p(A + B|C) = p(A|C) + p(B|C) - p(AB|C).$$ (4.4)

Two hypotheses, A and B, are *independent* if knowledge of the value of B does not affect our knowledge of A :

$$p(A|B) = p(A).$$ (4.5)

Two hypotheses A and B are *conditionally independent* if knowledge of the value of a third hypothesis C along with knowledge of B does not constrain A :

$$p(A|BC) = p(A|C).$$ (4.6)

Conditional independence does not imply unconditional independence.

In the context of probabilistic reasoning, we always need to ask: against which specific alternatives are we testing a model or hypothesis? Probability theory cannot invent alternative hypotheses for the biologist. Given some previously established set of *prior* probabilities, $p(A|X)$, where A is a hypothesis and X represents prior data, if we obtain some new data D, we can use the product rule to compute *posterior* probabilities:

$$p(A|DX) = \frac{p(AD|X)}{p(D|X)} = p(A|X)\frac{p(D|AX)}{p(D|X)},$$ (4.7)

thus updating our estimation of the plausibility of our hypothesis in light of the new data D. This is usually referred to as Bayes' rule. This can be written more symmetrically as

$$\frac{p(D|AX)}{p(D|X)} = \frac{p(A|DX)}{p(A|X)},$$ (4.8)

provided that the denominators do not vanish. This rule expresses exactly the fact that the proportion by which the data D affects the probability of the hypothesis A is the same proportion by which the hypothesis A affects the likelihood of the data D. Notice that other hypotheses are implicit in the update rules since

$$p(D|X) = \sum_A p(DA|X) = \sum_A p(A|X)p(D|AX),$$ (4.9)

summed over all hypotheses considered. In some cases, there may be an implicit alternative hypothesis in the problem, but in no case can one carry out probabilistic reasoning without a comparison of alternatives.

A note on terminology: We will use the terms *prior* and *posterior* fairly often, and it is important to emphasize here that "logical implication" is not the same as "biological causation"—in other words, we can infer a probability for a biologically earlier event from knowledge of a temporally later event. Thus, prior information is not necessarily about temporally prior events. For example, snow on the road in the morning may lead to a plausible inference that snow fell during the night, even though the causal connection goes in the opposite direction.

Probabilistic reasoning requires no optimization over unknown parameters. This would be akin to eliminating hypotheses explicitly by choosing only certain specific hypotheses based on non-probabilistic reasons and would render the logical consistency of the entire process suspect. The logically correct approach is to sum or integrate over unknown quantities, a process known as *marginalization*, so that the effects of unknown quantities, sometimes referred to as *nuisance variables* are averaged over all plausible values, weighted by their degree of plausibility as embodied in their probabilities.

What if the data comes out to have low probability with respect to the chosen prior distribution? This is not a disaster, nor does it imply that the reasoning process has broken down. Rather, it implies that the hypotheses encoded in the prior distribution are inadequate, and that new biology is needed to explain the data.

The flexibility of the probabilistic framework is daunting, since the biologist is required to think about known biology in order to formulate quantitative hypotheses for analyzing the data. The payoff is that the biology is front-and-center in the whole process. Prior knowledge is the input to mathematical models of biological systems. It is the biologist who is responsible for the connection between mathematics and reality. In particular, the expectation that enough data collection will automatically lead to emergent realistic models is a fallacy. Modeling and data collection cannot be separated: it is the analysis of new data that leads to posterior probabilities for alternative models, and it is the plausible models that must be used to guide the acquisition of new data. A guide to the choice of a sufficient number of plausible hypotheses is the value one obtains for $p(D|A)$. The key is to pick a set of hypotheses that are sufficient to explain the data, without making the set of hypotheses so general that the data is implausible.

4.2 An Example

A simple example (Skilling, 1998) should help clarify the reasoning process. Assume that we are given a liquid, known to be water or ethanol, and a thermometer, accurate to $\pm 2.5K$. We need to determine the probability that the liquid is water, given the temperature reading T on the thermometer. Let X be the true temperature of the liquid. We start by noting the a priori probabilities: $p(\text{water}) = p(\text{ethanol}) = 0.5$, given our lack of further information, and $p(X|\text{water}) = 1/100$ for $273K < X < 373K$, and $p(X|\text{ethanol}) = 1/160$ for $193K < X < 353K$. The likelihood function, given the uncertainty in the instrument, might be modeled as $p(T|X, \text{water or ethanol}) = 0.2$ for T between $X - 2.5$

and $X + 2.5$. In this particular case, we are assuming that the measured temperature uncertainty is independent of the liquid. We first note that

$$p(\text{water}, X) = p(X|\text{water})p(\text{water}) = 0.5 \, p(X|\text{water}), \qquad (4.10)$$

so

$$p(\text{water}, X|T) = p(\text{water}, X)\frac{p(T|\text{water}, X)}{p(T)} = 0.5 \, p(X|\text{water})\frac{p(T|\text{water}, X)}{p(T)}. \qquad (4.11)$$

Now, suppose we measure $T = 271K$. $p(T)$ is obtained by summing over the hypotheses, since its role is to normalize the probabilities:

$$p(T) = 0.5 \int dX \left[p(T|\text{water}, X)p(X|\text{water}) + p(T|\text{ethanol}, X)p(X|\text{ethanol}) \right]. \qquad (4.12)$$

Taking into account the values of X for which the probabilities in the integral are non-vanishing, we find $p(T = 271K) = 0.5(0.5/100 + 5/160) = 0.018125$. It follows that, *marginalizing* over X since we are interested in the classification of the liquid, not the nuisance variable X that we needed to introduce to formulate our hypothesis quantitatively,

$$p(\text{water}|T) = \int dX p(\text{water}, X|T) = \frac{0.0025}{0.01825} = 0.14. \qquad (4.13)$$

Thus, in this example, we find that the *odds ratio* for water is $0.14/0.86 = 0.16$, and the odds ratio for ethanol is $0.86/0.14 = 6.14$. It would seem that the hypothesis that the liquid is ethanol has much better odds than its alternative. The posterior distribution $p(\text{liquid}|T)$ is quite different from the prior distribution $p(\text{liquid})$.

This example has a set of hypotheses labelled by both a discrete variable (water or ethanol) and a continuous variable (the true temperature X), a common circumstance in biological inference where there are structurally different models and continuous rates and concentrations that all need to be part of the set of hypotheses considered. Furthermore, very often in biology we are not as interested in the most likely values of rates and concentrations as we are in finding the probable qualitative structure of the model, even though it isn't possible to formulate the model mathematically without the introduction of numerical rates. This example also shows the importance of averaging over the so-called nuisance variables, *marginalization*.

4.3   Information Theory

Probability and information are intimately related (Welsh, 1988; Cover and Thomas, 1991). If we have an observed variable X, for example the concentration of leptin, which takes the values $x_i, i = 1, \ldots, n$, with probabilities $p(X = x_i) = p_i : \sum_i p_i = 1$, the entropy of X is defined by

$$H(X) = -\sum_i p_i \log_2 p_i. \tag{4.14}$$

The logarithm to the base 2 is a normalization convention and leads to a unit entropy $H(X) = 1$ for a variable X that takes the values 0 or 1 with equal probabilities. The variable X is said to require one *bit* of information to describe it. The entropy is maximized when all the probabilities p_i are equal. An intuitive way of thinking about this maximum is that in such a case, we have no reason to prefer any of the n alternatives over the others. In other words, we are maximally uncertain about the n alternatives, and the entropy measures the uncertainty in our knowledge.

Any specific measurement/observation of a variable is an *event*. The *information* of an event E with non-zero probability is defined as

$$I(E) = -\log_2 p(E). \tag{4.15}$$

If X is an observed variable, each of the values it takes has an associated information $-\log_2 p_i$, so the *mean value of the information* associated with the observations of X $\sum_i p_i I(x_i)$ is in fact the entropy of X. This is the fundamental relation between the entropy and information of observed variables. The intuition for this relation is simply that the obtaining information is simply the removal of uncertainty, both of which are measured by the entropy.

The *conditional entropy* of X given an observation E is defined as

$$H(X|E) = -\sum_i p(X = x_i|E) \log_2 p(X = x_i|E). \tag{4.16}$$

Similarly, if Y is some other measured variable, with associated values $y_j, j = 1, \ldots, m$ the conditional entropy of X given Y is

$$H(X|Y) = \sum_j p(Y = y_j) H(X|Y = y_j). \tag{4.17}$$

What does the conditional entropy measure? Notice that $H(X|X) = 0$, since $p(X = x_i|X = x_j) = \delta_{ij}$. Extending this, $H(X|Y) = 0$ if and only if $X = f(Y)$ for some function f. In words, the conditional entropy vanishes if the observed value of X is completely predicted by the observed value of Y. On the other hand, if X and Y are independent, $H(X|Y) = H(X)$. We also note that the joint entropy of X and Y, $H(X, Y)$ satisfies

$$H(X, Y) \leq H(X) + H(Y). \tag{4.18}$$

In fact, it is not difficult to show that

$$H(X, Y) = H(Y) + H(X|Y), \tag{4.19}$$

showing that the conditional entropy exactly measures the uncertainty remaining in our knowledge of X, given our knowledge of Y.

The *relative entropy*, sometimes called the Kullback-Leibler divergence, of a set of probabilities p_i for a measurement X and another set of probabilities q_i for the same measurement (for example, these could be the prior and posterior probabilities) is defined by

$$D(p|q) = \sum_i p_i \log_2(p_i/q_i). \qquad (4.20)$$

D is always non-negative and only vanishes if $p_i = q_i$ for all i. In terms of information theory, the information about X contained in Y is

$$I(X|Y) = H(X) - H(X|Y) = I(Y|X) = D(p(XY)|p(X)p(Y)), \qquad (4.21)$$

which is symmetric in X and Y. This information is often called the *mutual information* of X and Y. If X and Y are independent, the mutual information vanishes. If the value of X is predicted by the value of Y, $H(X|Y) = 0$, and the mutual information is just the information in X. Mutual information is often used as a similarity measure in expression array (Butte et al., 2000) clustering of genes, but it is not a "distance measure" in the sense that it does not satisfy the triangle inequality

$$d(x,y) + d(y,z) \geq d(x,z), \qquad (4.22)$$

which holds for any three points x, y, z in Euclidean space. This inequality expresses the intuition that the length of any side of a triangle is less than the sum of the lengths of the other two sides. However,

$$m(X,Y) \equiv H(X|Y) + H(Y|X) \qquad (4.23)$$

is symmetric in X and Y and *does* satisfy the triangle inequality: $m(X,Y) + m(Y,Z) \geq m(X,Z)$. If X and Y are independent, $m(X,Y) = H(X) + H(Y)$, and if $X = f(Y)$ and $Y = f^{-1}(X)$, then $m(X,Y) = 0$.

In even a compressed account of information theory it is necessary to mention the connection of entropy with coding theory. Briefly, if we think of compressing our measurements $D = \{d_i, i = 1, \ldots, n\}$ into words made from an alphabet of N symbols, the average length of the words will be at least $H(D)/\log_2 N$. This result makes it possible to estimate the entropy $H(D)$ by finding an encoding of the data, in situations where, due to a lack of data or prior knowledge, we are unable to compute the entropy directly. If we use an encoding in terms of an alphabet consisting of 0 and 1, the average length of the words will be an upper bound on the entropy. Suppose each d_i is the result of an expression array measurement in a particular condition. We discretize d_i into bins defined by our expected uncertainty in the measurements. (The choice of binning can also be incorporated into our description of the information, but we do not do so here for the sake of simplicity.) A partitioning of the G genes into $N(\leq G)$ subsets with $n_\alpha >$

$0(\alpha = 1, \ldots, N)$ elements has a probability $p(N, G) = N!(G - N + 1)!N^{(N-G)}/G!$, so its information is $H_{\text{cluster}}(N, G) = - \log_2 p(N, G)$. (Prior information can be used in this step to reduce this information by taking into account information from the literature on known interactions between genes—the effect of this is to reduce the total number of independent genes G to some smaller number, using prior biological information to place certain genes together in a cluster.) In terms of these putative clusters, we only need N numbers to describe d_i instead of the original G. However, we now have to contend with the inaccuracy of our compression as well, in other words, with the information in the residuals $\epsilon_{i,\text{gene}} = d_{i,\text{gene}} - d_{i,\text{cluster}}$. The original information is

$$H_{\text{orig}} = \sum_{\text{genes}} H_{\text{gene}}(d) \tag{4.24}$$

where we compute $H_{\text{gene}}(d)$ by considering how we could encode the data. If the range of values that the gene takes over the n experiments is m, we need about $(\log_2 m)^n$ bits to encode the values. We also need to encode the information specifying the range of values for each gene, or use the same range for all the genes, and avoid specifying the range for every gene. Taking into account the information required to specify the clustering, the new information is

$$H_{\text{final}}(N) = H_{cluster}(N, G) + \sum_{\text{cluster}} H_{\text{cluster}} + \sum_{\text{genes}-\text{clusters}} H_{\text{gene}}(\epsilon), \tag{4.25}$$

where we also restrict the sum in the residual information to be over the genes that are not cluster centers (where we define the cluster center in a variety of ways, for example as the gene that exhibits the least deviation from the median of the cluster over all the n experiments). The cluster center will, by definition, show no deviation from the value accorded to the cluster. The term H_{cluster} is the term that favors model simplicity. At one extreme, there is a unique clustering of one cluster of G genes, which amounts to the original data expressed as residuals, and at the other extreme G clusters of one gene each, in which the residuals vanish. At both these extremes, the $H_{\text{cluster}}(N = 1, G) = H_{\text{cluster}}(N = G, G) = 0$. We expect a good clustering to reduce the amount of information in the residuals, because many of the entries in the residuals should vanish, and this should be balanced by the amount of information required to specify the clustering. For example, if the residual matrix is a sparse matrix, the coding required to specify it is just the gene name or index, the experiment index i, and the true value for every non-zero entry. This encoding will obviously take a lot less information to describe than the entire matrix, provided that the clustering is an accurate description of the correlations between gene expression values. Evaluating $H_{\text{final}}(N)$ for different values of N (minimizing over different choices of $\{n_\alpha\}$ for each N) gives us a criterion for picking the number of clusters. We can also use this approach to cluster the experiments.

4.4   Another Example: Probabilities Are Not Frequencies

Suppose we have expression data from several samples, normalized to message counts per cell. This is not the normalization commonly used for expression data, but the example will show that this normalization is helpful for certain considerations. The problem is to figure out the message counts in each cell, given the expression measurements. Abstractly posed, the problem is that there are a variety of colored balls in different jars, each jar corresponding to a sample. We have taken a handful of balls from each jar, corresponding to the expression data. We want to find the probable contents of each jar (Jaynes, 2003).

Let's focus on just one color of ball, red. We have drawn n balls out of a jar, and r of them have been red. What is the probability that we would draw r red balls in n tries if the total number of balls in the jar is N and the number of red balls in the jar is R? Since we are drawing the balls without replacement, this probability is easily computed. The probability of the first ball drawn being red is

$$p(r = 1|N, R, n = 1) = \frac{R}{N}, \qquad \text{with } p(r = 0|N, R, n = 1) = \frac{N - R}{N}. \qquad (4.26)$$

The probability of the second ball being red is

$$p(r = 2|N, R, n = 2) = p(r = 1|N, R, n = 1) \times p(r = 1|(N - 1), (R - 1), n = 1), \qquad (4.27)$$

and so on. Therefore, the probability of one of the two balls being red, $p(r = 1|N, R, n = 2)$, is

$$p(r = 1|N, R, n = 1) \times p(r = 0|(N - 1), (R - 1), n = 1) \qquad (4.28)$$

$$+p(r = 0|N, R, n = 1) \times p(r = 1|(N - 1), (R - 1), n = 1) \qquad (4.29)$$

$$= \tfrac{2R(N-R)}{N(N-1)}. \qquad (4.30)$$

A little further calculation shows that

$$p(r|N, R, n) = \binom{N}{n}^{-1} \binom{R}{r} \binom{N - R}{n - r} \qquad (4.31)$$

in general.

Now, N and R are unknown. Having drawn r red balls out of n balls, we know of course that $N \geq n$ and $R \geq r$. According to the rules of probability theory,

$$p(N, R|n, r) = p(N)p(R|N)\frac{p(n, r|N, R)}{p(n, r)}, \qquad (4.32)$$

since $p(N, R) = p(N)p(R|N)$. What is $p(n, r)$? It is a normalization constant given by

$$p(n, r) = \sum_{N=0}^{\infty} \sum_{R=0}^{N} p(N)p(R|N)p(n, r|N, R), \qquad (4.33)$$

where $p(n, r|N, R)$ obviously vanishes if $N < n$ or $R < r$ or $N - R < n - r$. Where is the biology in all this? It is in the probabilities we assign to $p(N)$ and $p(R|N)$ based on our biological knowledge. For example, the samples may be tissues taken from cancer or normal tissue. In this case, we might expect that cancerous cells are proliferating rapidly and may have an overexpression of genes involved in cell division, compared to normal samples. However, this proliferation has a metabolic cost as well, so these cells may also have an overexpression of messages corresponding to, for example, glucose transport. There may, on the other hand, be genes that are expressed at the same level as in normal tissue. Thus, if R corresponded to any of these classes of messages, we would have different prior distributions for $p(N, R)$. We might even choose to use the product rule differently, and factor $p(N, R) = p(R)p(N|R)$, if, for example, we knew the approximate rate of proliferation of the cells, and R was a message encoding for a mitotic spindle protein. The central point is that known biology dictates the choice of prior distributions.

We can also employ this logic in an *exploratory* mode, assuming that we do not know what form $p(N, R)$ should take, and compute $p(N, R|D)$ conditioned on all the data D we have available. Label the different samples with an index $\alpha = 1, 2 \ldots$. Suppose that the red balls correspond to a particular message, and we wish to ascertain if R scales as N^α for some non zero power α or if R is independent of N, based on our samples. For each α, we computed $p(N_\alpha, R_\alpha|n_\alpha, r_\alpha)$ as described in the previous paragraph, using some prior distribution $p_0(N, R)$. Since we do not know what to expect, other than the fact that $R \leq N$, we should choose $p_0(N, R)$ to reflect this ignorance, which is sometimes referred to as choosing a *maximally uninformative prior*. If we do have some biological knowledge to guide our choice, we need to incorporate it into the prior. There is no point in using uninformative priors when information is available. To this end, we can iterate through the samples, computing successively

$$p(N, R|n_2, r_2, n_1, r_1) = p(N, R|n_1, r_1) \frac{p(n_2, r_2|N, R, n_1, r_1)}{p(n_2, r_2|n_1, r_1)}, \qquad (4.34)$$

and so on, until we finally arrive at $p(N, R|D)$ where D stands for the entire data set $\{n_1, r_1, \ldots\}$. We can now compute $H(R|N) = H(N) - H(R, N)$ from $p(N, R|D)$, and answer our question: If $H(R|N) \approx 0$ then R is a function of N, conditioned on the given data D.

It should be noted here that there was no assumption in our considerations that the different samples were repetitions of some experiment. The probabilities that we calculate are not some measure of frequency of occurence in some idealized set of infinite numbers of trial experiments. In general, there is no logical consistency to assuming that probabilities are frequencies. Probabilities are nothing more or less than quantitative expressions of our state of knowledge. For experiments where the results are *exchangeable sequences* (for example, the identical experiment is performed n times), the expectation of the frequency of a particular result is numerically equal to the probability: $E(f_i) = p_i$. So the probability is an estimate of the frequency, but to understand the uncertainty in this estimate, we need to compute the covariance of f_i and f_j, which leads to

$$E(f_i f_j) - E(f_i)E(f_j) = (p_{ij} - p_i p_j) + \frac{1}{n}(\delta_{ij} p_i - p_{ij}). \tag{4.35}$$

Here p_{ij} is the joint probability of outcomes i and j at two different repetitions of the experiment. It is clear then that there is a finite n correction, and a non-zero $p_{ij} - p_i p_j$ correction to the probability assessment of the frequency which does not vanish even for infinite n. For the small numbers of repetitions available due to resource constraints in expression array measurements, for example, it is important to keep the finite-size correction in mind. In the particular case $i = j$ and $p_{ii} = p_i^2$ we get

$$E(f_i^2) - E(f_i)^2 = \frac{1}{n} p_i (1 - p_i). \tag{4.36}$$

If we are, conversely, attempting to assess probabilities by studying observed frequencies, these relations are again relevant.

4.5   Search Theory, or "Use the Information, Stupid"

We have a search space of possible models. We have finite resources available to find a correct model in the search space. How should this search be conducted? (Jaynes, 1985) Suppose we divide the space to be searched into n subsets, with search parameters $m_i, i = 1, \ldots, n$ for each subset. The search parameters measure the fractional "size" of the subset in terms of search difficulty, and satisfy $\sum_i m_i = 1$. For example, the m_i parameters could be the fractional volumes of the subsets: Larger subsets would take longer to search and therefore would have larger m_i values. We also have probabilities p_i assigned to each subset, also adding up to unity. These probabilities are our assessments of the presence of a correct model in a given subset. If a correct model is present in subset i, the probability that a search effort z will lead to finding it is

$$p_i(\text{discovery}|z) = (1 - \exp(-z/m_i)). \tag{4.37}$$

Search effort, which might be computer time or expression level measurements, is limited to be C. If we start with some prior probabilities $p_i^{(0)}$, and expend z_i of our search effort on subset i, the posterior probabilities will be

$$p_i^{(1)} = \frac{p_i^{(0)} \exp(-z_i/m_i)}{\sum_j p_j^{(0)} \exp(-z_j/m_j)} \equiv \frac{p_i^{(0)} \exp(-z_i/m_i)}{1 - p_D}, \tag{4.38}$$

if a correct model is not located, where p_D is the probability of finding a correct model with the given search efforts z_i. Intuitively, this shows that if we search a subset and do not find a model fitting the data in that subset, then the probability for the model being in that subset decreases. In this situation, how should we expend our search effort?

The information we possess about the subsets is measured in two relative entropies, which measure the sizes of the subsets versus the probabilities that a correct model is present in a subset. We define

$$I \equiv \sum_{i=1}^{n} p_i \log(p_i/m_i), \tag{4.39}$$

and

$$J \equiv \sum_{i=1}^{n} m_i \log(m_i/p_i). \tag{4.40}$$

Clearly, $I \geq 0$ and $J \geq 0$, with $I = J = 0$ only when $p_i = m_i$ for all $i = 1, \ldots, n$. As we search, we are expending our search effort z continuously, up to a maximum value C, so we can think of p_i, I, and J as functions of z, starting from initial values $p_i(0), I(0), J(0)$. Why are I and J relevant for search? We calculate

$$J(z) = J(0) + \log(1 - p_D) + z, \qquad z \equiv \sum_i z_i, \tag{4.41}$$

which can be rewritten as

$$p_D = 1 - \exp(-(z + \hat{z})), \qquad \hat{z} \equiv J(0) - J(z). \tag{4.42}$$

Thus, the detection probability p_D decreases if $J(z)$ increases in the course of the search. The best we can do is to make $J(z)$ decrease, but since $J \geq 0$, the optimal strategy is to reach $J = 0$ and to conduct further search so as to maintain $J(z) = 0$. In other words, we need to use up all the information we possess about the subsets by allocating search efforts z_i among the subsets to reach the $J = 0$ state.

Let us suppose that we want our expenditure of search effort to be optimal at all steps. We may not know how much computer time we will have available before an abstract needs to be submitted, for example. How should we allocate our next infinitesimal bit of search effort, δz? Notice that $(p_j/m_j)_{\max} \geq 1$, since $\sum_i p_i = \sum_i m_i = 1$, and $\delta J = (1 - p_j/m_j)\delta z$ if the search effort is expended in subset j. It therefore follows that we should search the subset with the highest value of p_j/m_j at any given step. We order the subsets so that $(p_1/m_1) \geq (p_2/m_2) \geq (p_3/m_3)\ldots$. We search subset 1 until $p_2/m_2 = p_1/m_1$ (note that all p_i are functions of the search effort expended, z). We then treat subsets 1 and 2 as one large subset and search it keeping $p_2/m_2 = p_1/m_1$ until $p_1/m_1 = p_2/m_2 = p_3/m_3$ at which point we treat the subsets 1, 2, and 3 as one big subset and proceed as before. This part of the search process continually decreases J and I and increases p_D until all the ratios p_i/m_i are equal. This is the state of maximum uncertainty characterized by $I = J = 0$. Having used up all our information, the best the rest of the search can do is to maintain this state until all the search effort available has been expended. An interesting point about this strategy is that it can be stopped at any given step, and we can be certain that we have done the best that we could have, given the information that we had available. Such finite-resource optimal strategies are likely to be very important in large-scale biological inference, given computational and experimental limitations.

4.6   Computational Techniques

While systems biology is generally associated with large-scale data collection, when it comes to inference of biological processes as a complex system, the scale of data collection is meager, and computational resources to analyze the data are limited. If our space of hypotheses has more than a few components, the entire set of probabilities $p(X|D)$ cannot be exhaustively computed, since there is a combinatorial explosion in the computational cost. The computational problems can be overcome with variants of Markov chain Monte Carlo (MCMC) methods (Skilling, 1998; D'Agostini, 2003). In general, Markov processes are processes where the next step only depends on the present location, not on the previous history of the process. A Monte Carlo method refers to a stochastic method for evaluating a quantity, for example estimating the value of an integral. An MCMC method marries the two, using a stochastic Markov process to generate new data points for the Monte Carlo estimation of the quantities of interest.

One of the key points is to consider biologically interesting questions. For example: we may want to know the probability that a certain hypothesis x is supported by the data D. In other words, of all the models that we can generate to fit the data with our full set of hypotheses $\mathcal{X}$, we want to ask how likely is it that x is used in the models that fit the data. We construct a function on the space of models, $I_x(Y)$, which takes the value 1 if x is used in the model Y and 0 if x is not. An example of x might be (a quantitative version of) "IKK activates NF-κB translocation to the nucleus." We can now evaluate the expectation E of $I_x(Y)$ over the space of models by computing

$$E(I_x(Y)) \equiv \sum_Y p(Y, D) I_x(Y). \tag{4.43}$$

This is, typically, a huge (or infinite) summation, and impossible to compute exactly. The trick is to approximate this summation using MCMC methods.

The Metropolis algorithm is a particular implementation of MCMC computations: We start from some initial model Y_0, and compute $p(Y_0|D)$ and $I_x(Y_0)$. We modify the model by changing the hypotheses incorporated or by changing the rate constants or kinetic parameters, generating a new model Y_1, for which we also compute $p(Y_1|D)$. (Since $p(Y, D) = p(D)p(Y|D)$, we are free to neglect the constant factor $p(D)$.) If $p(Y_1|D) \geq p(Y_0|D)$, we accept the change and add $I_x(Y_1)$ to our previous computation of $I_x(Y_0)$. However, if $p(Y_1|D) < p(Y_0|D)$ we compute a random number r between 0 and 1 and accept Y_1 only if $p(Y_1|D)/p(Y_0|D) \geq r$. If we do not accept Y_1, we accept Y_0 again and try a new change in the model, and repeat the process. After n accepted models, we compute an approximation to $E(I_x(Y))$:

$$E(I_x(Y)) \approx \frac{1}{n} \sum_{i=0}^{n-1} I_x(Y_i). \qquad (4.44)$$

The convergence of this approximation to the exact value scales as $n^{-1/2}$, which is slow but not impossible to compute. In the event that the probability of generating Y_1 from Y_0 is not symmetric, in other words, the probability of generating Y_1 from Y_0 is different from that for generating Y_0 from Y_1, a *Hastings ratio* $p(Y_1 \to Y_0)/p(Y_0 \to Y_1)$ multiplies the ratio $p(Y_1|D)/p(Y_0|D)$. This factor is significant in Markov chains on spaces of models.

Another approach to MCMC computations is to use *genetic algorithms*. The interesting point about these algorithms for the purposes of model selection is that they are better suited to multimodal problems and problems with discrete variables (common in testing collections of hypotheses, for example). The main point is to start with a family of P sample models and generate new models by two means: mutations (changes in a single model) and crossovers (exchanging hypotheses between two distinct models). Other means of generating new models can be used as well, as long as the method is reversible. For example, changing one of the models in the population based on the difference between two other models in the population is a possible way to exploit the population as a whole, and not just one or a pair of the models in the population. It is particularly useful for biological applications if the allowed mutation and crossover transformations are actually biologically feasible alternative mechanisms for implementing a given biological function. Having implemented the genetic algorithm, it remains to explain how this fits in with the expectation computation of interest. For this we just have to go back to the Metropolis algorithm, described above, and think of each genetic algorithm step as a step on the P-fold product of our space of models. We now apply the same acceptance or rejection criterion to each step, except that the probabilities that we use are computed as the product $\prod_{\alpha=0}^{P-1} p(Y_\alpha|D)$. The expectation is computed by picking a random selection Y_i out of the P models in the population at each step, and again using

$$E(I_x(Y)) \approx \frac{1}{n} \sum_{i=0}^{n-1} I_x(Y_i). \qquad (4.45)$$

Asymptotically, this will again converge to the true expectation.

In this way, by computing expectations, we can assign plausibilities to our set of hypothetical interactions $\mathcal{X}$. A note of caution: Taking the most plausible hypotheses in $\mathcal{X}$ and putting them together in a model does not necessarily result in a model that has a high probability $p(Y|D)$ since there may be correlations or anticorrelations between hypotheses. In other words, there may be alternative explanations for a phenomenon which may be antagonistic. To figure out which hypothetical interactions combine well to match the data, we could, for example, compute the pair-wise expectations $E(I_x(Y)I_z(Y))$ where x and z are hypotheses in $\mathcal{X}$. If we pick a threshold for the mutual information t, these pair-wise expectations lead to a graph where the hypotheses are vertices and the edges are links between hypotheses with mutual information $I(x|y) \equiv H(x) + H(y) - H(x,y) > t$ (using $H(x) = -E(I_x(Y)) \log_2 E(I_x(Y)) - (1 - E(I_x(Y))) \log_2 (1 - E(I_x(Y)))$ and similarly for $I_z(Y)$ and $I_x(Y)I_z(Y)$, up to a finite sample size correction). The connected components of this graph (or a subset of them) would be an interesting starting point for more detailed model inference.

An important technical point to speed up the computations is to use *simulated annealing.* In this numerical method, based on analogies with statistical mechanics, we replace $p(Y, D) = p(Y)p(D|Y)$ with $p(Y)p(D|Y)^\lambda$ where $0 < \lambda \leq 1$. $1/\lambda$ plays the role of temperature, so when we approach $\lambda = 1$ from higher values of the temperature (lower values of λ), gradually the peaks and valleys in $\log p(D|Y)$ get more pronounced, and therefore make it less likely that a step that would decrease the probability is accepted. Thus, at small values of λ it is easier for the MCMC update algorithm to find acceptable steps resulting in a wider coverage of the population of models. As λ is brought closer to the true value 1, the MCMC update steps will stay in the vicinity of the optimal model found at lower values of λ. It is the process of reaching the optimal model that is shortened by the cooling-down phase of simulated annealing. Care must be taken in the multi-modal case to find the expectations around each locally modal value. This is usually accomplished by repeating the calculation with different starting points.

4.7 Three Applications

There is a wide range of applications of probabilistic inference in systems biology, indicative of the universality and flexibility of the methodology expounded in this chapter. In this section, we review briefly three examples from the literature:

1. The use of multiple types of experimental data to organize genes in modules (Lee et al., 2004)

2. Model selection on the basis of a Bayesian comparison of models (Sachs et al., 2005)

3. Studying the sensitivity and specificity of Bayesian inference of genetic regulatory interactions (Husmeier, 2003)

A specific application of probabilistic techniques that is used in two of these examples is the concept of a *Bayesian network*: If we have a set of measured quantities, and the probabilities of the values observed of some of the quantities are conditional on the observed values of some of the other quantities, we can draw a graph of dependencies, with the measured quantities represented as nodes and directed arrows going into every measured quantity from the measured quantities upon which the probabilities of its values are conditional. An example of a Bayesian network is

$$
\begin{array}{ccc}
 & A & \\
\swarrow & & \searrow \\
B & & C \\
\searrow & & \swarrow \\
 & D &
\end{array}
\qquad (4.46)
$$

while

$$
\begin{array}{ccc}
 & A & \\
\swarrow & & \nwarrow \\
B & & C \\
\searrow & & \nearrow \\
 & D &
\end{array}
\qquad (4.47)
$$

is *not* a Bayesian network. The *parents* of a node are the tail ends of the arrows pointing to that node. If the graph obtained in this manner has no directed cycles (in other words, no closed loops with arrows all connected head to tail), then it is called a Bayesian network. This special case is computationally much more tractable than the general case (usually referred to as a *graphical model*). The calculational tractability arises from the fact that variables can be easily marginalized in Bayesian networks, since the probability distribution factorizes:

$$
p(X_1, X_2 \ldots X_n) = \prod_{i=1}^{n} p(X_i | \text{parents of } X_i). \qquad (4.48)
$$

Feedback loops in a graphical representation of dependencies between different quantities correspond to networks that are not Bayesian networks, according to the definition above. The general Bayesian logic expounded in previous sections is still applicable, of course, but the analytical simplifications that go along with the factoring of the probability distribution do not hold. A better way to understand the consequences of feedback loops, in any event, is to think of the probabilities in a time dependent context, which amounts to taking a particular graphical representation and unfolding the arrows in a new time direction. As an example, the graph of probabilistic dependence

$$
\begin{array}{ccc}
A & \rightarrow & B \\
\searrow & & \swarrow \\
 & C &
\end{array}
\qquad (4.49)
$$

unfolds to

$$
\begin{array}{ccc}
\downarrow & \downarrow & \downarrow \\
A_t & B_t & C_t \\
\downarrow & \downarrow & \downarrow \\
B_{t+1} & C_{t+1} & A_{t+1} \\
\downarrow & \downarrow & \downarrow \\
C_{t+2} & A_{t+2} & B_{t+2} \\
\downarrow & \downarrow & \downarrow
\end{array}
\tag{4.50}
$$

Thus variables gain an additional time label and the arrows point from variables at one time-slice to the next. The resulting graph is certainly acyclic, and is therefore a Bayesian network. Such unfolded networks are referred to as *dynamic Bayesian networks*.

Lee et al. (2004) considered several different sources for deriving gene-gene interaction information: mRNA coexpression across microarrays, gene fusions, phylogenetic profiles, co-citations, and protein interaction experiments. They calibrated the likelihood that any given one of these sources was reliable by picking KEGG pathway database annotations and computing the ratio of the frequency with which the source linkage operated in the same pathway as the KEGG annotation to the frequency with which the source linkage operated in different KEGG pathways. They normalized this ratio by picking random pairs of genes and computing the ratio of the frequency with which the genes operated operated in the same KEGG pathway to the frequency with which the pair of genes operated in different KEGG pathways. They use the logarithm of this normalized ratio as a score for the accuracy of the source. In the context of this chapter, their likelihoods for the accuracy of any given source of information was determined by their data for that source, conditioned on the KEGG database. They then used these probabilities to score gene linkages that were not in the KEGG database, but were predicted by the source. Since the probabilities for each source were independently obtained, they could produce a cumulative log likelihood score for each linkage by adding up each individual score. Thus the framework of probability theory allowed an integrated use of all available data to predict the reliability of a given gene-gene interaction linkage, placed on a common scoring basis.

Learning the probability distributions $p(X_i|\text{parents of } X_i)$ for a given Bayesian network is a major part of determining the probability that the network is a likely description of the data. These probability distributions are specific to each hypothesized model since the dependencies between the entities in the network may differ between models. kSachs et al. (2005) applied this procedure to infer the most likely protein signaling network from multi-parameter flow cytometry data, emphasizing the importance of data in the presence of different perturbations in network inference. The role of a given perturbation is to fix the measured values of certain variables in the network, and therefore constrain the possible dependencies in the set of probability distributions.

The requirements for successful reverse engineering of genetic regulatory networks using dynamic Bayesian networks are considered in Husmeier (2003), who shows the importance both of known biology in the form of possible interactions and of time series data in disequilibirum after a perturbation as the system relaxes in inferring the network structure. This work also shows the promise (and limitations) of MCMC methods in inferring local structures of the genetic network. Rather than try to infer a single most likely network, the importance of marginalizing over models is apparent in these results, since the posterior probability distribution on the space of models is diffuse for sparse data sets.

4.8 Summary

The logical analysis of biological data has many advantages, indicated in the introduction. It has only one "disadvantage": Known biology must be incorporated in the analysis from the beginning, and thought must be expended on the translation of this knowledge into quantitative models. The consistency, optimality and uniqueness properties of logical inference imply that one cannot "do better" in extracting knowledge from new data.

The key steps are:

(A) Encode known knowledge into prior probabilities for models that are plausible explanations for the new data.

(B) Compute the likelihood of the new data for these models.

(C) Compute the posterior probabilities for the models using the likelihoods (B) and the prior probabilities (A).

(D) Examine the likely models using these posterior probabilities and ask what experiment would differentiate best between these likely models.

(E) With new data (D), go to step (A) with the posterior probabilities now serving as the prior probabilities.

This general scheme of inference applies to sequence analysis on one end to reverse engineering on the other end with no change. A consistent application of the simple rules of probability theory is all that is needed.

5 Stoichiometric and Constraint-based Modeling

Steffen Klamt and Jörg Stelling

A major current challenge in systems biology is to clarify the relationship between structure, function, and regulation in complex networks that can be reconstructed from genomic or biochemical data. However, dynamic mathematical modeling of large-scale networks meets difficulties as the necessary mechanistic detail and kinetic parameters are rarely available. In contrast, structural (topological) analyses require only reaction stoichiometries and reversibilities, which are often well-known. This chapter introduces the main concepts of stoichiometric network analysis, a special class of structural analysis methods. We emphasize practical applications for obtaining a system-wide understanding of metabolic networks, including functional and regulatory aspects. In particular, we aim at providing a critical evaluation of the different theoretical approaches available regarding their prerequisites, predictive power, and inherent differences. This approach should finally enable the audience to make critical judgments on the applicability of stoichiometric network analysis for their special problems in systems biology.

5.1 Overview and Applications

One of the most important challenges in systems biology is to understand the functionality of cellular networks that can be reconstructed from genomic and biochemical data for a wide variety of organisms. Current theories have different strengths and shortcomings in providing an integrated, predictive description of complex networks. For dynamic mathematical modeling of large–scale systems, often the necessary mechanistic detail and kinetic parameters are not available. In contrast, structure-oriented analyses only require the usually well-characterized network topology. Graph theory uses the scheme of network connectivities, which is a simplified representation of real reaction networks (see chapter 7). Here we introduce a class of analysis methods that consider network stoichiometry explicitly and potentially other constraints such as maximal pathway capacities as well. These approaches can be subsumed under the term *stoichiometric network analysis* (SNA) (Heinrich and Schuster, 1996; Simpson et al., 1999).

Stoichiometric modeling has become a particularly important approach for understanding the function of metabolic networks. Hence, we focus on metabolism, and we discuss extensions to cellular regulation. One aim of this chapter is to critically review virtues and limitations of the approaches with respect to their potential applications for realistic biological networks. For this comparison of altogether four major approaches to stoichiometric network analysis, we will address the following issues:

- **Network consistency:** Blocked reactions and missing network elements can compromise the validity of reconstructed networks. They should be detectable by analytical methods.

- **Functional pathways and cycles:** Pathways should be sets of connected reactions, but establishing a theoretically sound notion of "meaningful" pathways is difficult. Pathway analysis may suggest new hypothetical routes between specific inputs and outputs that only emerge in the context of a complex network. Identifying "futile cycles" that involve only a net consumption of energy can help to recognize potential energy-wasting routes. Cycles without any net energy consumption point to thermodynamic inconsistencies.

- **Network capabilities:** The evaluation of, for instance, maximal product yields in terms of the moles of product generated per mole of substrate has clear relevance for biotechnological applications. Stoichiometrically derived yields may give indicators of the maximal efficiency of engineered organisms. The identification of alternative optimal pathways, or of sub-optimal pathways can, however, be of equal importance with regard to the feasibility of genetic engineering approaches.

- **Importance of reactions:** A prominent application of network analysis is to determine the importance of single reactions for the overall systems performance, in particular, by studying knockout mutations. Predicting the effects of enzyme deficiencies that cause human diseases is of clear medical relevance. Estimates of the relative importance of a reaction may differentiate between essential and nonessential genes under specific (environmental) conditions.

- **Correlated reactions:** Reactions that always have to operate together are likely to be coregulated. This applies to many unbranched linear pathways in biosynthesis. Reactions that never appear together point to differential regulation, for instance, to establish qualitatively different network operation modes depending on the environmental conditions. Hence, such groups of reactions help to understand, and possibly predict, features of regulatory networks.

- **Network design:** Studying of the effects of adding reactions to or deleting reactions from a given network is closely related to analyzing the importance of single reactions. In addition, it can unravel how (additional) constraints on reaction reversibilities influence the set of possible pathways in the network. Assessing the effect of newly introduced genes with respect to functional capabilities and potential, unanticipated side effects *in silico* could help to identify targets for the addition or removal of genes *in vivo*.

- **Network flexibility and robustness:** Robustness is generally defined as the (relative) insensitivity of a system to changes in its parameters (Csete and Doyle, 2002) (see chapter 2). Flexibility means the capacity to switch between different functional modes. Here we regard both concepts as equivalent because a metabolic system should tolerate changes in its set of enzymes once it provides for alternative pathways when a specific reaction is not functional. Hence, one has to investigate the set of all possible behaviors of a system.

5.2 Stoichiometric Networks

Before we turn to theoretical approaches for stoichiometric network analysis (SNA), we first need the introduction of some (formal) terms related to the stoichiometry and structure of biochemical reaction networks. A biochemical reaction is usually characterized by the following properties:

- **Stoichiometry:** The stoichiometry specifies the reactants (educts or products) participating in a reaction as well as the molar ratios in which they are consumed or produced. The stoichiometric coefficient of a metabolite, by convention, is positive if it is produced when the reaction proceeds in its forward direction, and negative otherwise.

- **Reaction directionality:** In principle, all chemical reactions are thermodynamically reversible. Certain reactions in biochemical networks, however, can be considered to be practically irreversible because they (nearly) exclusively proceed in one direction. Examples include the irreversible fixation of carbon dioxide by the most abundant enzyme in nature, namely *Rubisco*. Knowledge on the reversibility of reactions, as will be seen in the next sections, allows to constrain the number of possible pathways in a network, since pathways that would involve reactions proceeding in the "wrong" direction can be excluded from the analysis.

- **Catalyzing enzyme:** Many biochemical, in particular metabolic, reactions are characterized by the participation of an enzyme that facilitates or even enables a reaction to proceed. The connections between reactions and enzymes do not have to be unique, because several enzymes (isoenzymes) may catalyze the same reaction, whereas multifunctional enzymes have the ability to catalyze different reactions. Specification of the catalyzing enzyme, however, allows one to directly relate structural network properties to features of the genome encoding those enzymes.

- **Reaction kinetics:** Reaction kinetics describe the dynamics of the reaction based on the reaction mechanism and the enzyme properties. In many cases, these characteristics of a reaction are unknown. However, the metabolism is characterized by usually fast reactions and high turnover of substances when compared to regulatory events. Then, at least for certain modeling aspects, dynamics may be neglected (see below).

In structural analyses of biochemical networks, only the first three properties are considered. A formal description of the structure and stoichiometry of a reaction network can be given as follows:

- m: number of (internal) species

- q: number of reactions; if desired, the catalyzing enzyme(s) and the corresponding gene(s) can be assigned to each reaction.

- $\mathbf{N}$: $q \times m$ stoichiometric matrix—each row corresponds to one species and each column to one of the reactions; the matrix element n_{ij} represents the stoichiometric coefficients of species i in reaction j.

- **rev**: the set of the reversible reactions

- **irrev**: the set of the irreversible reactions ($\mathbf{rev} \cap \mathbf{irrev} = \emptyset$)

The structure of any reaction network can be captured by this formalism. In the following we will focus on metabolic networks because stoichiometric network analysis is especially suited for them. Note that in metabolic networks, biomass synthesis may be considered as a pseudo reaction whose (cumulative) stoichiometry can accordingly be collected in one of the columns of $\mathbf{N}$.

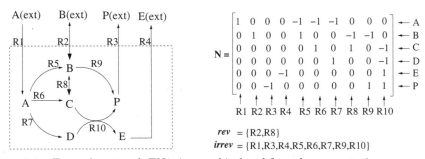

Figure 5.1   Example network EN1: its graphical and formal representation.

Fig. 5.1 shows the map and the corresponding variables of a simple example network, called EN1 throughout the paper. This network comprises 6 (internal) metabolites and 10 reactions, two of which are reversible. Characteristics of any network are its boundaries and its connections to "the rest of the world." Related to this issue is the notion of *internal* and *external* species. *Internal* species are those which are explicitly considered in the network model (and, hence, in $\mathbf{N}$). In contrast, *external* species are thought to be sinks or sources (Heinrich and Schuster, 1996), which can lie physically outside the system (for example, substrates or products as the four external compounds in figure 5.1), but might also be located inside the cell. If, for example, metabolite E in figure 5.1 represented a metabolite in great excess, such as water, we would probably not consider this metabolite as part of our model and neglect it as is done in many stoichiometric (and also dynamical) studies. External metabolites are the reason for why the stoichiometric coefficients of a reaction in $\mathbf{N}$ may have only positive (R1, R2 in EN1) or only negative (R3, R4) signs.

The stoichiometric matrix $\mathbf{N}$ is fundamental, not only for SNA. First, the underlying graph of a given reaction network—needed for graph-theoretical studies (see chapter 7)—can easily be derived from the stoichiometric matrix. Furthermore, $\mathbf{N}$ is essential also for dynamic modeling of metabolic processes. The changes of the species concentrations over time can be described by a system of differential equations (see chapter 6 for details of the approach) as follows (Heinrich and Schuster, 1996):

$$\frac{d\mathbf{c}(t)}{dt} = \mathbf{N} \cdot \mathbf{r}(t) \tag{5.1}$$

The $m \times 1$ - vector $\mathbf{c}(t)$ represents the current metabolite concentrations and the $q \times 1$ - vector $\mathbf{r}(t)$ represents a *flux distribution* in the network, that is, it contains the q reaction rates. Vector $\mathbf{r}(t)$ is given by a—often approximated—function of the current metabolite concentrations and of many—often unknown or uncertain—parameters (contained in vector $\mathbf{p}$), that is,

$$\mathbf{r}(t) = \mathbf{f}(\mathbf{c}(t), \mathbf{p}, t) . \tag{5.2}$$

Hence, as already mentioned above, the uncertainties in describing a metabolic system dynamically lie within the kinetic description of the reaction rates. However, the other part of equation 5.1 is given by $\mathbf{N}$, which in most cases is well-known and represents an *invariant* of the system. $\mathbf{N}$ is invariant against time, kinetics, and concentrations (although, under certain conditions, only subnetworks, that is, submatrices of $\mathbf{N}$ may be active). $\mathbf{N}$ describes the structural relationships between the network components which are of eminent importance for the overall function and behavior of the network. Therefore, results obtained by stoichiometric network analysis do often have direct implications also for the dynamic behavior. Of course, since equation 5.2 is practically neglected, only some of the major characteristics of a metabolic system can be extracted by SNA.

5.3 Conservation Relations

Conservation relations (CR) characterize weighted sums of metabolite concentrations which remain constant in the system. Here, concentrations are denoted by brackets. A typical example occurring frequently in studies on metabolic networks is [NADH] + [NAD] = S = *CONST*. When one of these cosubstrates is consumed, then the other is produced, keeping the sum of both concentrations constant. For

NADH and NAD, this is reflected by the phenomenon that the corresponding row of NAD in the stoichiometric matrix $\mathbf{N}$ is exactly the same as for NADH, except that it is multiplied by –1. This actually means that the rows are *linearly dependent*.

It is a general property of any conservation relation $\mathbf{y}$ that it represents a combination of rows (species) of $\mathbf{N}$ that are linearly dependent (Heinrich and Schuster, 1996). Linear combinations of rows of $\mathbf{N}$ can be represented by $\mathbf{N}^T\mathbf{y}$ ($\mathbf{N}^T$ = transpose of $\mathbf{N}$), and finding linearly dependent rows means that $\mathbf{y}$ must fulfill:

$$\mathbf{N}^T\mathbf{y} = \mathbf{0} \qquad (5.3)$$

($\mathbf{0}$ is the $m \times 1$ zero vector). This means that a CR $\mathbf{y}$ must lie in the *null-space* of the transpose of $\mathbf{N}$. One also says that $\mathbf{y}$ lies in the *left null-space* of $\mathbf{N}$ (Strang, 1980), since equation 5.3 is equivalent to $\mathbf{y}^T\mathbf{N} = \mathbf{0}^T$. The dimension of the left null-space is m-rank($\mathbf{N}$), that is, conservation relations only exist if rank($\mathbf{N}$)$< m$. Then, m-rank($\mathbf{N}$) linearly independent CRs can be found, which can be arranged in a matrix $\mathbf{Y}$. For terms related to a null-space see also section 5.4, where the null-space of $\mathbf{N}$ is analyzed.

Network EN1 (figure 5.1) does not contain any CR since rank($\mathbf{N}$)=m=6. A simple example would be a network that contains the four metabolites A,B,C,D and only one reaction: A + B → C + D. In this case,

$$\mathbf{N} = \begin{pmatrix} -1 \\ -1 \\ 1 \\ 1 \end{pmatrix} \qquad (5.4)$$

and, hence, three linearly independent CRs exist (because m-rank($\mathbf{N}$) $= 4-1 = 3$). They can be found by searching for linearly independent solutions $\mathbf{y}$ for

$$\mathbf{N}^T\mathbf{y} = [-1 \ -1 \ 1 \ 1]\,\mathbf{y} = 0 \qquad (5.5)$$

Three selected independent solutions for $\mathbf{y}$ are arranged as columns in the matrix $\mathbf{Y}$:

$$\mathbf{Y} = \begin{pmatrix} 1 & 1 & 0 \\ -1 & 0 & 1 \\ 0 & 1 & 0 \\ 0 & 0 & 1 \end{pmatrix} \qquad (5.6)$$

In the order of the columns this means (i) [A] – [B] = S1 = CONST.; (ii) [A] + [C] = S2 = CONST.; (iii) [B] + [D] = S3 = CONST. Furthermore, each linear combination of these CRs is also a CR, e.g. (i) + (ii) = 2 [A] – [B] + [C] = CONST.

Identifying the CRs is a simple task, but brings important benefits. First, CRs are helpful for detecting conserved moieties by searching only for those CRs that are composed of a positive sum of metabolite concentrations (Heinrich and Schuster, 1996; Cornish-Bowden and Hofmeyr, 2002). Algorithms for this task (similar to those for computing elementary modes, see section 5.6.3) exist (Heinrich and Schuster, 1996). Secondly, CRs are a nice example for how stoichiometric relations affect systems dynamics: CRs *shrink the possible dynamic behavior* (equation 5.1) of a given network. If at a given time point, say at the beginning of the simulation or experiment, the value of a CR is known, then it will be constant for all the time. In our small example above, this would mean that if [A] - [B] is 6 at any time point then there will never be a state of the system where the difference between [A] and [B] will be unequal to 6. For this reason, CRs express redundancies with respect to the considered states of the systems. It is therefore possible to remove m-rank($\mathbf{N}$) states from the set of (modeled) system variables without losing information. In our example above, we might, thus, remove the three metabolites B, C, D and model only A explicitly. Using the CRs and the initial concentrations, we can then derive the concentrations of B, C, and D from the current concentration of A at any time point (Heinrich and Schuster, 1996; Reder, 1986).

5.4 Balanced Networks: The Quasi Steady State Assumption

5.4.1 Metabolite Balancing Equation and Null-Space of N

Metabolism usually involves fast reactions and high turnover of substances when compared to regulatory events. Therefore, analysis of metabolic networks is often based on the assumption that, on longer time scales, metabolite concentrations and reaction rates are constant. Applying this *quasi (pseudo) steady state assumption* to equation 5.1 leads to the fundamental *metabolite balancing equation* (Heinrich and Schuster, 1996)

$$0 = \mathbf{Nr} . \tag{5.7}$$

This homogeneous system of linear equations demands that the production (sum of positive fluxes) and consumption (sum of negative fluxes) of a metabolite must be equal, similar to Kirchhoff's first law for electric circuits. As we will see in section 5.5, the metabolite balancing equation is *the* main constraint in constraint-based modeling. Note that in oscillating systems, where the metabolite concentrations are not constant (Wolf et al., 2000), equation 5.7 is fulfilled at least for the averaged reaction rates.

The trivial solution $\mathbf{r} = \mathbf{0}$ always fulfills equation 5.7. However, this would represent thermodynamic equilibrium. We are, for obvious reasons, only interested in other solutions, and the cell should (must) have degrees of freedom. Indeed, as the number of reactions q in real networks mostly is much larger than the number m of internal metabolites, an infinite number of flux distributions $\mathbf{r}$ usually complies with the system of equations (5.7). From linear algebra, it is known that all possible solutions are contained in a vector space called the null-space (or kernel) of $\mathbf{N}$ (cf. with left null-space in equation 5.3) (Strang, 1980). The dimension of the null-space is q-rank$(\mathbf{N})$, which equals the number of linearly independent solutions for equation 5.7. A set of $q - \text{rank}(\mathbf{N})$ linearly independent solutions can easily be found and is arranged in a *kernel matrix* $\mathbf{K}$. Then, all flux distributions $\mathbf{r}$ fulfilling equation 5.7, that is, which lie in the null-space of $\mathbf{N}$, can be constructed by a linear combination $\mathbf{b}$ of the columns of $\mathbf{K}$:

$$\mathbf{r} = \mathbf{Kb} \; . \tag{5.8}$$

For illustration, figure 5.2 shows the map and formal representation of a very simple network called EN2. The null-space has dimension q-rank$(\mathbf{N})$=4-2=2.

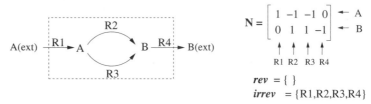

Figure 5.2   Example network EN2.

A kernel matrix for this system, which accordingly must have two columns, reads:

$$\mathbf{K} = \begin{pmatrix} 1 & 0 \\ 0 & -1 \\ 1 & 1 \\ 1 & 0 \end{pmatrix} \tag{5.9}$$

A special balanced flux distribution in this network is $\mathbf{r} = (2, 1, 1, 2)^T$ which can be constructed from $\mathbf{K}$ by using $\mathbf{b} = (2, -1)^T$ in equation 5.9.

Note that the kernel matrix is, in general, not unique. For example, we may substitute one of the columns of $\mathbf{K}$ in equation 5.9 by the vector $\mathbf{r}$ given above. Therefore, usually not all qualitatively different flux distribution in the network are captured. An even more problematic point for analyzing the null-space by the kernel matrix is that neither sign (reversibility) nor other capacity restrictions of the reactions are considered. For example, as all reactions in EN2 are irreversible, the second column of $\mathbf{K}$ is not a valid flux distribution in this network because for R2 a negative sign occurs. It can even happen that a null-space has many dimensions (that is, many columns in $\mathbf{K}$), although no other steady state flux than the trivial one is feasible in the network. Hence, the "real" degrees of freedom (possibilities for distributing metabolic fluxes) can only roughly be estimated from the dimension of $\mathbf{K}$. These shortcomings are overcome by constraint-based approaches (section 5.5). Some important steady state properties of the system can, nevertheless, be derived from $\mathbf{K}$.

5.4.2 Analysis of the Kernel Matrix

It may happen that a reaction can only have a zero rate in steady state. This applies, but is not restricted to, whenever an internal "dead-end" metabolite participates only in this reaction. If this reaction carried a non-zero flux, then this metabolite could not be in steady state. Because of equation 5.8, many blocked reactions (BRs, also called "strictly detailed balanced reactions" (Heinrich and Schuster, 1996)) can easily be identified from (any) $\mathbf{K}$ if their corresponding row in $\mathbf{K}$ is a zero row. Checking a network for BRs is especially useful in reconstructed networks, since BRs can hardly perform any function and therefore often indicate missing network elements. For any further network analysis involving the steady state assumption (sections 5.5–5.6), they can for practical reasons be removed.

An enzyme subset (ES), or coupled/correlated reaction set, is a set of reactions that must always operate together with a fixed ratio in their rates (Pfeiffer et al., 2001). Typical examples are reactions in a linear pathway, such as {R4,R7,R10} in EN1 (figure 5.2). The rates of these reactions will be equal in *any* steady state flux distribution. Consequently, if one reaction is removed from the network (for example, by a gene deletion), then the others cannot work properly and their flux will be zero in steady state. Since the reactions of an ES are structurally so strongly coupled, they are often commonly regulated (Schuster et al., 2002c). ESs are not restricted to linear pathways as shown by EN2 (figure 5.2), where {R1,R4} is the only ES. ESs can be verified by the null space matrix because the corresponding rows in $\mathbf{K}$ of two reactions of the same ES can only differ by a (scalar) factor. In equation 5.2, the factor for the corresponding rows for R1 and R4 in $\mathbf{K}$ (first and fourth row, respectively) is even unity, which means that the reactions operate always with the same stationary rate.

Other important conclusions can be drawn if $\mathbf{K}$ *is block-diagonisable*. Then, certain sub-networks can be identified in the system that are either completely disconnected or whose steady state fluxes are independent from the fluxes in the rest of the network (Heinrich and Schuster, 1996).

5.4.3 Metabolic Flux Analysis

By applying metabolic flux analysis (MFA), one tries to shrink the possible solution space of equation 5.7 by measuring some of the reaction rates (such as uptake or excretion rates) in a certain steady state experiment (Stephanopoulos et al., 1998). Ideally, one unique solution (a point in the null space of $\mathbf{N}$) remains for the actual flux distribution in the respective experiment. The procedure is straightforward: one divides equation 5.7 into the measured (index m) and unknown part (u), possibly after rearranging the columns in $\mathbf{N}$ and components in $\mathbf{r}$:

$$0 = \mathbf{N}\mathbf{r} = \mathbf{N}_u\mathbf{r}_u + \mathbf{N}_m\mathbf{r}_m \quad \Rightarrow \mathbf{N}_u\mathbf{r}_u = -\mathbf{N}_m\mathbf{r}_m \; . \tag{5.10}$$

The right part of equation 5.10 is the central equation for MFA and characterizes a flux scenario. The ideal case with only one unique and exact solution occurs if N_u is a square matrix and invertible because then all unknown rates in r_u can be determined. However, in general, from the rank of N_u, a scenario can be classified with respect to determinacy (determined or underdetermined) and redundancy (redundant or non-redundant). If a scenario is underdetermined, then only some or even none of the unknown rates can be determined. In redundant systems, a consistency check can be performed, which is useful for detecting gross measurement or modeling errors. The basic techniques for MFA are extensively described in Stephanopoulos et al. (1998); van der Heijden et al. (1994); and Klamt et al. (2002). In larger networks, despite a number of measurements, many or all rates in the system often remain completely unobservable. Then, only isotopic tracer experiments may deliver further constraints (Wiechert, 2001).

To give a small example, we assume that in EN1 we measured the rates R1=R3=2 and R4=1 (figure 5.3). We could then calculate R2=R7=R9=R10=1. The other three rates remain unknown.

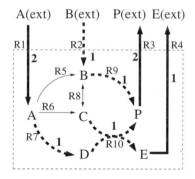

Figure 5.3   Example for metabolic flux analysis: stationary rates of R1, R3, and R4 are measured (bold arrows). Using this information, one can determine the fluxes of R2, R7, R9, and R10 (dashed arrows). The other rates remain unknown (thin arrows).

In general, MFA is useful for analyzing specific flux distributions, but it is not able to characterize the complete admissible steady state solution space.

5.5   Constraint-Based Modeling

5.5.1   Principles of Constraint-Based Modeling

In the previous section we introduced the metabolite balancing equation which resulted from the assumption of quasi steady state. As a consequence of this constraint, the space of possible flux distributions in a reaction network reduces from "everything is possible" to the null-space of N. The basic idea of the constraint-based approach, mainly developed by B.O. Palsson and colleagues, is to incorporate further well-defined physicochemical and biological constraints that limit the network overall behavior *with respect to the possible flux patterns* (Varma and Palsson, 1993; Reed and Palsson, 2003; Price et al., 2003, 2004). As a result, the solution space, encompassing all flux distributions satisfying the imposed constraints, shrinks. Different types of constraints can be involved and they can all be expressed by linear equations or inequalities:

C1) Quasi steady state: $0 = N r$

C2) Capacity/Reversibility: $\alpha_i \leq r_i \leq \beta_i$

For all irreversible reactions one usually sets $\alpha_i = 0$ in C2. Flux capacity constraints are often known for exchange (uptake/excretion) reactions. If capacity constraints, for internal reactions normally given by the v_{max} value of the enzyme, are unknown then the boundary values of the reaction rates are set to $\pm\infty$. C2 can be simplified to a pure reversibility constraint when no capacity values are known/considered:

C2') Reversibility: $r_i \geq 0$ (for all irreversible reactions i)

C3) Measurements: $r_i = m_i$ (for measured/known rates i)

C4) Optimality: $s^T r = s_1 r_1 + s_2 r_2 + ... + s_q r_q = Max!$

Note that null-space and metabolic flux analysis can be seen as special constraint-based methods which take into account the constraints C1 (+partially C2) and C1+C3, respectively.

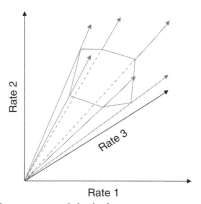

Figure 5.4 Example of a convex polyhedral cone.

Constraints C1 and C2' are in practice often well-known in a given network. The set F of all flux vectors r obeying these constraints

$$F = \{r \in R^q : 0 = Nr, \ r_i \geq 0 \ \forall \ i \in \mathbf{irrev}\} \tag{5.11}$$

represents, mathematically, a *convex polyhedral cone* (Rockafellar, 1970; Bertsimas and Tsitsiklis, 1997). In stoichiometric studies, it is often referred to as *flux cone*. According to C1 and C2', this cone is an intersection of the null-space with the positive halfspaces of the irreversible reactions. An example of a three-dimensional polyhedral cone is given in figure 5.4. As suggested by this picture, the edges of such a cone are of eminent importance; they are the subject of pathway analysis (section 5.6).

The constraints C2–C4 further restrict the cone to a smaller subset of flux vectors representing then, in general, a polyhedron. Note that the optimality condition C4 is not always considered as a constraint. However, one may treat it as the same since it reduces the space of flux vectors of interest as the other constraints do. The optimality condition C4 is central to the approach of flux balance analysis, which is introduced next.

5.5.2   Flux Balance Analysis

Flux balance analysis (FBA) seeks to identify extreme patterns of flux distributions that keep the network balanced (constraint C1), are thermodynamically feasible (C2) and maximize a linear objective function (C4). Thus, the characteristic and necessary assumption of FBA is the optimal function of the network expressed by the optimality constraint C4. The three constraints C1, C2, and C4, in mathematical terms, represent a linear optimization problem (Kauffman et al., 2003a; Bertsimas and Tsitsiklis, 1997), which may be optionally extended by measurements (C3). In most cases, the (linear) objective function is the maximization of growth or product yield. The vector $\mathbf{s}$ in the linear objective function used in C4 represents the optimization criteria and weights the reaction rates. For maximizing the growth rate, for example, only the coefficient corresponding to the growth rate is set to one and all others to zero. As an example, assume we want to maximize the yield of P (reaction R3) for growth on substrate A in network EN1 (figure 5.1). The variables for the constraints then read:

- Stoichiometry (for C1): $\mathbf{N}$ as given in figure 5.1
- Boundaries (for C2): $\alpha = (0, -\infty, 0, 0, 0, 0, 0, -\infty, 0, 0)$; $\beta = (1, 0, \infty, \infty, \infty, \infty, \infty, \infty, \infty, \infty)$
- Linear objective function (for C4): $\mathbf{s} = (0, 0, 1, 0, 0, 0, 0, 0, 0, 0)$

Note that only α_2 and α_8 are $-\infty$ because only R2 and R8 are reversible. Furthermore, we set $\beta_2 = 0$ (B cannot be taken up), because exclusive growth on substrate A is considered. Only c_3 is non-zero as we want to optimize R3. Finally, we assume that the maximal uptake rate of A is 1 ($\beta_1 = 1$). Using available computer routines like the simplex algorithm (Bertsimas and Tsitsiklis, 1997), one can easily solve such a linear optimization problem. In our example, one might get a solution as shown in figure 5.5 with an optimal yield of P/A = 1.

The following main applications of FBA became attractive for metabolic engineering, but also for systems biology:

- **Predicting optimal yield and optimal behavior**: FBA enables one to predict production capabilities of a micro-organism (Varma and Palsson, 1993). This is of

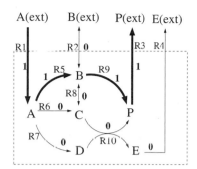

Figure 5.5 Optimal flux distribution for producing P from A in EN1.

high interest for industrial applications (Stephanopoulos et al., 1998; Nielsen, 1998) and FBA can also be used to search for optimal knock-outs (Burgard et al., 2003) with respect to certain criteria. Furthermore, bacteria such as *E. coli* have been shown to behave (stoichiometrically) optimal with respect to biomass yield, at least under selective pressure (Edwards et al., 2001a; Ibarra et al., 2002). Thus, at least for certain conditions, the quantity of this optimal behavior can be calculated *in silico*.

- **Predicting functionality and phenotypes (after gene deletions)**: A very useful application of FBA is to investigate whether a certain function can be performed *at all* in a network, especially after removal of network elements (simulating gene deletions): if a reaction is removed in the network (that is, from $\mathbf{N}$), then one may optimize the network again. If the optimal value, for example, for the growth rate, now becomes zero, then one definitely knows that this function (growth) is not possible anymore. This procedure has been applied, for example, in (Edwards and Palsson, 2000; Förster et al., 2003), and it could be shown that the prediction "growth is/is not possible" agrees well with the real phenotype. Especially when a function is possible, although the *in silico* analysis of the network predicts the opposite (false negative prediction), there must be an error or something missing in the considered network.

- **Flux coupling**: FBA can be used to analyze flux couplings in a network (Burgard et al., 2004). Similar to investigating the null-space matrix, blocked and fully coupled reactions may be identified, but reversibility constraints are explicitly considered. Additionally, weaker couplings may also be identified, for instance, where one reaction is used when another reaction is active, but not automatically vice versa (as for R1 and R5 in EN1, for example). The results of such investigations can help inferring the underlying regulatory rules.

The usefulness of FBA has been proven in many applications, in particular for microbial model organisms (Price et al., 2004), but there are also limitations one should be aware of. FBA critically depends on the optimality criterion applied. Not all cells, and bacterial cells not under all circumstances, will behave stoichiometrically optimal. This means that, in general, network capabilities but not the real phenotype can be predicted. Moreover, the optimal value of the objective function is unique and an optimal solution will usually be found. However, especially in large (genome-scale) networks, the calculated optimal flux distribution itself may be not unique. Look at our optimal solution in figure 5.5. It is easy to find another optimal flux distribution that also realizes optimal yield (P/A =1), such as the left one in figure 5.6. We can even (linearly) combine this solution with the one in figure 5.5 (here with a factor of 0.5 for both) yielding the right flux map in figure 5.6. Thus, actually, infinitely many optimal flux distributions exist even in this small network. Therefore, in most cases, albeit the additional constraints C2 and C4 of FBA shrink the solution space considerably, infinitely many solutions can remain. FBA delivers one particular optimal solution. Thus, even if optimality is assumed, it may happen that only little can be said about the internal behavior, that is, how the fluxes are distributed inside the cell (Mahadevan and Schilling, 2003).

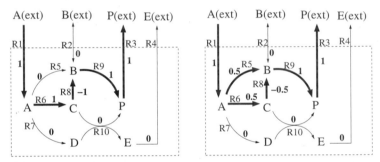

Figure 5.6 Two further optimal flux distributions for producing P from A in EN1.

However, often one can specify some reaction rates that are fixed in any optimal solution. For the example optimization problem in EN1, we could derive that R4, R7, and R10 must be zero during optimal behavior (because they are involved in side production of E) and that R1, R3, and R9 carry a fixed flux of unity. Thus, only R5, R6, and R8 remain variable. Fixed rates in optimal flux patterns can easily be identified (Mahadevan and Schilling, 2003). Moreover, one may also determine the qualitatively distinct optimal solutions (as the two in figs. 5.5 and 5.6 (left)) for a given FBA problem, for example, by mixed-integer linear programming (Lee et al., 2000), or—in smaller networks—by elementary modes as described in a later section.

5.5.3   Minimization of Metabolic Adjustment (MoMA)

The analysis of the stoichiometric implications of gene deletions is one important application of FBA, because FBA can find a new (optimal) flux distribution. Even if the wild type grows optimally, mutants may not necessarily behave optimally with respect to their retained resources. Instead they could adjust their metabolism with minimal effort (Segre et al., 2002). This assumption suggests that the cell searches for the nearest solution in the new feasible space of steady state flux distributions, which is part of the wild type solution space. Formally, this leads again to a constraint-based problem, where $\mathbf{w}$ represents the optimal solution of the wild type and d the index of the deleted reaction whose rate is set to zero:

$$\mathbf{Nr} = \mathbf{0} \qquad (5.12)$$

$$\alpha_i \leq r_i \leq \beta_i$$

$$r_d = 0$$

$$(\mathbf{r} - \mathbf{w})^T(\mathbf{r} - \mathbf{w}) = \text{Min!}$$

The first three terms correspond to C1–C3 in the usual FBA, whereas the fourth term leads to a quadratic programming problem whose handling, however, is mathematically straightforward (Segre et al., 2002).

For *E. coli* mutants, this approach lead to better predictions than FBA (Segre et al., 2002). However, MoMA at first needs the flux distribution from the wild type, which is also assumed to be optimal and, hence, determined by FBA. Therefore, MoMA also faces the problem of non-unique optimal flux distributions in the wild type. It can, thus, also result in non-unique solutions for the mutant (Mahadevan and Schilling, 2003). Hence, for MoMA it is essential to identify the real flux distribution in the wild type under a given environment.

5.6 Pathway Analysis

Pathway analysis deals with the discovery and analysis of meaningful routes in (primarily) metabolic networks using the concepts of extreme pathways (EPs) and of elementary flux modes (EFMs) (Papin et al., 2003). In contrast to FBA or MFA, it characterizes the complete space of admissible steady-state flux distributions by particular flux vectors.

5.6.1 Principles of Pathway Analysis

Extreme pathways/elementary flux modes are structural elements that are unique for a given network and can be considered as the smallest functional entities (Schuster and Hilgetag, 1994; Schilling et al., 2000). They both are defined by a flux vector $\mathbf{e}$ composed of q elements $(e_1, e_2, ...e_q)$, each describing the net rate of the corresponding reaction. The pathway represented by $\mathbf{e}$ can be identified by the utilized reactions. We denote this by

$$P(\mathbf{e}) = \{i : e_i \neq 0\}. \qquad (5.13)$$

In other words, the pathway representation $P(\mathbf{e})$ specifies all reactions that participate in the EP or EFM $\mathbf{e}$. If $\mathbf{e}$ is an EFM or EP, it fulfills the following three conditions (Schuster et al., 1999, 2000; Schuster and Hilgetag, 1994; Schilling et al., 2000; Schuster et al., 2002b):

(C1) Pseudo steady state: None of the metabolites is consumed or produced in the overall stoichiometry according to equation 5.7. Hence, EP or EFM **e** is in the null-space of **N** and **Ne** = **0** holds.

(C2) Feasibility: All fluxes in an EP or EFM have to be thermodynamically feasible, that is, irreversible reactions have to proceed in the "right" direction. Formally, this requires that all rates $e_i \geq 0$ if reaction $i \in$ **irrev**.

(C3) Non-decomposability: The central property of EPs and EFMs is that they represent the *minimal* functional units in a network. No reaction from an EP or EFM can be deleted, still resulting in a valid (non-trivial) steady state flux distribution. Formally, there exists no vector **v** unequal to the zero vector and to **e** fulfilling C1, C2, and that $P(\mathbf{v})$ is a proper subset of $P(\mathbf{e})$. This feature is also called genetic independence because C3 implies that the participating enzymes in one pathway are not a subset of the enzymes in another pathway. C1 and C3 together ensure that the sub-network spanned by the reactions in pathway **e** is connected.

Conditions C1–C3 completely define an EFM up to a scaling factor for each pathway. Note that C1 and C2 are identical to C1 and C2' used in the constraint-based approach (section 5.5). For an EP, two additional conditions have to be satisfied (see section 5.6.2). Importantly, both approaches provide a unique decomposition of a given network structure into EPs or EFMs, respectively. Hence, they unambiguously represent a particular network. The small example network EN2 illustrates these basic properties of EFMs (figure 5.7). Only two EFMs can occur, namely one using the upper branch of the central reaction couple, and the other one using the lower branch. The third flux distribution is not an EFM because the irreversible reaction R2 operates in the backward direction, and thereby violates feasibility condition C2. The rightmost flux distribution violates condition C3; it can be decomposed into EM1 scaled by a factor of two and EM2.

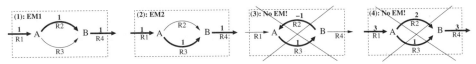

Figure 5.7 Elementary flux modes in the example network EN2 (left), and two flux distributions that do not constitute EFMs (right). Bold face denotes participating reactions and their (normalized) rates.

The last example referred to a particular property of EFMs and EPs, namely convexity, which is of paramount importance for pathway analysis. The basic conditions C1–C3 imply that all feasible steady state flux distributions **v** can be described by a nonnegative superposition of all EFMs or all EPs, respectively.

With the complete set of EFMs in a network denoted by S_{EFM} and, in analogy the complete set of EPs being S_{EP}, this feature is formally represented by

$$\mathbf{v} = \sum_j \alpha_j \mathbf{e}^j . \qquad (5.14)$$

Here, α_j is a positive scaling factor ($\alpha_j \geq 0$), and $\mathbf{e}^j$ denotes the j=th EFM $\in S_{EFM}$ or the j-th EP $\in S_{EP}$, respectively. The pattern of superposition does not necessarily have to be unique for a given flux distribution, that is, different combinations of EFMs or EPs may lead to an identical flux pattern. Hence, in most cases a direct decomposition of a flux distribution into the underlying EFMs and EPs is not possible.

Importantly, all edges—the so-called extreme rays—of the convex flux cone (section 5.5) are contained in the sets of EPs and EFMs, respectively (Schuster et al., 2002b), which directly follows from equation 5.14. In convex analysis, EPs and EFMs are called generating vectors of the convex cone. The concepts of EPs and EFMs were derived from a more general convex analysis approach to stoichiometric networks. There, pathways have been called extreme currents, but they were restricted to irreversible reactions (Clarke, 1988). EFMs permit all reactions to be reversible, while for EPs, this is allowed for certain fluxes (see below).

5.6.2 Elementary Flux Modes and Extreme Pathways

The conditions C1–C3 already uniquely determine the complete set of EFMs in a network (up to a scaling factor for each pathway vector). Two additional conditions delimit the EPs from the EFMs (Schilling et al., 2000):

(C4$_{EP}$) Network configuration: Reactions have to be classified either as exchange fluxes, which allow a metabolite to enter or to exit the system, or as internal reactions. All reversible internal reactions must be described by two separate irreversible reactions for the forward and the backward direction, respectively. Exchange fluxes can be reversible and each metabolite may only participate at most in one exchange flux.

(C5$_{EP}$) Systemic independence: The set of EPs in a network configured according to condition C4$_{EP}$ is the *minimal* set of generating vectors, allowing to describe all feasible steady state flux distributions by equation 5.14. The network configuration (C4$_{EP}$) ensures that the set of EPs is unique for a given network.

Thus, extreme pathways are only defined in a particular representation of a given network.

Reconfiguration and the particular conditions for EPs lead to the following consequences, which can be exemplified by EN1 (figure 5.8 and table 5.1): (i) Each split reversible reaction leads to a "two-cycle" constituted by the forward and backward branches, for instance, EM9' in EN1 for R8. This type of pathway, however, has no practical meaning and is usually not further considered (Papin et al., 2002). (ii) Except for these two-cycles, the EFMs in the original and the reconfigured network are equivalent. (iii) In a reconfigured network, the set of EPs is always a (proper or non-proper) subset of the EFMs because each EP obeys conditions C1–C3, that is, $S_{EP} \subseteq S_{EM'}$. Each EP can be mapped onto a corresponding EFM, while the inverse is not true. For instance, EFMs1'–3' (table 5.1) can be represented by non-negative linear combinations of EPs and, hence, are not systemically independent. (iv) Systemically dependent EFMs that are not EPs occur only when a network contains reversible exchange fluxes (Klamt and Stelling, 2003; Papin et al., 2004b) such as R2 in EN1. There, the direct pathway (EM1) can formally be decomposed into two pathways that rely on the reversible exchange flux of metabolite B (EM5',8' = EP2,5).

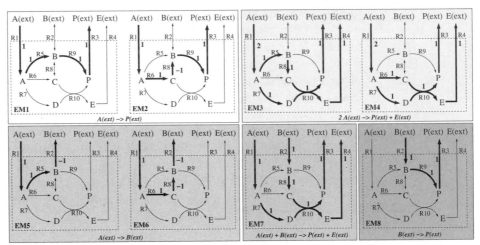

Figure 5.8 Elementary flux modes EM1–8 in the example network EN1. EFMs were grouped by the net conversion of external metabolites (bottom of each box) as indicated by different gray background levels.

As the above examples indicate, the set of EFMs related to a network shows certain conservation properties. When a reversible reaction is changed to irreversible, a new pathway set is obtained by excluding those EFMs from the original set that use the specific reaction in the forbidden direction (Schuster et al., 2002b). Hence, one can calculate the EFMs separately for forward and backward direction and then assemble the complete set of EFMs for the original network by uniting the two sub-sets. Likewise, if a reaction is deleted, the subset of EFMs not involving this reaction is the complete set of EFMs in the reduced network (Schuster et al., 2000). In contrast, the set of EPs needs to be recalculated whenever a (partial) reaction is removed.

Table 5.1   Relations between elementary flux modes in the original EN1 (EFMs; cf. figure 5.8), in EN1 after reconfiguration (EFMs'), and extreme pathways (EPs).

EFMs	EFMs'	EPs	Sum of EPs
EM1	EM1'	—	EP2 + EP5
EM2	EM2'	—	EP3 + EP5
EM3	EM3'	—	EP2 + EP4
EM4	EM4'	EP1	—
EM5	EM5'	EP2	—
EM6	EM6'	EP3	—
EM7	EM7'	EP4	—
EM8	EM8'	EP5	—
—	EM9'	EP6	—

5.6.3   Calculation of Pathway Sets

Several algorithms have been proposed for the enumeration of pathways (Schuster et al., 2000; Wagner, 2004). They contain a common core (Gagneur and Klamt, 2004) shown as pseudo-code in figure 5.9. Pathway sets, stored in matrices $\mathbf{M}^i$, are built iteratively by successively processing the imposed equality (C1), inequality (C2), and elementarity (C3) constraints . An initial matrix $\mathbf{M}^0$ can be derived from $\mathbf{N}$, for example, using a special kernel matrix $\mathbf{K}$ of $\mathbf{N}$ (Wagner, 2004); in this aspect, existing algorithms differ most. Until all constraints are satisfied, the rows in $\mathbf{M}^i$, which represent preliminary pathways, have to be processed for compliance with conditions C1–C2. Thereby, new candidate pathways are generated by Gaussian combination of pairs of rows in $\mathbf{M}^i$. Additionally, computationally expensive tests have to be performed to comply with C3.

Figure 5.9   Pseudo-code for pathway calculation.

Construct initial matrix $\mathbf{M}^0$ from $\mathbf{N}$
for all constraints of C1/C2 not satisfied

$$\mathbf{M}^{i+1} = \left\{ \begin{array}{l} \text{Process current constraint for all rows in } \mathbf{M}^i \\ \text{Pairwise Gaussian combinations of rows of } \mathbf{M}^i \\ \text{Test for elementarity of all candidate pathways (C3)} \end{array} \right.$$

$\qquad \mathbf{M}^i \leftarrow \mathbf{M}^{i+1}$
end
EFMs $= \mathbf{M}^{i+1}$

These requirements render the combinatorial problem of pathway identification NP-hard. With increasing network size, the number of pathways and the associated computational costs are likely to grow more than linearly (Klamt and Stelling, 2002). Therefore, pathway analysis has mainly been applied to networks of small or moderate size. With algorithmic improvements, however, the networks investigated became increasingly more complex (Stelling et al., 2002; Förster et al., 2002; Wiback and Palsson, 2002). We will discuss prospects for EFM/EP computation in genome-scale networks in section 5.7.

5.6.4 Applications of Pathway Analysis

Pathway analysis, which comprises the approaches of elementary flux modes and of extreme pathways, per se aims at the dissection of complex networks into smaller functional units. Here, we consider how these entities help in understanding metabolic networks by focusing on EFM analysis because for most applications, the sets of EPs and EFMs are identical. However, the slight differences in the methods may have important consequences for specific applications (Klamt and Stelling, 2003).

EFM analysis can be used to identify all routes that enable a cell to convert a certain substrate into a product. In EN1, for instance, four genetically independent routes (EM1–4) produce P from A as sole substrate (figure 5.8). Purely internal reaction cycles without net energy consumption, in contrast, would point to thermodynamic inconsistencies (Beard et al., 2002). Since all possible steady-state flux distributions are linear combinations of EFMs, the pathway(s) that are optimal or sub-optimal regarding the ratio of two reaction rates have to be among these units. In the example network, the two EFMs with highest P:A yield (R3/R1) of one (EM1,2) correspond to the two qualitatively different optimal routes in section 5.5.2, figs. 5.5 and 5.6). However, FBA allows including additional constraints such as maximal reaction capacities when searching for optimal flux vectors. Pathway analysis uncovers *all* qualitatively different (potentially optimal) pathways, the superposition of which gives the actual flux distribution observed *in vivo*. Potential contributions of individual pathways to this flux distribution may be analyzed through the spectrum of α-values in equation 5.14 (Wiback et al., 2003).

As pathway analysis yields all possible routes, the importance of single reactions for the network behavior in a certain context can be analyzed. For instance, reaction R9 in EN1 is indispensable for the production of P from B alone, but several alternative routes without R9 exist for the conversion of A to P. Similarly, correlated reactions (see sections 5.4.2 and 5.5.2) can be dealt with. The number of reactions in a pathway might be of interest because it indicates the amount of cellular resources that is needed to establish the pathway, for instance, to provide for the necessary enzymes. Moreover, the distribution of pathway lengths can characterize the complexity of a given network or differences between seemingly similar networks in two organisms.

The analysis of network functionality directly relates to the conservation properties of EFMs. When a reaction is removed from a network, the new set of EFMs contains all those EFMs of the original network in which the specific reaction does not participate. An empty set for the perturbed network, hence, indicates that the organism is structurally unable to achieve a steady-state flux distribution. This is a reliable predictor of inviability for the corresponding gene deletion mutant (Edwards et al., 2001a; Stelling et al., 2002). The concept of "minimal cut sets," the smallest sets of reactions the inactivation of, which will guarantee network failure, systematically extends these analyses; it allows to search for optimal intervention strategies (Klamt and Gilles, 2004). Likewise, the introduction of new genes can be assessed.

The analysis of network robustness/flexibility may be performed by assessing the effects of all possible mutations. Here, however, it may be important to analyze the complete (reduced) sets of EFMs in order to investigate the effects of pathway redundancy, or the sensitivity of network performance in terms of yields upon perturbations. For instance, such an analysis would show that production of P from substrate A alone (four alternative pathways EM1–4) likely is less affected by random mutations than the production of P from exclusively B (one route EM8; figure 5.8). Hence, MPA represents a suitable approach for extracting a large number of structural features from a given network, but it is limited by the increasing combinatorial complexity in larger networks.

5.7 Advanced Topics and Future Directions

The most challenging fields in stoichiometric network analysis concern predominantly (i) analyzing networks of increasing complexity, (ii) decomposing networks into modules and hierarchies, and (iii) incorporating and predicting cellular regulation (Price et al., 2004; Stelling, 2004).

5.7.1 Genome-Scale Network Analysis

The first task in genome-scale network analysis is network reconstruction from genomic, biochemical, and physiological data. Stoichiometry, directionality, and catalyzing enzymes (and their genes) of organism-specific metabolic (sub-)networks can now be obtained from databases such as KEGG (www.genome.ad.jp, Goto et al. (2002)) or MetaCyc (www.biocyc.org, Karp et al. (2002)). Unknown reactions and the necessary validation of database entries, however, pose challenges for model development. To date, genome-scale stoichiometric models have been established mainly for microbial model organisms. With up to ~1,200 reactions and ~700 metabolites, they belong to the largest models of cellular systems known so far (Price et al., 2004).

The analysis of such complex networks is straightforward for FBA, which requires only linear optimization (Edwards et al., 2001a). Pathway analysis, however, has to deal with a combinatorial explosion of possible routes with increasing network complexity. EFM analysis in a model of *E. coli* central metabolism that comprised only 89 metabolites and 110 reactions yielded up to half a million pathways (Stelling et al., 2002) (see chapter 2). However, this number is far below the theoretical upper bound of $4.39 \cdot 10^{21}$ for that network size, which is typical for highly structured cellular networks (Klamt and Stelling, 2002). Despite the progress in pathway computation (section 5.6.3), it seems still impossible to calculate all pathways directly in genome-scale networks. Restricting the number of simultaneously active system inputs (substrates) and outputs (product pathways) (Papin et al., 2002; Price et al., 2002) allows to describe many situations in practice, but it does not provide an assessment of all organismic capabilities. Alternative approaches aim at decomposing complex networks into biologically meaningful modules and hierarchies, which is closely related to the study of general design principles in cellular networks.

5.7.2   Modularity and Hierarchies

Modules are semi-autonomous entities that show dense internal functional connections, but looser connections with their environment. They occur at all levels of cellular organization (and beyond), for instance, as metabolic pathways, or as anabolism and catabolism at higher levels (see chapter 3). Modularity and hierarchies are directly linked because in general, smaller modules combine into larger modules of the next layer. Their biological relevance lies in the possibility to evolve, maintain, and coordinate cellular functions effectively because changes in one module primarily affect this entity and do not (unintentionally) spread through the network (Lauffenburger, 2000; Oltvai and Barabási, 2002).

Graph theory has been the method of choice for uncovering modules and hierarchies in genome-scale networks in various organisms. For metabolic and transcriptional networks, several studies yielded a surprising overlap of the identified modules with "classical" biochemical entities, but also divergences (Ravasz et al., 2002; Holme et al., 2003; Gagneur et al., 2003; Ihmels et al., 2004b). Consequently, formal approaches have been proposed for graph-based network decomposition and subsequent stoichiometric analysis (Schilling and Palsson, 2000; Schuster et al., 2002a). For example, metabolites with higher connectivity numbers can be considered as external to obtain "local" EFMs in small subnetworks (Schuster et al., 2002b). Alternatively, one may consider subnetworks by neglecting reactions; the resulting EFMs will be valid for the complete network and, thus, approximate its capabilities. However, because graph-theoretical approaches use only little biological knowledge and, consequently, roughly represent reality, it would be desirable to employ other structural approaches for this type of analysis. First attempts into this direction rely on correlated reaction sets, which correspond to enzyme subsets for perfectly coupled reactions (Papin et al., 2004a). Pathway analysis could provide starting points for future methods because per se it aims at identifying functional subunits in complex networks. In fact, EPs and EFMs may represent overlapping modules. Sound theoretical criteria for the demarcation of modules from pathway structures, however, still have to be developed.

5.7.3 Network Structure and Cellular Regulation

Analysis of cellular regulation in the SNA framework is at an early stage. Different objectives distinguish three broad classes of approaches. Starting from regulation as an additional constraint on network function, known regulatory interactions were incorporated into FBA (Covert et al., 2001) and into EP analysis (Covert and Palsson, 2003). Only those flux distributions were allowed that complied with regulatory rules superimposed onto the stoichiometric model. A simple, yet realistic control model uses logical rules, for instance "*IF* the favored substrate is available *THEN* uptake of less preferred substrates is suppressed." This "dynamic" FBA improved predictions of mutant phenotypes of *E. coli* for a large variety of conditions (Covert et al., 2004). Even such coarse descriptions of regulation may serve as powerful constraints because they eliminate the majority of structurally possible pathways (Covert and Palsson, 2003).

A second class of approaches aims at inferring regulatory features from network structure. It assumes that evolution established regulatory circuits that are adapted to the network they control. Inference of regulation from network structure may be possible because the underlying regulatory logic could be relatively simple compared to the networks' complexity (Lauffenburger, 2000). Enzyme subsets and correlated reactions help to qualitatively predict the relative control of fluxes. Singular value decomposition (SVD) of pathway matrices has been proposed to define the most important "eigenpathways" that could approximate the functionalities of a network and thereby unravel potential key control points. First evidence from human red blood cell metabolism supports this claim (Price et al., 2003). The pathway-based concept of "control-effective fluxes" allows to estimate gene expression ratios under different growth conditions solely from network structure (Stelling et al., 2002; Cakir et al., 2004). Analysis of *E. coli* central metabolism pointed to a different control logic of gene expression (long-term flexibility) versus regulation at the enzyme level (fine-tuning of fluxes in a specific situation) (Stelling et al., 2002).

Finally, the direct application of SNA approaches to regulatory networks has just begun. Examples include the formulation of stoichiometric models for gene expression (Allen et al., 2003). Extreme pathway analysis was extended to characterize information flows in signal transduction (Papin and Palsson, 2004) and in gene regulatory networks (Xiong et al., 2004). Approaches like these may become important for elucidating crosstalk between signaling systems or for yielding functional cycles involved in signal propagation and resetting. It has to be noted, however, that regulatory processes are often characterized by their dynamics, which cannot easily be captured by SNA methods. Here, as in the other fields described above, the establishment of new approaches and the testing of existing ones against biological data are necessary for the further development of stoichiometric network analysis.

5.8   Conclusions

Stoichiometric and constraint-based modeling provides powerful methods for the characterization of, in particular, metabolic networks. It can form a basis for more detailed dynamical modeling of such systems. However, none of the approaches we discussed is able to adequately address all the potential applications of SNA (table 5.2).

Table 5.2   Approaches for stoichiometric network analysis, their requirements, and fields of application. Parentheses denote partial applicability.

Approach	Constraints incorporated					Computational costs	Flux distribution(s)
	Stoichiometry	Thermodynamics	Quasi steady state	Reaction capacities	Optimality		
Graph theory	$(-)^a$	$(+)$	$-$	$-$	$-$	Low	None
CRs	$+$	$-$	$-$	$-$	$-$	Low	None
Kernel matrix	$+$	$-$	$+$	$-$	$-$	Low	All
MFA	$+$	$(+)$	$+$	$(+)$	$-$	Low	Single
FBA	$+$	$+$	$+$	$+$	$+$	Low	Single
MOMA	$+$	$+$	$+$	$+$	$+$	Medium	Single
EFMs/EPs	$+$	$+$	$+$	$-$	$-$	High	All

Approach	Applications						
	Functional pathways	Optimal operation	Reaction importance	Reaction correlations	Pathway length	Network function	Robustness
Graph theory	$-$	$-$	$(+)$	$(-)$	$(+)$	$(+)$	$(+)$
CRs	$-$	$-$	$-$	$-$	$-$	$-$	$-$
Kernel matrix	$-$	$-$	$-$	$(+)$	$-$	$(+)$	$-$
MFA	$-$	$-$	$(+)$	$-$	$-$	$-$	$-$
FBA	$-$	$+$	$(+)$	$-$	$-$	$+$	$(+)$
MOMA	$-$	$+$	$(+)$	$-$	$-$	$+$	$(+)$
EFMs/EPs^b	$+$	$+$	$+$	$+$	$+$	$+$	$+$

^{a} Graph-theoretical methods use only connectivities and, possibly, directions.

^{b} For the realistic case of equivalent sets of EPs and EFMs.

Hence, the methods for tackling a specific problem have to be carefully selected. More specifically, FBA and related approaches are most suitable for finding particular flux solutions even in genome-scale networks. Pathway analysis delivers a multitude of structural and functional aspects but is, in very large networks, hampered by combinatorial complexity. Despite such limitations, we expect the importance of SNA for systems biology to increase, particularly for an effective initial characterization of large-scale systems. We anticipate that the field will move towards a closer connection of the analysis of network structures in metabolism and regulation, which requires the development of new or modified theoretical methods.

6 Modeling Molecular Interaction Networks with Nonlinear Ordinary Differential Equations

Emery D. Conrad and John J. Tyson

Cellular processes, like growth, division, motility, and death, are controlled by complex networks of interacting macromolecules (genes, mRNAs, and proteins). These networks are sets of chemical reactions that convert reactant species into product species at rates that depend on reactant concentrations and, often, on the concentrations of other molecules (enzymes, inhibitors, transcription factors). To a first approximation, a reaction network can be described mathematically by a set of nonlinear ordinary differential equations that track the effects of these simultaneously occurring reactions. To gain some insight into the dynamical possibilities of such networks, we explore a set of increasingly more complicated network motifs, describing their effects in terms of signal-response curves. From our collection of simple functional motifs (buzzers, fuses, toggle switches, and a variety of oscillators) we can create realistic models of control systems actually employed by cells. As an example, we discuss the DNA-damage response pathway in mammalian cells.

6.1 Introduction

Molecular biologists often rely on suggestive cartoons to capture the complex interactions between many molecular components in functional networks of genes, proteins, and metabolites. In such cartoons (for example, figure 6.1), icons represent the interacting molecules and solid arrows their chemical transformations, for example, synthesis, degradation, phosphorylation, dephosphorylation, binding, and dissociation. Enzymatic and other indirect effects (such as allosteric activations or inhibition) are often represented by dashed arrows. These cartoons (or "wiring diagrams") are useful in summarizing many experimental observations, in capturing the way biologists think about molecular mechanisms, and in suggesting new experiments to test or extend this molecular understanding of cell physiology.

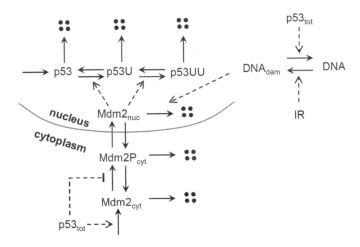

Figure 6.1 Example wiring diagram (reproduced from (Ciliberto et al., 2005), with permission). Intracellular proteins (p53, Mdm2, etc.) participate in a network of chemical reactions (solid arrows), such as synthesis, degradation, phosphorylation, and ubiquitination. Dashed arrows represent catalytic or regulatory effects on a reaction. This wiring diagram is a hypothetic mechanism (Ciliberto et al., 2005) for the interactions between p53, a transcriptional activator involved in cell cycle arrest and apoptosis, and Mdm2, a protein involved in degradation of p53. Mdm2 catalyzes the ubiquitination of p53, and polyubiquitinated p53 is rapidly degraded. Two feedback signals govern the behavior of the reaction network: (1) p53 stimulates the synthesis of Mdm2 in the cytoplasm, and (2) p53 indirectly inhibits the transport of Mdm2 into the nucleus. In response to DNA damage, the degradation of Mdm2 in the nucleus is upregulated. (IR = ionizing radiation)

Although most cell biologists use molecular wiring diagrams in these informal ways, we would like to pursue the idea that a reaction network is fundamentally a complex dynamical system and that its wiring diagram instructs how the concentrations of all the interacting components will change over time as the chemical reactions play out within the cell. From this point of view, the next question is how best to capture the dynamics of the network in mathematical form, in order to analyze and simulate its behaviors and ultimately to use the model to answer real physiological questions. For the purposes of this chapter, we will use nonlinear ordinary differential equations (ODEs) to represent the dynamical properties of reaction networks.

Realizing a reaction network as a system of ODEs is based on two assumptions. First, that our system is a "well-stirred" chemical reactor, so that component concentrations don't vary with respect to space. This is a reasonable assumption for cell-free extracts, but it hardly seems appropriate for an intact cell. Whether it is a good approximation or not depends on the time and space scales involved.

In box 6.1, we show that molecular diffusion is sufficiently fast to mix proteins throughout a yeast-sized cell in less than a minute. If we are interested in cell cycle processes (time scale = hours) or circadian rhythms (period = 1 day), then the "well-stirred" assumption is justified. If we are interested in membrane oscillations (time scale = seconds), then the "well-stirred" assumption would not be advisable.

When spatial information is required, then partial differential equations (PDEs) would be indicated. We will not discuss modeling by PDEs in this chapter, but note that, before one can appreciate the special properties of nonlinear PDE models (see chapter 9 of this book and Murray (2002b)), one must first master the principles in this chapter.

Box 6.1: How fast is molecular diffusion?
Given the typical diameter of a cell to be 10^{-3} cm and a typical diffusion constant for a protein in aqueous solution to be D=10^{-7}cm^2/s, we can calculate the average time for a protein to diffuse across a cell to be: $t = \frac{(10^{-3}cm)^2}{2 \times 10^{-7}cm^2/s} = 5s$. If diffusion is 10-fold slower in cytoplasm, then the average time to cross a cell is roughly 1 min. These are expected "mixing times" for macromolecules in cells. Metabolites (small molecules) will mix on a faster time scale.

The second basic assumption is that the variables (chemical concentrations) are continuous functions of time. This assumption is valid if the number of molecules of each species in the reaction volume (the cell or subcellular compartment) are sufficiently large (say, thousands of molecules each, at least). For concentrations greater than about 10 nM, we are safe using ODEs (see box 6.2).

Box 6.2: How many molecules of a regulatory protein in a cell?
A spherical cell of diameter 10^{-5} m has a volume of roughly 0.5×10^{-15} m^3 = 5×10^{-13} L. Given a typical concentration of a specific regulatory protein to be 10 nM, we calculate $10^{-8}\frac{mol}{L} \times 6 \times 10^{23}\frac{molecules}{mol} \times 5 \times 10^{-13} \frac{L}{cell}$=3,000 $\frac{molecules}{cell}$. For a reaction volume containing 3,000 molecules, we are justified in using ordinary differential equations to describe changes in a continuous variable $X(t)$ = concentration of species X. Were the concentration to drop below 1 nM, we would need to reformulate the model in terms of stochastic variables to capture the effects of molecular noise in the dynamical system.

If the total number of molecules of any particular substance, say, a transcription factor, is less than 1,000, then a stochastic differential equation or a Monte Carlo model would be more appropriate (Rao et al., 2002; McAdams and Arkin, 1999). Stochastic modeling is much more difficult than ODEs and requires a preliminary understanding of the deterministic dynamical system. For this reason, it makes sense to limit this chapter to ODE modeling and leave the harder stuff to chapter 8.

Granted these two simplifying assumptions, then ordinary differential equations are a very useful language in which to express mathematically the dynamical consequences of a molecular interaction network. By applying a set of simple rules, we can express an arbitrarily complex reaction network as a set of coupled differential equations. The computer can then keep track of all the complex, interweaving interactions in the network and tell us with great precision what are the consequences of the mechanism that purports to describe some aspect of cell physiology. In this sense, kinetic modeling by differential equations is a tool for hypothesis testing (see chapter 1). If the mathematical consequences of the mechanism do not agree with observations, we must search for the problems in our hypotheses. If the consequences agree with the observations, then we can have some confidence that we are on the right track to understanding the mechanism.

We assume the reader has no familiarity with how to do kinetic modeling of chemical reactions beyond some vague (and possibly regretful) memories of the Michaelis-Menten equation. We start with the basic idea of using a simple rate law to describe how fast a chemical reaction proceeds and show how to estimate kinetic rate constants for isolated reactions from data. Then we assemble a few simple reactions (for protein synthesis, degradation, phosphorylation, and dephosphorylation) into modules for chemical buzzers, switches, and oscillators. These reaction motifs can then be linked together to form more complicated and realistic control systems. Writing the differential equations describing these systems can be largely automated, and solving the equations can be fully automated (see chapter 16). Fitting the results to experimental data and estimating rate constants are difficult tasks, which are the subjects of active research (chapter 11). We shall touch on all these issues in what follows.

6.2   Basic Building Blocks

6.2.1   From a Wiring Diagram to a Set of ODEs

To get from a wiring diagram to a set of ODEs, we must think about a network as a dynamical system whose state is changing from one moment of time to the next. We assign to each species (or icon) in the diagram a single *state variable*, $X(t)$ = the concentration of species X. The collection of values of all these variables $\{X_1(t), X_2(t), X_3(t), ...\}$ at any point in time constitutes the *state of the system*. Then, for each molecular species, we write a differential equation that describes how its concentration changes over time due to its interactions with the other species in the network. For example, for species X, we write

$$\frac{dX}{dt} = \text{synthesis} - \text{degradation} - \text{phosphorylation}$$
$$+ \text{dephosphorylation} - \text{binding} + \text{release, etc.} \qquad (6.1)$$

The rate of each reaction (synthesis, degradation, etc.) must be represented by a *kinetic rate law*, which will have one or more *rate constants* associated with it. By assigning specific values to these rate constants, we fine-tune general rate laws to particular reactions. The set of all rate constants needed to describe the reactions in a molecular interaction network is called the *parameter set* $\{p_1, p_2, \ldots p_m\}$ of the model.

In this paradigm, the dynamical consequences of a reaction network are determined by a system of nonlinear ordinary differential equations,

$$\frac{dX_i}{dt} = F_i(X_1, X_2, ..., X_n; p_1, p_2, ..., p_m), \quad i = 1, 2, ..., n \tag{6.2}$$

The ODEs are nonlinear because the rate laws on the right-hand sides of equation 6.2 are often nonlinear functions of the state variables. Notice that the ODEs tell us how each state variable is *changing* with respect to time; they do not tell us the value of X at any specific time t. To solve the differential equations is to find these functions, $X_i(t)$, for each species (i) in the network. Each function corresponds to a measurable property of the system, the *time course* of species i. In order to solve equation 6.2 for the time courses $X_i(t)$, we must first prescribe a set of initial conditions $\{X_1(0), X_2(0), ..., X_n(0)\}$. The combination of rate equations, initial conditions, and parameter values is called a well-posed initial value problem (IVP), and its solution is guaranteed by a famous theorem stated informally in box 6.3.

Box 6.3: Existence and uniqueness theorem
Given very weak conditions on the smoothness of the rate laws on the right-hand side of equation 6.2, conditions that are usually satisfied by realistic models of reaction networks, the initial value problem has one and only one solution $\{X_1(t), X_2(t), ..., X_n(t)\}$ for all $0 \leq t < \infty$. By running time backwards, we can also find a unique prehistory of the system (for $-\infty < t \leq 0$).

Box 6.4: Linear and nonlinear differential equations
If the F_i's in equation 6.2 are linear functions of the variables, $X_1, X_2, \ldots, X_n$, then much can be said about the dynamical characteristics of the reaction system. The good news is that the solution can be expressed analytically in terms of exponential functions, $\exp(\lambda_i t)$, and harmonic functions, $\sin(\omega_i t + \phi_i)$. The bad news is that the dynamical possibilities of a linear system are very impoverished. In general, there can be only a single steady state solution, and all other solutions either approach this steady state as $t \rightarrow \infty$ or they blow up (some $X_j \rightarrow \infty$ as $t \rightarrow \infty$). Linear systems show none of the interesting dynamical behaviors (multiple steady states, limit cycle oscillations) to be described later in this chapter. The interesting dynamical features depend crucially on nonlinear dependencies of the F_i's on the X_j's.

We can imagine three types of "solutions" of a system of ODEs.

1. Analytical. Under very special circumstances (see, for example, box 6.4), it is possible to write the solution of a set of ODEs in terms of elementary functions,

such as $X_1(t) = X_1(0)e^{-k_a t}$, $X_2(t) = X_2(0)[1 + 0.5 \sin(k_b t)]$, $X_3(t) = ...$, where $k_a, k_b, ...$ are rate constants and $X_1(0), X_2(0), ...$ are initial values.

2. Numerical. It is always possible to solve a well-posed IVP numerically on a computer. In principle, we can write

$$X_i(t + \Delta t) = X_i(t) + F_i(X_1(t), X_2(t), ..., X_n(t)) \cdot \Delta t \qquad (6.3)$$

for each i. By starting at $X_i(0)$ and taking sufficiently small steps, Δt, we can "walk along" the time course to any time t in the future (or in the past). In practice, there are much more sophisticated, efficient, and accurate numerical schemes for walking along the time course (see chapter 16).

3. Qualitative. Whereas numerical integration of the ODEs gives us quantitative information about the solution (which is necessary if we are trying to account for quantitative experimental data), sometimes we are more interested in answers to qualitative questions, like, "What will the network do if I wait for a sufficiently long time?" (that is, characterize the solutions—the "stable attractors"—of the ODEs as $t \to \infty$) or "How will the long-term behavior of the network change if I double the rate of synthesis of protein X?" (that is, characterize the dependence of the stable attractors on any parameter in the ODEs).

To explore the examples that we will present, we suggest that the reader download XPPAUT or one of the other tools for simulating dynamical systems listed in the appendix.

Many of our qualitative methods depend on identifying and characterizing the steady state solutions of equation 6.2. A steady state solution is a set of constants $\{X_1^*, X_2^*, ..., X_n^*\}$ for which the net rate of change of every variable is zero, that is, $F_i(X_1^*, X_2^*, ..., X_n^*) = 0$ for all $i = 1, 2, ..., n$. A steady state is a special time-invariant solution of the ODEs, where the reactions producing and consuming each species perfectly cancel each other. Steady states can be either stable or unstable. Stable steady states attract all nearby solutions, whereas unstable steady states repel some nearby solutions as time increases.

6.2.2　Constant Synthesis

For starters, let's consider a constant rate of synthesis of some macromolecule, which can be described by the initial value problem $\frac{dX}{dt} = k_1$, $X(0) = X_0$. In this case, the differential equation is simple enough that we can guess the solution of the initial value problem: $X(t) = X_0 + k_1 t$. The numerical value of the rate constant must be estimated from experimental data. For example, from observations of accumulating cyclin in a frog egg extract (figure 6.2), we estimate that $k_1 = 1\text{nM/min}$.

$X(t) = X_0 + k_1 t$ is an example of an explicit, analytical solution. The uniqueness part of the theorem in box 6.3 assures us that once we have guessed a solution to the initial value problem, it is the only solution. We can sleep soundly at night, assured that we have not overlooked some other solution of this dynamical system.

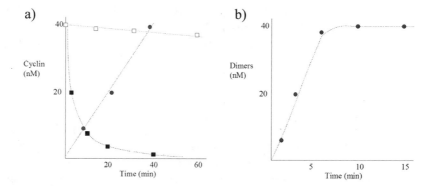

Figure 6.2 Experimental data used to estimate kinetic rate constants. a) Accumulation of cyclin (filled circles) in a frog egg extract; degradation of cyclin in interphase cells (open squares; (Felix et al., 1990)) and in metaphase cells (filled squares, from (Tang et al., 1993)). b) Formation of dimers of Cdk1 and cyclin B in an extract for which the initial concentration of Cdk1 monomers was approximately 100nM (Kumagai and Dunphy, 1995).

Once we have the solution, we can ask, "What happens to $X(t)$ as $t \to \infty$?" Well, it appears that the concentration of X grows without bound. We get this undesirable result because there is no term to counteract the growth rate in the differential equation.

6.2.3 Linear Degradation

Biochemical molecules naturally experience decay or degradation, and the rate at which this happens depends on how much of the molecule is present. In mathematical terms, $\frac{dX}{dt} = -k_2 X$, $X(0) = X_0$. The unique solution to this initial value problem is $X(t) = X_0 e^{-k_2 t}$. An interesting property of exponential decay is that X disappears with a constant half-life, $t_{1/2}$, defined by $X(t_{1/2}) = \frac{1}{2}X_0$. For linear degradation, $t_{1/2} = \frac{\ln 2}{k_2}$. From the data on cyclin degradation in figure 6.2, we see that cyclin is disappearing with a half-life of about 10 minutes, hence $k_2 \cong 0.07 \text{ min}^{-1}$.

At this point, the reader should consider what happens when we combine a constant rate of synthesis with linear degradation. That is, what is the analytical solution of the initial value problem: $\frac{dX}{dt} = k_1 - k_2 X$, $X(0) = X_0$? From the exact solution, show that $X(t) \to k_1/k_2$ as $t \to \infty$, for any $X_0 \geq 0$.

6.2.4 Autocatalytic Production

Autocatalysis is a process whereby a molecule activates its own production, either directly or indirectly through intermediates. In molecular biology, important examples include DNA synthesis and ribosome biogenesis. The simplest equation expressing autocatalysis is $\frac{dX}{dt} = k_2 X$. This is identical to the equation of the previous subsection, except for a difference of sign. The solution is $X(t) = X_0 e^{k_2 t}$. In this case, the solution grows with a constant *doubling time*, $t_2 = \frac{\ln 2}{k_2}$. We'll see more complex, indirect autocatalytic effects when we discuss feedback, later in the chapter.

6.2.5   Dimerization

Another fundamental reaction in biochemical networks is *dimerization*, where two species combine to form a complex. Examples include enzymes binding substrates, and the successive steps in the formation of hemoglobin, a four subunit heteromer ($\alpha_2 \beta_2$). According to the Law of Mass Action, dimerization proceeds at a rate proportional to the product of the concentrations of the two binding species. Hence, we can express C binding X, forming the complex M, by the following scheme

reaction	C	+ X	→ M
initial concentrations	C_0	X_0	0
extent of reaction	-M	-M	M
concentrations at a later time	$C_0 - M$	$X_0 - M$	M

$$\frac{dM}{dt} = k_3 C X = k_3 (C_0 - M)(X_0 - M) \ , \ [k_3] = \frac{1}{\text{nM} \cdot \text{min}} \tag{6.4}$$

where we've chosen to write $C(t)$ and $X(t)$ in terms of $M(t)$ so that we have a single, solvable equation for the unknown function $M(t)$. The notation $[k_3]$ means "the units of k_3."

Now, guessing a solution to this equation requires a bit more imagination. Let's suppose that we receive a mysterious letter claiming that $M(t) = \frac{C_0 X_0 (1 - e^{-\alpha t})}{C_0 - X_0 e^{-\alpha t}}$, where $\alpha = k_3 (C_0 - X_0)$, solves the initial value problem, when $M(0) = 0$, as in the scheme above. We can verify this claim by differentiating and doing a bit of algebra:

$$\frac{d}{dt} \left(\frac{C_0 X_0 (1 - e^{-\alpha t})}{C_0 - X_0 e^{-\alpha t}} \right) = \frac{\alpha C_0 X_0 (C_0 - X_0) e^{-\alpha t}}{(C_0 - X_0 e^{-\alpha t})^2} \tag{6.5}$$

and

$$
\begin{aligned}
k_3 (C_0 - M)(X_0 - M) &= k_3 C_0 (1 - \frac{X_0 (1 - e^{-\alpha t})}{C_0 - X_0 e^{-\alpha t}}) X_0 (1 - \frac{C_0 (1 - e^{-\alpha t})}{C_0 - X_0 e^{-\alpha t}}) \\
&= \frac{\alpha C_0 X_0 (C_0 - X_0) e^{-\alpha t}}{(C_0 - X_0 e^{-\alpha t})^2}
\end{aligned}
\tag{6.6}
$$

Remember that once we have a solution (even if it comes in the mail), it is the only solution we ever need (thanks to the existence and uniqueness theorem in box 6.3).

Notice from the analytical solution, $M(t) = \frac{C_0 X_0 (1 - e^{-\alpha t})}{C_0 - X_0 e^{-\alpha t}}$, where $\alpha = k_3 (C_0 - X_0)$, that, if $C_0 > X_0$, then $\alpha > 0$ and $M(t) \to X_0$ as $t \to \infty$. On the other hand, if $C_0 < X_0$, then $\alpha < 0$ and $M(t) \to C_0$ as $t \to \infty$. In either case, the asymptotic concentration of the complex is the initial concentration of the subunit in short supply. In principle, this conclusion is incorrect, because we have neglected dissociation of the complex (M → C + X, with some rate constant k_{-3}).

In order to estimate the rate constant from the data in figure 6.2b, we notice that $C_0 = 100$ nM and $X_0 \cong 40$ nM (why?). Considering that it takes about 3 minutes for $M(t)$ to reach 20 nM, we can solve $M(3) = 20$ for α:

$$M(3) = \frac{4000(1 - e^{-3\alpha})}{100 - 40e^{-3\alpha}} = 20 \tag{6.7}$$

$$\Rightarrow \quad 200 - 200e^{-3\alpha} = 100 - 40e^{-3\alpha} \tag{6.8}$$

$$\Rightarrow \quad \frac{5}{8} = e^{-3\alpha} \Rightarrow \alpha = \frac{1}{3}\ln\frac{8}{5} \text{ min}^{-1} \tag{6.9}$$

Therefore, we estimate that $k_3 = \frac{1}{180}\ln\frac{8}{5}\text{nM}^{-1}\text{min}^{-1} = 2.6 \times 10^{-3}\text{nM}^{-1}\text{min}^{-1}$.

6.2.6 Michaelis-Menten Kinetics

The diagram in figure 6.3a represents the enzymatic transformation of substrate X into product P. Michaelis and Menten (1913) and Briggs and Haldane (1925) first explored the elementary reaction mechanism (figure 6.3b) for this process. Assuming that the total enzyme concentration E_T is much less than the initial substrate concentration, X_0, they showed that the rate of the enzyme-catalyzed reaction can be written as: $\frac{dX}{dt} = -\frac{dP}{dt} = \frac{-k_2 E_T X}{K_m + X}$, where $K_m = \frac{k_{-1} + k_2}{k_1}$ is called the Michaelis constant. Note that $[K_m] = $ nM. A rigorous derivation of the Michaelis-Menten rate law can be found in (Murray, 2002a), and in (Segel, 1988).

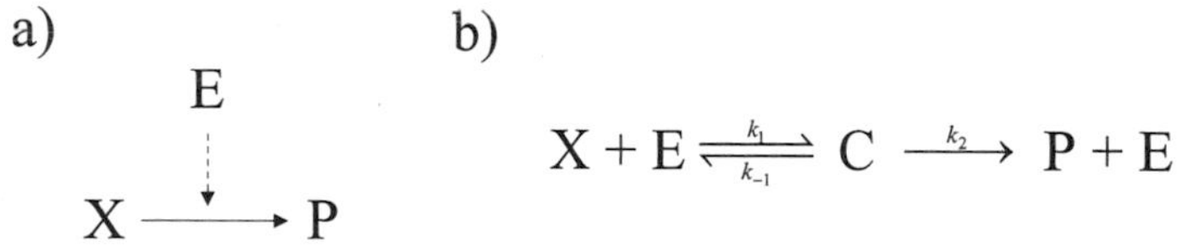

Figure 6.3 Michaelis-Menten kinetics. a) Enzyme E catalyzes the conversion of substrate X into product P. b) Michaelis-Menten mechanism for an enzyme-catalyzed reaction: E binds the substrate X to form a complex C; in the complex, E converts X to P; once the conversion is done, E dissociates from P and is free to bind another molecule of substrate.

Among other things, the Michaelis-Menten rate law can be used to reduce the number of variables which describe a typical enzymatic conversion process, such as phosphorylation or dephosphorylation. This reduction is often useful when trying to understand the dynamic possibilities of a network using analytical and qualitative methods. On the other hand, one must keep in mind the assumption ($E_T \ll X_0$) so that the rate law is applied in a consistent fashion.

6.3 Simple Networks and Signal-Response Curves

The basic rate laws just described can be combined to form reaction motifs that are commonly found in biochemical networks. These motifs have specific characteristics that dominate their behavior within larger networks. In order to build some dynamical intuition that may be useful in understanding larger, more realistic macromolecular networks, we first explore the properties of some common network motifs.

6.3.1 Synthesis and Degradation

Our first motif is simultaneous synthesis and degradation (figure 6.4a), described by $\frac{dX}{dt} = k_1 S - k_2 X$, with $X(0) = 0$. In this equation, we might think of S ("signal") as the concentration of mRNA encoding protein X. Notice that $[k_1] = [k_2] = \text{min}^{-1}$. The solution of this ODE, which was posed as a problem earlier in the chapter, is $X(t) = \frac{k_1 S}{k_2}(1 - e^{-k_2 t})$. Notice that as $t \to \infty$, $e^{-k_2 t} \to 0$, and our solution tends towards the value $X_{ss} = \frac{k_1 S}{k_2}$. Notice also that $k_1 S - k_2 X_{ss} = 0$, so X_{ss} is the steady state solution of the differential equation, as described earlier.

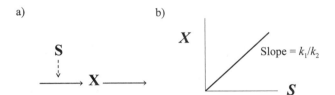

Figure 6.4 A signal-response relationship. a) Signal S stimulates the synthesis of protein X. b) Linear response of steady state protein concentration to signal strength.

If we think of S as an input signal (mRNA concentration) and X as the response (protein concentration), then this motif at steady state generates a *linear signal-response curve*, as depicted in figure 6.4b.

6.3.2 Phosphorylation and Dephosphorylation

Now suppose X is phosphorylated and dephosphorylated as depicted in figure 6.5a. Choosing to model both the forward and reverse steps using simple linear kinetics, we write $\frac{dX_P}{dt} = k_1 S(X_T - X_P) - k_2 X_P$, where X_T is the total concentration of both phosphorylated and unphosphorylated forms of X (so that $X_T - X_P = X$), and S is the concentration of the protein kinase. (The concentration of the protein phosphatase is absorbed into the value of k_2.) Notice that $[k_1] = \text{nM}^{-1}\,\text{min}^{-1}$, $[k_2] = \text{min}^{-1}$. Solving $\frac{dX_P}{dt} = 0$ results in a single steady state solution, $X_{P,ss} = \frac{X_T S}{(k_2/k_1)+S}$, which is a hyperbolic function of S(see figure 6.5b). This is called a *hyperbolic signal-response curve*.

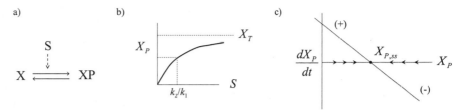

Figure 6.5 Hyperbolic signal-response curve (see text).

We can determine the stability of the steady state graphically by plotting $\frac{dX_P}{dt}$ as a function of X_P. Noting that trajectories lie along the x-axis, we see that for $\frac{dX_P}{dt} > 0$ (that is, wherever the curve is above the x-axis), the solution, $X_P(t)$, moves to the right along the x-axis and for $\frac{dX_P}{dt} < 0$ (where the curve is below the x-axis), the solution moves to the left. The curve crosses the x-axis at $X_{P,ss}$, the steady state. The stability of the steady state is then obvious because $X_P(t)$ moves towards $X_{P,ss}$ along the x-axis (figure 6.5c). This method of determining stability can be applied to any single-variable system.

Our assumption of linear kinetic rate laws implies that X_T is much less than the Michaelis constants of both the kinase and the phosphatase. If this is not the case, then we should use Michaelis-Menten rate laws.

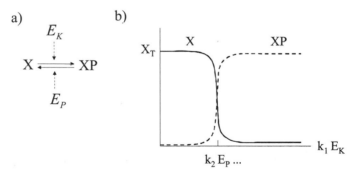

Figure 6.6 Sigmoidal signal-response curve (see text).

In this case (figure 6.6a), the governing ODE is

$$\frac{dX}{dt} = -\frac{k_1 E_K X}{K_{m1} + X} + \frac{k_2 E_P(X_T - X)}{K_{m2} + X_T - X} \,, \tag{6.10}$$

where $X_T - X = X_P$, E_K and E_P are the total concentrations of kinase and phosphatase (taken to be constant in this equation), and K_{m1} and K_{m2} are the Michaelis constants. At steady state, we have

$$\frac{k_1 E_K X}{K_{m1} + X} = \frac{k_2 E_P(X_T - X)}{K_{m2} + X_T - X} \tag{6.11}$$

or, after simplifying and scaling the relevant variables,

$$u_1 x (J_2 + 1 - x) = u_2 (1 - x)(J_1 + x) \tag{6.12}$$

where $x = X/X_T$, $u_1 = k_1 E_K$, $u_2 = k_2 E_P$, $J_1 = K_{m1}/X_T$, $J_2 = K_{m2}/X_T$. Using the quadratic formula, we can solve this equation for x as a function of u_1, u_2, J_1, and J_2. We get $x = G(u_1, u_2, J_1, J_2)$, where the *Goldbeter-Koshland* function, G, is defined as

$$G(u_1, u_2, J_1, J_2) = \frac{2 u_1 J_2}{B + \sqrt{B^2 - 4(u_2 - u_1) u_1 J_2}} \tag{6.13}$$

where $B = u_2 - u_1 + u_2 J_1 + u_1 J_2$ (see Goldbeter and Koshland (1981)). In terms of the original variables, X_{ss} is a sigmoidal function of the input signal E_K (see figure 6.6b), and so we call this a *sigmoidal signal-response curve*. The sigmoid becomes more and more switch-like as J_1 and J_2 become much less than 1.

To confirm the sigmoidal character of the Goldbeter-Koshland function, it is easier to think of u_1 as a function of x than x as a function of u_1. Rearranging equation 6.12, we find that $u_1 = u_2 \frac{J_1 + x}{J_2 + 1 - x} \cdot \frac{1 - x}{x}$. As a function of x, this curve crosses the x-axis at $x = 1$ and $x = -J_1$ and has vertical asymptotes at $x = 0$ and $x = 1 + J_2$. For $0 < J_1, J_2 \ll 1$, the curve must have the shape illustrated in figure 6.6b.

We can prove the stability of the steady state by the same graphical methods used for the case of linear reaction kinetics, but we omit the details.

6.4   Networks with Feedback

6.4.1   What Is Feedback?

Biochemical reaction networks commonly contain feedback loops, for which the output of one reaction affects the progress of an upstream reaction. Feedback can be characterized as positive or negative, depending on the net effect of the interactions. When reaction networks have intertwined feedback loops, their dynamical properties can be exceedingly complex (see chapter 1 and chapter 2).

We start our investigation of feedback loops with two-component interactions (figure 6.7), which can be categorized as negative feedback (6.7a and b), positive feedback (6.7c), or mutual antagonism (6.7d). Mathematically speaking, the effect of species X_j on the rate of change of another species X_i, $\frac{dX_i}{dt} = F_i(X_1, ..., X_n)$, is the partial derivative $\frac{\partial F_i}{\partial X_j}$. The sign of this derivative determines whether the feedback is positive or negative. Naturally, this partial derivate need not be constant and may change sign based on the state and on parameter values, so classifying the effect isn't always unambiguous. A chain of such effects makes a feedback loop if it starts and ends with the same species.

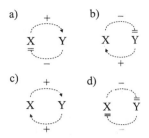

Figure 6.7 Three types of feedback are possible between two components: a) and b) are negative feedback, c) positive feedback, d) mutual antagonism.

6.4.2 Negative Feedback

We start with a simple example of negative feedback (figure 6.8). The phosphorylated form of Y activates the degradation of X, and X is the kinase that phosphorylates Y. In this case, we need at least two differential equations to characterize the system:

$$\frac{dX}{dt} = k_1 S - k_2 Y_P X \tag{6.14}$$

$$\frac{dY_P}{dt} = \frac{k_3 X (Y_T - Y_P)}{K_{m3} + Y_T - Y_P} - \frac{k_4 E Y_P}{K_{m4} + Y_P} \tag{6.15}$$

where $Y = Y_T - Y_P$ is the concentration of the unphosphorylated form of Y, $[X] = [Y_P] - [S] = [E] = $ nM, $[k_1] = [k_3] = [k_4] = $ min^{-1}, $[k_2] = $ nM $\cdot$ min^{-1}, and $[K_{m3}] = [K_{m4}] = $ nM. The equation for X is constant synthesis (proportional to S) minus degradation (proportional to $Y_P \cdot X$). The equation for Y_P is just the case studied in the subsection 6.3.2. We know how each of these differential equations behaves in isolation, but what happens when they are coupled together?

6.4.3 Phase Planes, Vector Fields, and Nullclines

As described earlier, at any point in time t_0, the network must reside in a particular state, $(X(t_0), Y(t_0))$, which is just a point in the two-dimensional *state space* of the system of ODEs. For the case of a two-species network, the state space is called the *phase plane*. At each point in the phase plane, the differential equations define a vector that tells us which direction and how far the dynamical system will move over the next small increment of time, Δt. We can think of the phase plane as covered with little vectors, like the hair on the head of a new military recruit. This collection of vectors is called the *vector field*. A solution to the ODEs is just a curve that starts at some initial point and follows the vector field.

The vector field in the phase plane is conveniently characterized by the X- and Y-nullclines, the curves for which the corresponding species' time derivative is exactly zero. Along the X-nullcline, the vector field points north (N) or south (S) because $\frac{dX}{dt} = 0$ (that is, no change in the east-west direction). Along the Y-nullcline,

a)   b)

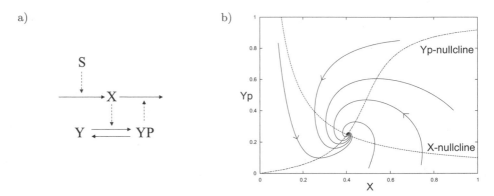

Figure 6.8 Example of negative feedback. a) Wiring diagram. b) Phase portrait. Dashed curves: nullclines given by equations 6.16 and 6.17; solid curves: trajectories of equations 6.14 and 6.15. Parameter values are $k_1 = k_2 = k_3 = k_4 = 1$, $S = K_{m3} = K_{m4} = 0.1$, $Y_T = 1$, and $E = 0.5$. In b), trajectories spiral into a stable steady state at the intersection of the nullclines.

the vector field points east (E) or west (W) because $\frac{dY}{dt} = 0$ (no change in the north-south direction). In the region between the nullclines, the vector field adopts one of four characteristic compass directions (NE, SE, SW, or NW). Wherever the nullclines intersect, the pair of ODEs has a steady state solution (both $\frac{dX}{dt} = 0$ and $\frac{dY}{dt} = 0$).

In the above example for negative feedback, the nullclines are:

$$\text{X-nullcline:} \qquad k_1 S = k_2 Y_P X \Rightarrow Y_P = \frac{k_1 S}{k_2 X} \qquad (6.16)$$

$$\text{Yp-nullcline:} \qquad \frac{k_3 X (Y_T - Y_P)}{K_{m3} + Y_T - Y_P} = \frac{k_4 E Y_P}{K_{m4} + Y_P}$$

$$\Rightarrow \qquad Y_P = Y_T \cdot G(k_3 X, k_4 E, \frac{K_{m3}}{Y_T}, \frac{K_{m4}}{Y_T}) \qquad (6.17)$$

where G is the Golbeter-Koshland function defined by equation 6.13. These curves are easily plotted on the phase plane (figure 6.8b) along with representative trajectories that point out how the system evolves with time given several different initial conditions. The X-nullcline is a hyperbola, while the Y_P-nullcline is a sigmoidal curve with the switch point at $X = \frac{k_4 E}{k_3} \cdot \frac{Y_T + 2 K_{m3}}{Y_T + 2 K_{m4}}$. Of particular note is how all trajectories seem to be sucked into the steady state. When this is the case, we call the steady state locally and globally stable. It is possible to be locally stable but not globally stable or to be locally unstable, as we shall soon see.

6.4.4   Positive Feedback

Figure 6.9 presents a simple example of positive feedback, where species X activates species Y (via phosphorylation) and the phosphorylated form of Y promotes the synthesis of X.

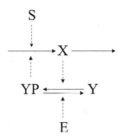

Figure 6.9 Example of positive feedback. Wiring diagram.

One possible set of equations to describe this network is

$$\frac{dX}{dt} = k_1 S + k_2 Y_P - k_3 X \tag{6.18}$$

$$\frac{dY_P}{dt} = \frac{k_4 X (Y_T - Y_P)}{K_{m4} + Y_T - Y_P} - \frac{k_5 E Y_P}{K_{m5} + Y_P} \tag{6.19}$$

where $[X] = [Y_T] - [S] = [E] = $ nM, $[k_1] = [k_2] = [k_3] = [k_4] = [k_5] = $ min^{-1}, and $[K_{m4}] = [K_{m5}] = $ nM. For this system of equations, the X-nullcline is $Y_P = (k_3/k_2)X - k_1 S$ and the Y_P-nullcline is $Y_P = Y_T \cdot G(k_4 X, k_5 E, K_{m4}/Y_T, K_{m5}/Y_T)$ (plotted in figure 6.10a). Notice that as we increase or decrease S, the X-nullcline moves down or up, and there is a range of S values, $S \in (S_{c1}, S_{c2})$, for which the nullclines intersect in three places. The points at the end of this range, where the system changes from one to three steady states, are called saddle-node (SN) bifurcation points. For $S_{c1} < S < S_{c2}$, we say that the system is *bistable*.

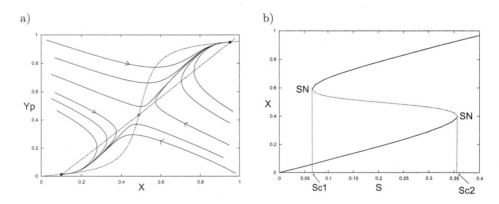

Figure 6.10 Example of positive feedback. a) Phase portrait. b) One parameter bifurcation diagram. Solid curves: stable steady states; dashed curve in between: unstable steady states. For $S_{c1} < S < S_{c2}$, the control system is bistable. Parameter values are $k_1 = k_4 = 1$, $k_2 = 0.8$, $k_3 = 1.2$, $S = 0.2$, $K_{m3} = K_{m4} = 0.05$, $Y_T = 1$, and $E = 0.5$. In a), trajectories move away from the unstable steady state (in the center) to one of two stable steady states.

A good way to visualize this bifurcation behavior is to plot a one parameter bifurcation diagram with S on the abscissa and either the X or YP concentration for each steady state on the ordinate, as in figure 6.10b. In general, this is a compact way to visualize how the dynamics of a system depend on its parameters. In this particular case, the system exhibits hysteresis as the parameter S passes back and forth through the region of bistability. That is, for low S, the system is at rest in the lower steady state (which is globally attracting). As S increases, the control system remains at this lower steady state, even after passing into the region of bistability because the lower steady state is stable with respect to small perturbations. Finally, as S increases past the upper bifurcation point (S_{c2}), the system abruptly shifts to the upper stable steady state. Now, if S were to decrease, the control system would remain in the upper steady state until S falls below the lower critical value, S_{c1}. Only then will the system switch back to the lower steady state. This non-reversibility is called *hysteresis*.

6.4.5 Mutual Antagonism

Mutual antagonism is a situation where an increase in either species means a decrease in the other, as in figure 6.11. Here, X phosphorylates Y, so more X implies less Y. Further, Y degrades X, so more Y means less X. The equations for this module are:

$$\frac{dX}{dt} = k_1 S - (k_2' + k_2 Y)X \tag{6.20}$$

$$\frac{dY}{dt} = \frac{k_3 E(Y_T - Y)}{K_{m3} + Y_T - Y} - \frac{k_4 XY}{K_{m4} + Y} \tag{6.21}$$

where $Y_T = Y + Y_P$ is constant, and the dimensions of the variables and rate constants are as before. In this case, the X-nullcline is now a hyperbola $Y = \frac{k_1 S - k_2' X}{k_2 X}$, which is similar to the negative feedback case. The Y-nullcline is $Y = Y_T \cdot G(k_3 E, k_4 X, K_{m3}/Y_T, K_{m4}/Y_T)$, which is a switch function that turns off as X increases. As in the case of positive feedback, there may be multiple intersections of the nullclines and a region of bistability for the parameter S (see figure 6.12a). Figure 6.12b shows a one-parameter bifurcation diagram for this system.

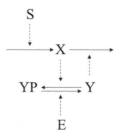

Figure 6.11   Example of mutual antagonism. Wiring diagram.

a)   b)

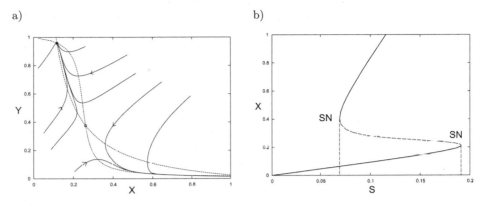

Figure 6.12 Example of mutual antagonism. a) Phase portrait. b) One parameter bifurcation diagram. Parameter values are $k_1 = k_2 = k_3 = k_4 = 1$, $k_2' = 0.1$, $S = 0.125$, $K_{m3} = K_{m4} = 0.05$, $Y_T = 1$, and $E = 0.25$. In a), trajectories move away from the unstable steady state (in the center) to one of two stable steady states.

Recently there have appeared a number of interesting experimental studies of bistability in macromolecular regulatory networks: in the MAP kinase signaling pathway of frog eggs (Ferrell Jr. and Machleder, 1998; Xiong and Ferrell Jr., 2003), in the activation of MPF in frog egg extracts (Sha et al., 2003; Pomerening et al., 2003), in the lactose utilization network of bacteria (Ozbudak et al., 2004), and in artificial genetic networks (Gardner et al., 2000).

6.5 Networks That Oscillate

There are three simple motifs that generate oscillatory behavior: activator-inhibitor, substrate-depletion, and delayed negative feedback.

6.5.1 Activator-Inhibitor

In figure 6.13, R stimulates its own production by phosphorylating E, and E_P also stimulates the production of X. (Think of E_P as the active form of a transcription factor.) As X increases, it promotes degradation of R. This negative feedback loop between X and R can cause oscillation (figure 6.14a). The equations for this system are

$$\frac{dR}{dt} = k_0 E_p + k_1 S - k_2 X R \tag{6.22}$$

$$\frac{dX}{dt} = k_3 E_p - k_4 X \tag{6.23}$$

where $E_p = E_T \cdot G(k_5 R, k_6, \frac{K_{m5}}{E_T}, \frac{K_{m6}}{E_T})$. The X-nullcline is $X = (k_3/k_4)E_p$ and the R-nullcline is $X = (k_0 E_p + k_1 S)/k_2 R$. The phase portrait (figure 6.14a) clearly shows the tendency of the vector field to drive trajectories in a circulatory pattern. For appropriate values of the parameters, the control system exhibits a closed trajectory, called a stable limit cycle. As the system rotates around the limit cycle, $R(t)$ and $X(t)$ oscillate periodically in time. The classic example of activator-inhibitor oscillations in cell biology is the cyclic AMP signaling system of the cellular slime mold, *Dictyostelium discoideum* (Martiel and Goldbeter, 1987); see box 6.5.

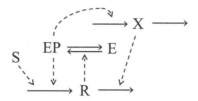

Figure 6.13   An activator-inhibitor oscillator. Wiring diagram.

a)   b)

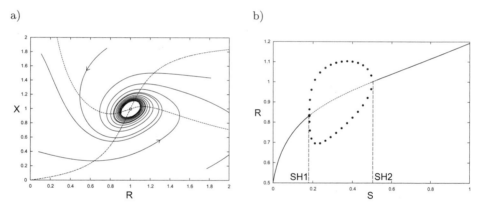

Figure 6.14   An activator-inhibitor oscillator. a) Phase portrait, b) One parameter bifurcation diagram. Parameter values are $k_0 = k_1 = k_2 = k_3 = k_5 = k_6 = 1$, $k_4 = 0.5$, $S = 0.5$, $K_{m5} = K_{m6} = 0.1$, and $E_T = 1$. In a), trajectories spiral in towards a limit cycle surrounding the unique unstable steady state. In b), the min and max values of R on the limit cycle oscillation are plotted in the region between the two Hopf bifurcations, S_{H1} and S_{H2}.

As we increase or decrease the signal strength, S, the R-nullcline shifts up or down, and though there is always only one steady state (one intersection of the nullclines), the stability of the steady state changes as we change S. For $S_{H1} < S < S_{H2}$, the steady state is unstable and surrounded by a limit cycle. The boundary points, S_{H1} and S_{H2}, are called Hopf bifurcation points. Figure 6.14b plots the one-parameter bifurcation diagram for this system, along with the amplitude of the oscillatory solution where it exists.

Box 6.5: Cyclic AMP oscillations in *Dictyostelium*

Cyclic AMP binds to a membrane receptor, which activates adenylate cyclase, the enzyme that catalyzes the synthesis of cyclic AMP from ATP (this is the positive feedback loop promoting the autocatalytic production of cyclic AMP). Meanwhile, cyclic AMP binding to the receptor promotes phosphorylation and desensitization of the receptor (this is the negative feedback loop, the desensitized receptor being the "inhibitor" that shuts off autocatalytic production of cyclic AMP). Next, cyclic AMP is hydrolyzed to 5'-AMP, which allows the receptor to slowly regain its sensitivity. Only then can there be a new burst of cyclic AMP synthesis.

6.5.2   Substrate-Depletion

In the substrate-depletion motif (figure 6.15), substrate X is converted by enzyme E into product R in a process which is autocatalytically amplified by R-dependent phosphorylation of E. This positive feedback loop leads to an explosive production of R which depletes the pool of the substrate, X. Naturally, once X is depleted, the production of R ceases and the degradation of R reduces its concentration below the level necessary to sustain the positive feedback loop. At this point, the pool of X begins to replenish. When X builds up sufficiently high, the positive feedback loop reengages, and a new burst of R synthesis commences.

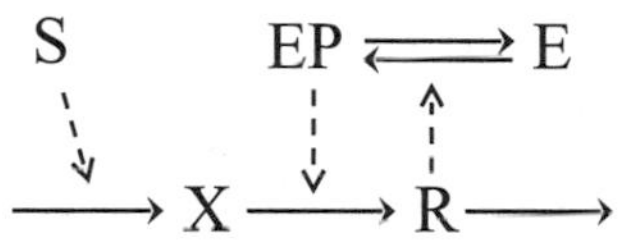

Figure 6.15   A substrate-depletion oscillator. Wiring diagram.

The differential equations for the model in figure 6.15 are

$$\frac{dX}{dt} = k_1 S - (k_0' + k_0 E_p)X \qquad (6.24)$$

$$\frac{dR}{dt} = (k_0' + k_0 E_p)X - k_2 R \qquad (6.25)$$

where $E_p = E_T \cdot G(k_3 R, k_4, \frac{K_{m3}}{E_T}, \frac{K_{m4}}{E_T})$. The X-nullcline is $X = \frac{k_1 S}{(k_0' + k_0)E_p}$ and the R-nullcline is $X = \frac{k_2 R}{(k_0' + k_0)E_p}$. Again, the phase portrait (figure 6.16a) shows a circulatory pattern around the steady state, and for a suitable choice of parameters, the system executes a stable limit cycle oscillation. In this case, the X-nullcline shifts upward (downward) as S increases (decreases). As before, the one-parameter bifurcation diagram shows two Hopf bifurcations and oscillatory solutions in between (figure 6.16b). Substrate-depletion oscillations are common in biochemical networks (see table 6.1).

a)   b)

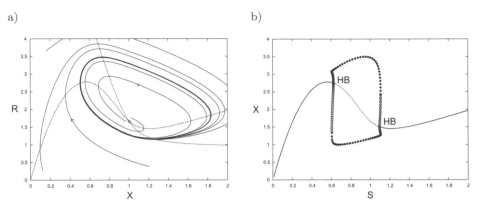

Figure 6.16 A substrate-depletion oscillator. a) Phase portrait. b) One parameter bifurcation diagram. Parameter values are $k_0 = k_1 = k_2 = k_3 = k_4 = 1$, $k_0' = 0.1$, $S = 1$, $K_{m3} = K_{m4} = 0.1$, and $E_T = 1$. In a), trajectories spiral in towards a limit cycle surrounding the unique unstable steady state. In b), the min and max values of X on the limit cycle oscillation are plotted in the region between the two Hopf bifurcations.

Table 6.1 Examples of substrate-depletion oscillators.

Example	Substrate	Activator	Reference
Frog egg	Cyclin B	MPF	(Tyson, 1991)
Glycolysis	F6P+ATP	FDP+ADP	(Selkov, 1968)
Calcium	Ca^{2+} in ER^a	Ca^{2+} in cytosol	(Dupont et al., 1991)
Ecosystem	Prey	Predator	(Maynard-Smith, 1974)

[a] ER = endoplasmic reticulum.

6.5.3   Delayed Negative Feedback

In figure 6.17, we present an example of delayed negative feedback. In this scheme, R phosphorylates E, which then binds to C to form X, and X is the active complex that degrades R itself (closing the negative feedback loop). This motif is derived from components of the cell cycle regulatory mechanism in eukaryotes, where R is MPF (mitosis promoting factor), E is APC (anaphase promoting complex), C is Cdc20, and X is a complex of APC and Cdc20.

The corresponding set of equations is

$$\frac{dR}{dt} = k_1 S - k_2 X R \qquad (6.26)$$

$$\frac{dE_P}{dt} = \frac{k_3 R (E_T - E_P)}{K_{mk} + E_T - E_P} - \frac{k_4 Q E_P}{K_{mp} + E_P}$$
$$\qquad\qquad - k_5 [E_P (C_T - X) - K_d X] \qquad (6.27)$$

$$\frac{dX}{dt} = k_5 [E_P (C_T - X) - K_d X] \qquad (6.28)$$

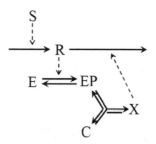

Figure 6.17   A negative feedback oscillator. Wiring diagram.

where $E_T = E + E_P$ is the total concentration of the APC, Q is the (fixed) concentration of a phosphatase, and $C_T = C + X$ is the total concentration of Cdc20. Having left the familiar territory of two-variable systems and phase plane portraits, we must now rely on numerical and qualitative results.

Table 6.2   Parameter values for the delayed negative feedback oscillator.

Parameter	Description	Value	Units
k_1	1st-order rate const	1	min^{-1}
k_2	2nd-order rate const	1	$nM^{-1}min^{-1}$
k_3	1st-order rate const	1	min^{-1}
k_4	1st-order rate const	1	min^{-1}
k_5	2nd-order rate const	0.01	$nM^{-1}min^{-1}$
K_{mk}	Michaelis constant	1	nM
K_{mp}	Michaelis constant	1	nM
K_d	Equilibrium constant	50	nM
S	Signal	0.3	nM
Q	Phosphatase concen.	100	nM
E_T	Total APC concen.	100	nM
C_T	Total Cdc20 concen.	1	nM

Using the parameter values in table 6.2 and S as the control parameter, we can compute a one-parameter bifurcation diagram (figure 6.18a) using numerical tools. In this case, there are two critical values of S at which the system undergoes Hopf bifurcations, with oscillatory solutions in between, $0.2 < S < 0.4$ (roughly). A typical oscillation for S in this range is plotted in figure 6.18b.

Small amplitude oscillations due to a "pure" negative feedback loop have recently been observed by Pomerening et al. (2005) in frog egg extracts (see box 6.6). A long negative feedback loop on PER-protein synthesis seems to play a major role in circadian rhythms, as described in chapter 2.

a)   b)

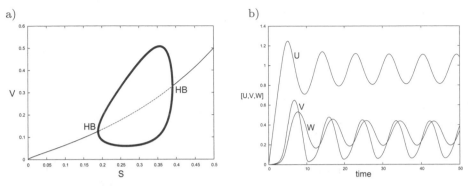

Figure 6.18 A negative feedback oscillator. a) One parameter bifurcation diagram. b) Simulation for $S = 0.3$. See table 6.2 for parameter values. Definitions: $u = R/Q$, $v = E_P/E_T$, $w = X/C_T$. In a), the min and max values of v on the limit cycle oscillation are plotted in the region between the two Hopf bifurcations.

Box 6.6: Negative feedback oscillations in frog egg extracts

Frog egg extracts are convenient preparations in which to observe the negative feedback loop involving MPF and APC, although the native regulatory system also includes a substrate-depletion oscillator involving phosphorylation of MPF (see table 6.1). By clever experimental techniques, Pomerening et al. (2005) have knocked out the substrate-depletion oscillator in a frog egg extract, revealing the negative feedback oscillator in its (presumably) unadulterated state. They observed "pure" negative feedback oscillations in their preparations. In the absence of the self-amplification of MPF activity provided by the substrate-depletion motif, the pure negative feedback oscillations are of considerably smaller amplitude and drive ambiguous transitions into and out of mitosis. It seems that the positive feedback mechanism is important to amplify the negative feedback oscillations and give unambiguous signals to nuclei to enter and leave mitosis.

6.6   A Multiple-Feedback Network: p53 and Mdm2

Transcriptional activator p53 is involved in cell cycle arrest and *apoptosis* (programmed cell death). In normal cells, the level of p53 is kept low by Mdm2, which promotes degradation of p53. The transcription of Mdm2 is activated by p53, creating a negative feedback loop (p53 → Mdm2 —| p53). When a cell is subjected to environmental stress causing DNA damage or oncogene activation, the activity of Mdm2 is weakened, allowing accumulation of p53 in the nucleus. Recently, it has been observed (Lahav et al., 2004) that p53 and Mdm2 undergo one or more oscillations in response to ionizing radiation (which causes double-stranded breaks of DNA), in an apparent attempt to repair the damage. Ciliberto et al. (2005) have proposed a simple mechanism (figure 6.1), including both negative and positive feedback, which quantitatively reproduces this behavior. The equations for the network in figure 6.1 are

$$\frac{d[\text{p53}]}{dt} = k_{s53} - k'_{d53}[\text{p53}] - k_f[\text{Mdm2}_{\text{nuc}}][\text{p53}]$$
$$+ k_r[\text{p53U}] \tag{6.29}$$

$$\frac{d[\text{p53U}]}{dt} = k_f[\text{Mdm2}_{\text{nuc}}][\text{p53}] - k_r[\text{p53U}] - k'_{d53}[\text{p53U}]$$
$$- k_f[\text{Mdm2}_{\text{nuc}}][\text{p53U}] + k_r[\text{p53UU}] \tag{6.30}$$

$$\frac{d[\text{p53UU}]}{dt} = k_f[\text{Mdm2}_{\text{nuc}}][\text{p53U}] - k_r[\text{p53UU}]$$
$$- (k'_{d53} + k_{d53})[\text{p53UU}] \tag{6.31}$$

$$\frac{d[\text{Mdm2}_{\text{cyt}}]}{dt} = k'_{s2} + \frac{k_{s2}[\text{p53}_{\text{tot}}]^m}{J^m_{s2} + [\text{p53}_{\text{tot}}]^m} - k'_{d2}[\text{Mdm2}_{\text{cyt}}]$$
$$- \frac{k_{ph}}{J_{ph} + [\text{p53}_{\text{tot}}]}[\text{Mdm2}_{\text{cyt}}] + k_{deph}[\text{Mdm2P}_{\text{cyt}}] \tag{6.32}$$

$$\frac{d[\text{Mdm2P}_{\text{cyt}}]}{dt} = \frac{k_{ph}}{J_{ph} + [\text{p53}_{\text{tot}}]}[\text{Mdm2}_{\text{cyt}}] - k'_{d2}[\text{Mdm2P}_{\text{cyt}}]$$
$$- k_{deph}[\text{Mdm2P}_{\text{cyt}}] - k_i[\text{Mdm2P}_{\text{cyt}}]$$
$$+ k_o[\text{Mdm2}_{\text{nuc}}] \tag{6.33}$$

$$\frac{d[\text{Mdm2}_{\text{nuc}}]}{dt} = V_{ratio}(k_i[\text{Mdm2P}_{\text{cyt}}] - k_o[\text{Mdm2}_{\text{nuc}}])$$
$$- k_{d2}[\text{Mdm2}_{\text{nuc}}] \tag{6.34}$$

$$\frac{d[\text{DNA}_{\text{dam}}]}{dt} = k_{dam}[\text{IR}] - k_{rep}[\text{p53}_{\text{tot}}]\frac{[\text{DNA}_{\text{dam}}]}{J_{dna} + [\text{DNA}_{\text{dam}}]} \tag{6.35}$$

where

$$k_{d2} = k'_{d2} + \frac{[\text{DNA}_{\text{dam}}]}{J_{dam} + [\text{DNA}_{\text{dam}}]}k''_{d2} \tag{6.36}$$

$$[\text{p53}_{\text{tot}}] = [\text{p53}] + [\text{p53U}] + [\text{p53UU}] \tag{6.37}$$

$$[\text{Mdm2}_{\text{tot}}] = [\text{Mdm2}_{\text{cyt}}] + [\text{Mdm2P}_{\text{cyt}}] + \frac{1}{V_{ratio}}[\text{Mdm2}_{\text{nuc}}] \tag{6.38}$$

$$V_{ratio} = \frac{V_{cytoplasm}}{V_{nucleus}} \tag{6.39}$$

$$[\text{IR}] = \text{imposed dose of ionizing radiation} \tag{6.40}$$

The network contains a long negative feedback loop (p53 $\rightarrow$ Mdm2$_{\text{cyt}}$ $\rightarrow$ Mdm2P$_{\text{cyt}}$ $\rightarrow$ Mdm2$_{\text{nuc}}$ —| p53) and a long positive feedback loop (p53 $\rightarrow$ PTEN —| PIP3 $\rightarrow$ Akt $\rightarrow$ Mdm2P$_{\text{cyt}}$ $\rightarrow$ Mdm2$_{\text{nuc}}$ —| p53). The positive feedback loop is shortened to p53 —| Mdm2P$_{\text{cyt}}$ $\rightarrow$ Mdm2$_{\text{nuc}}$ —| p53.

A simulation of this network (figure 6.19) compares very favorably with the experimental observations of (Lahav et al., 2004). As the radiation dose increases (figure 6.19d), the number of pulses of p53 increases. The reason for this curious "digital" response of p53 to DNA damage is made clear by the one-parameter bifurcation diagram in figure 6.20, where we plot system response, $[\text{p53}_{\text{tot}}]$, as a function of the extent of DNA damage, measured by k_{d2}. The positive feedback

Table 6.3 Parameter values for p53-Mdm2 network

Rate Constants ($\min^{-1}$)		
$k_{s53} = 0.055$	$k'_{d53} = 0.0055$	$k_{d53} = 8$
$k_f = 8.8$	$k_r = 2.5$	$k'_{s2} = 0.0015$
$k_{s2} = 0.006$	$k'_{d2} = 0.01$	$k''_{d2} = 0.01$
$k_{ph} = 0.05$	$k_{deph} = 6$	$k_i = 14$
$k_0 = 0.5$	$k_{dam} = 0.18$	$k_{rep} = 0.017$

Other Constants (dimensionless)		
$J_{s2} = 1.2$	$J_{ph} = 0.01$	$J_{dna} = 1$
$J_{dam} = 0.2$	$V_{ratio} = 15$	$m = 3$

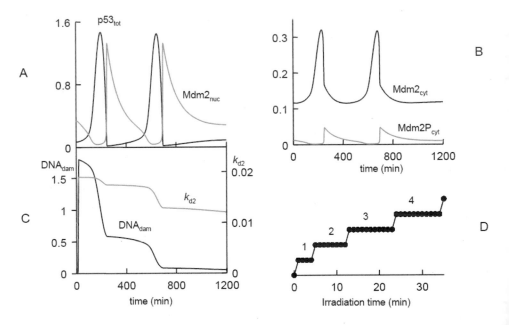

Figure 6.19 Simulation of gamma-irradiation experiment (reproduced from (Ciliberto et al., 2005), with permission). At the beginning of the simulation, the system is at steady state. (A) Between time 10 and 20, the control system is exposed to a transient damaging agent, which induces two large amplitude oscillations in p53$_{\text{tot}}$ and Mdm2nuc. (B) The oscillations of the two cytoplasmic forms of Mdm2 have a smaller amplitude compared to Mdm2$_{\text{nuc}}$ concentration in panel (A). (C) The oscillations are initiated as a consequence of k_{d2} increase, which is induced by irradiation. As the damage is repaired, k_{d2} decreases back to its basal value. (D) The number of pulses increases with the amount of damage. In the simulation, we count the number of oscillations as a function of the irradiation time. In panels A through C irradiation time = 10 min.

in the network creates multiple steady states (the S-shaped curve at $k_{d2} \cong 0.01$), but the negative feedback loop makes the upper steady state unstable. The unstable upper steady state is surrounded by stable limit cycle oscillations of $[\text{p53}_{\text{tot}}]$ and $[\text{Mdm2}_{\text{tot}}]$. The region of stable oscillation is bounded above (at $k_{d2} \cong 0.853$) by a Hopf bifurcation, and below (at $k_{d2} \cong 0.0135$) by a saddle-node-loop bifurcation. For a broad range of values of k_{d2}, that is, of DNA damage, the system responds with pulses of p53 and Mdm2 of fixed amplitude and period, exactly as observed by Lahav et al. (2004). As the damage is repaired, k_{d2} drops toward $k_{d2} \cong 0.01$, and the oscillations disappear abruptly as k_{d2} crosses the saddle-node-loop bifurcation point.

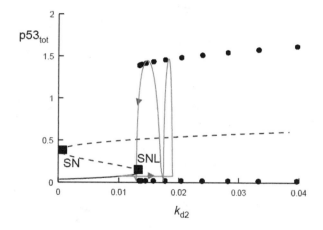

Figure 6.20   Bifurcation diagram (reproduced from (Ciliberto et al., 2005), with permission). Recurrent states (steady states and limit cycles) for p53_{tot} are plotted as functions of k_{d2}, the degradation rate of Mdm2_{nuc}. The solid line represents stable steady states, the dotted line unstable steady states. Black dots are the maxima and minima of the stable limit cycles. The grey solid line represents p53_{tot} as a function of k_{d2} from the simulation shown in figure 6.19. Notice that in figure 6.19 k_{d2} is a variable (see equations), while here it is a parameter (all other equations and parameter values as in table 6.3). When the qualitative behavior of the system changes, it is said to undergo a bifurcation. In the p53/Mdm2 model there is a saddle-node (SN) bifurcation at $k_{d2}=0.0018$ and a saddle-node-loop (SNL) bifurcation at $k_{d2}=0.0135$. Before the SNL bifurcation there is only one stable steady state, with low p53 ("p53 OFF"); after the SNL the steady state becomes unstable, surrounded by a stable limit cycle. The family of stable limit cycles disappears at a Hopf bifurcation at $k_{d2}=0.8532$ (not shown on the diagram).

A somewhat different model of p53/Mdm2 oscillations in response to ionizing radiation has recently been published by (Ma et al., 2005).

6.7   Conclusions

How are cell biologists to make reliable connections between molecular interaction networks and cell behaviors, when intuition fails in all but the simplest cases? In this chapter, we propose to make the connection by translating the reaction network into a set of nonlinear differential equations that describe how all the interacting species are changing with time (figure 6.21).

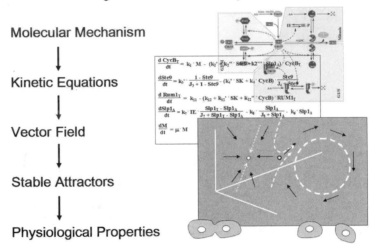

Figure 6.21   A dynamical perspective on molecular cell biology. To make a connection between molecular mechanisms and cell physiology, we convert the mechanism into a set of kinetic equations, by standard principles of biochemical kinetics, and view the kinetic equations as defining a vector field in the state space of the dynamical variables. The vector field has attracting solutions (steady states and oscillations) that correspond to characteristic physiological responses of the cell. The dependence of these attractors on kinetic constants (hence, on genetics and environment) are robustly captured in bifurcation diagrams.

The differential equations define a vector field in the state space of the network. The vector field points to certain stable attractors, which can be correlated with long-term, stable behavior of the network and of the cell it governs. Transitions from one stable attractor to another represent the responses of the cell to specific perturbations (signals). A natural way to describe the signal-response properties of a regulatory network is in terms of a one-parameter bifurcation diagram, which efficiently displays the stable attractors (steady states and oscillators) and transitions between attractors as signal strength (the "parameter") varies.

We have illustrated these ideas with simple examples of linear, hyperbolic, and sigmoidal signal-response curves, of bistable switches based on positive feedback or mutual inhibition, and of limit cycle oscillators based on substrate depletion, activator-inhibitor interactions, or time-delayed negative feedback. These fundamental motifs (switches and oscillations) can be coupled together into networks of increasing complexity and dynamical potential. Interested readers should now be ready to read and understand the growing body of literature that takes this dynamical perspective on interesting topics in cell physiology. Some nice examples include: cell-cycle control (Tyson et al., 2002), circadian rhythms (Leloup and Goldbeter, 1998), lysogenic viruses (Arkin et al., 1998), quorum sensing in bacteria (James et al., 2000; Usseglio Viretta and Fussenegger, 2004), NF-κB signaling (Hoffmann et al., 2002), and programmed cell death (Eissing et al., 2004).

7 Qualitative Approaches to the Analysis of Genetic Regulatory Networks

Hidde de Jong and Delphine Ropers

There is a growing demand for methods that can make predictions of qualitative properties of the dynamics of molecular interaction networks, that is, properties that are invariant for a range of reaction mechanisms and values of kinetic constants. On the one hand, precise and quantitative information on reaction mechanisms and kinetic constants is not available for most networks of biological interest. On the other hand, in many situations predictions of qualitative rather than quantitative dynamical properties are appropriate for gaining an understanding of the functioning of a molecular interaction network. This chapter discusses three examples of qualitative approaches for the analysis of genetic regulatory networks, allowing qualitative dynamical properties to be inferred from currently-available incomplete and non-quantitative data. The approaches are based on different formalisms, namely discrete abstractions of differential equations, Boolean networks, and graphs. We illustrate the approaches by means of a simple two-gene network and give an example of their application to real biological systems.

7.1 Motivation for Qualitative Approaches

Differential equations are the classical formalism for modeling the behavior of natural and man-made systems. Therefore not surprisingly, they form the most prominent approach for the modeling, analysis, and simulation of molecular interaction networks (chapter 6). The application of differential equations rests on a well-established theoretical framework for the deterministic modeling of the kinetics of biochemical reaction systems (Cornish-Bowden, 1995; Heinrich and Schuster, 1996). In addition, a variety of mathematical methods and computer tools is available for transforming the model equations into experimentally-testable predictions. Many excellent examples exist to demonstrate the capability of differential equations to help gain insight into the functioning of molecular interaction networks of biological importance, such as the control of circadian rhythms in *Drosophila* (Leloup and Goldbeter, 2003), the metabolism of the red-blood cell in humans (Joshi and Palsson, 1989), the regulation of the cell cycle in yeast and higher eukaryotes (Tyson et al., 2001), and the signaling pathway involved in the maturation of oocytes in *Xenopus laevis* (Ferrell Jr. and Machleder, 1998).

In principle, the use of differential equations allows precise numerical predictions of the behavior of molecular interaction networks to be made. However, for many networks of biological interest, such predictions are difficult or even impossible to obtain. In the first place, the biochemical reaction mechanisms underlying the interactions are usually not or incompletely known, which complicates the formulation of the differential equation models. In the second place, quantitative data on kinetic constants and molecular concentrations are generally absent, even for extensively studied textbook systems.

In addition to these practical difficulties, one can raise the question of whether quantitative information on reaction mechanisms and kinetic constants is essential for understanding the functioning of molecular interaction networks. In fact, it is reasonable to assume that many important dynamical properties of living systems do not depend on precise numerical values or a specific reaction mechanism (Barkai and Leibler (1997); Eldar et al. (2002); Rao et al. (2004); see also chapter 2). In other words, in many situations *qualitative dynamical properties*—dynamical properties that are invariant for a range of reaction mechanisms and values of kinetic constants—are more important than quantitative dynamical properties. The qualitative properties express the intimate connection between the behavior of the system and the structure of the network of molecular interactions, independently from the quantitative details of the latter.

For all of the above reasons, there is a growing interest in qualitative approaches for the modeling, analysis, and simulation of molecular interaction networks, capable of inferring qualitative properties of the system dynamics from currently-available incomplete and non-quantitative data. The aim of this chapter is to review existing qualitative approaches focusing on one particular type of molecular interaction network, *genetic regulatory networks*. These networks mainly involve interactions between proteins and nucleic acids, controlling the transcription and translation of genes. In the next sections, we first explain the notion of qualitative dynamical property and then discuss three representative examples of qualitative approaches. These approaches are based on formalisms increasingly remote from the differential equation models traditionally used: discrete abstractions of differential equations, Boolean networks, and graphs.

Of course, our review of qualitative approaches does not pretend to be exhaustive. Some of the more obvious omissions of this chapter are Petri nets (Koch et al., 2005; Reddy et al., 1996), constraint-based models (Covert et al., 2004; Edwards and Palsson, 2000; Edwards et al., 2001a; Stelling et al., 2002), and process algebras (Eker et al., 2002; Regev et al., 2001). On the one hand, these model formalisms partially overlap with the formalisms discussed here, or are reviewed at length in other chapters of the book (chapter 5). On the other hand, they seem to have been mostly used for metabolic and signal transduction networks, rather than for genetic regulatory networks. For further reference, the reader may consult other reviews of qualitative approaches in systems biology (de Jong, 2002; Gagneur and Casari, 2005). Another restriction of the scope of this chapter is that we focus on methods that derive behavior predictions from structural information on the network, thus leaving out of consideration methods that aim at inferring the network structure from observations on the behavior of the system. Such reverse engineering methods have been developed for Boolean networks and their relatives (Akutsu et al., 2000; Ideker et al., 2000; Laubenbacher and Stigler, 2004; Liang et al., 1998; Perkins et al., 2004), while Bayesian methods for inferring graph models from gene expression data are discussed in chapter 4.

7.2 Qualitative Properties of the Dynamics of Genetic Regulatory Networks

In order to develop the notion of qualitative dynamical property, we will consider a simple network of two genes (figure 7.1). Each of the genes encodes a regulatory protein that inhibits the expression of the other gene, by binding to a site overlapping the promoter of the gene. Simple as it is, this *cross-inhibition network* is a basic component of more complex, real networks and makes it possible to analyze some characteristic aspects of cellular differentiation (Monod and Jacob, 1961; Thomas and d'Ari, 1990). Moreover, its dynamical properties have been experimentally tested by Gardner et al., who have reconstructed the network in *Escherichia coli* cells by cloning the genes on a plasmid. The genes on the plasmid have been chosen such that the network functions independently from the rest of the cell and the activity of the corresponding proteins can be regulated by external signals (Gardner et al. (2000); see also chapter 13).

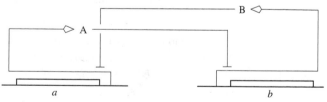

Figure 7.1 Example of a simple genetic regulatory network, composed of two genes *a* and *b*, the proteins A and B, and their regulatory interactions.

The cross-inhibition network can be modeled by means of differential equations. Generally speaking, a genetic regulatory network of n genes is conveniently described by a system of n ordinary differential equations:

$$\frac{dx_i}{dt} = f_i(\boldsymbol{x}), \; i \in \{1, \ldots, n\}, \tag{7.1}$$

where $\boldsymbol{x} = (x_1, \ldots, x_n)' \in \Omega$ is a vector of cellular protein concentrations, and $\Omega \subseteq \mathbb{R}^n_{\geq 0}$. The function $f_i : \ \Omega \to \mathbb{R}$ expresses how the rate of change of the concentration of the protein encoded by gene i depends on the concentrations $\boldsymbol{x}$ of the proteins in the cell. A differential equation model of the cross-inhibition network is shown in figure 7.2a. The variables x_a and x_b represent the concentration of the proteins A and B. The time derivative of x_a equals the difference between the rate of synthesis of A, given by $\kappa_a\, h^-(x_b, \theta_b, m_b)$, and the rate of degradation, given by $\gamma_a\, x_a$. The use of the sigmoidal *Hill function* h^-, shown in figure 7.2b, means that for low concentrations of the protein B, gene a is expressed at a rate close to its maximum rate κ_a, whereas for high concentrations of B, the expression of the gene is almost completely repressed. The shape of the Hill function is in agreement with experimental evidence (Ptashne, 1992). The rate of degradation of A is proportional to the concentration of the protein. The differential equation for x_b has an analogous interpretation.

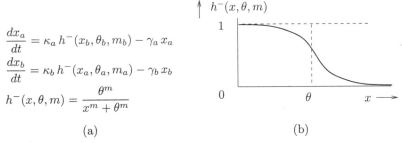

$$\frac{dx_a}{dt} = \kappa_a\, h^-(x_b, \theta_b, m_b) - \gamma_a\, x_a$$

$$\frac{dx_b}{dt} = \kappa_b\, h^-(x_a, \theta_a, m_a) - \gamma_b\, x_b$$

$$h^-(x, \theta, m) = \frac{\theta^m}{x^m + \theta^m}$$

(a)   (b)

Figure 7.2   (a) Nonlinear ordinary differential equation model of the cross-inhibition network (figure 7.1). The non-negative variables x_a and x_b correspond to the concentrations of proteins A and B, respectively, the parameters κ_a and κ_b to the synthesis rates of the proteins, the parameters γ_a and γ_b to degradation constants, the parameters θ_a and θ_b to the threshold concentrations, and the parameters m_a and m_b to the degree of cooperativity of the interactions. All parameters are positive. (b) Graphical representation of the characteristic sigmoidal form, for $m > 1$, of the Hill function $h^-(x, \theta, m)$.

The use of the nonlinear Hill function does not make it possible to analytically solve the system of differential equations. However, the dynamics of the two-gene network can be analyzed in the phase plane, by means of standard techniques developed in dynamical systems theory (Kaplan and Glass (1995); Strogatz (2000); see also chapter 6). The phase portrait in figure 7.3a shows that the system is *bistable*, in the sense that it possesses two asymptotically stable equilibrium points, at which either protein A or protein B is present at a high concentration. The third equilibrium point, characterized by intermediate concentrations for proteins A and B, is unstable and cannot be experimentally observed. The phase-plane analysis also reveals that the system exhibits *hysteresis*. If one strongly perturbs the system from one of its stable equilibria—for instance, by provoking the degradation of the protein present at a high concentration—the other equilibrium can be reached (figure 7.3b). From then onwards, even if the source of strong degradation has disappeared, the system will remain at the new equilibrium. In other words, the analysis suggests that a simple molecular mechanism may allow the system to switch from one functional state to another. Interestingly, this has been confirmed by the experiments of Gardner et al. mentioned above.

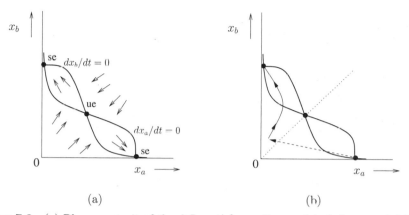

(a)   (b)

Figure 7.3 (a) Phase portrait of the differential equation model of the cross-inhibition network (figure 7.2). The system has two asymptotically stable equilibrium points (se) and one unstable equilibrium point (ue). The equilibria lie at the intersection of the nullclines of x_a and x_b (drawn curves annotated by $dx_a/dt = 0$ and $dx_b/dt = 0$). (b) Hysteresis effect, resulting from a transient perturbation of the system (dashed line with arrow).

The above-mentioned dynamical properties of the cross-inhibition network—bistability and hysteresis—are invariant for a range of parameter values and molecular mechanisms. That is, they are qualitative properties of the system. For instance, a moderate increase of the value of θ_b causes the nullcline of x_a to move upwards (figure 7.4a). This deforms the phase portrait, but does not lead to the loss of the bistability and hysteresis properties. For large changes in parameter values though, the qualitative properties may not be invariant. Figure 7.4b shows what happens for values of θ_b close to, or above, κ_b/γ_b. In this case one of the stable equilibria and the unstable equilibrium approach annihilate each other, so that the system is no longer bistable and no longer exhibits hysteresis. In the terminology of dynamical systems theory, a bifurcation has occurred (Kaplan and Glass (1995); Strogatz (2000); see also chapter 6).

The invariance of the dynamical properties of genetic regulatory networks for changes in the reaction mechanisms and the value of kinetic constants can be defined more generally and more rigorously than has been done here. A classical treatment is found in the book by Andronov et al., who define qualitative dynamical properties as those properties invariant under trajectory-preserving topological mappings of (a region of) the phase space (Andronov et al., 1973; Kalagnanam et al., 1991). What will interest us in this chapter are not the technicalities of the definition, but rather the practical question of how qualitative properties of the dynamics of

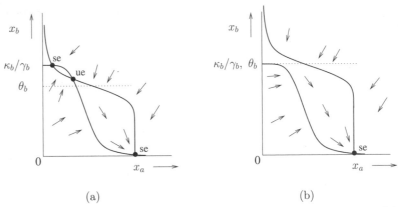

(a)   (b)

Figure 7.4 Changes in the phase portrait of the differential equation model of the cross-inhibition network (figure 7.2), following a change in the value of the parameter θ_b. The change in (a) preserves the bistability and hysteresis properties, whereas the more important change in (b) does not.

genetic regulatory networks can be inferred from weak information on the structure of the network and the type of the interactions.

The example of the two-gene network shows that dynamical systems theory provides concepts and techniques for the characterization of qualitative properties of dynamical systems, notably the construction and analysis of phase portraits. Unfortunately, the theoretical results become much weaker when studying higher-order systems. While higher-order systems can sometimes be reduced to second-order systems, by time-scale abstraction or other model simplifications, this is not always possible. More fundamentally, the insights to be gained from dynamical systems theory are to a large extent based on geometrical representations that are difficult to manipulate in higher dimensions.

The qualitative methods discussed in the remainder of this chapter also try to infer qualitative properties of the dynamics of genetic regulatory networks. However, they employ model formalisms and representations of the system dynamics that are more abstract than differential equations and phase portraits. Although the predictions that can be made by means of these qualitative methods are less precise, they are based on theoretical results and computational techniques that better scale up to large and complex systems. In the next sections, we will discuss three examples of qualitative methods, based on model formalisms that make increasingly stronger abstractions of the process of gene regulation.

7.3   Discrete Abstractions of the Dynamics of Differential Equations

In response to the problem that the use of phase portraits does not scale up well to higher dimensions, alternative representations of the qualitative dynamics could be proposed. A closer look at figure 7.3a, the phase portrait of the two-gene network, suggests one such alternative. In every region of the phase plane bounded by the nullclines of x_a and x_b, the system behaves in a qualitatively homogeneous way, in the sense that the derivatives of the concentration variables have the same sign everywhere. When solution trajectories leave one region and enter another, the sign of one or both derivatives changes. The partition of the phase space into regions suggests a *discrete abstraction* of the qualitative dynamics of the system, which will be formally developed in this section.

Consider again the system of differential equations 7.1, describing a network of n genes. The nullclines of the system are the hypersurfaces on which $f_i(x) = 0$. They define a *partition* $\mathcal{R}$ of the phase space Ω, consisting of regions R in which the time derivative of each of the concentration variables x_i has a unique sign. We introduce a function $\pi : \mathcal{R} \to \{-, 0, +\}^n$, associating a derivative sign pattern to each region $R \in \mathcal{R}$. Figure 7.5a shows the partition of the phase space obtained in the case of the model of the two-gene network discussed in the previous section, while figure 7.5c shows the derivative sign pattern for each region. Suppose that $R, R' \in \mathcal{R}$ are two contiguous regions in the phase space. If there is a solution of equation 7.1 that on a time interval T reaches R' from R, without leaving $R \cup R'$, then we say that there exists a *transition* from R to R', denoted $R \to R'$. Formally, $\to \subseteq \mathcal{R} \times \mathcal{R}$. For instance, by looking at the direction of the vector field in figure 7.3a, it can be easily inferred that $R^1 \to R^8$, $R^1 \to R^9$, and $R^1 \to R^1$ are possible transitions. Self-transitions like $R^1 \to R^1$ occur when solutions do not instantaneously cross a region, but remain in it for some time.

The above definitions underlie an abstract, discrete representation of the dynamics of the continuous differential equation system in the form of a *transition graph*:

$$\text{TG} = (\mathcal{R}, \to). \tag{7.2}$$

The vertices of the graph correspond to the regions of the phase space and the edges to the transitions between regions. Each of the regions can be seen as a discrete or qualitative state of the system, in which the derivatives of the concentration variables have a particular sign pattern. A sequence of regions σ is a *path* in the transition graph if and only if $\sigma = (R^0)$ or $\sigma = (R^0, \ldots, R^m)$, $m > 0$, and for all $i \in [0, \ldots, m-1]$, we have $R^i \to R^{i+1}$. A path in the transition graph gives a qualitative description of the behavior of the system, in the sense that it describes how the derivative sign pattern changes over time. The transition graph corresponding to the cross-inhibition network is shown in figure 7.5b.

This discrete representation of the dynamics of a continuous differential equation system facilitates the analysis of the behavior of genetic regulatory networks. In fact, equilibrium points correspond to regions $R \in \mathcal{R}$, such that $\pi(R) = (0, \ldots, 0)'$, while the stability of the equilibrium points can be inferred from the outgoing transitions of the contiguous regions. As expected, the transition graph in figure 7.5b contains three regions coinciding with equilibrium points. In the case of R^5 and R^{13}, the paths starting in the contiguous regions lead towards the equilibrium points, thus

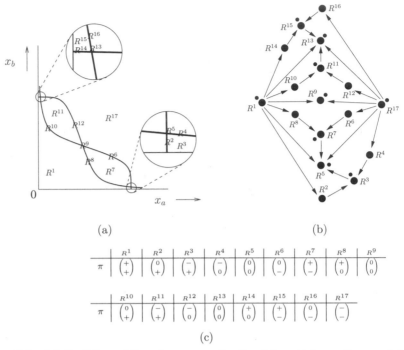

(a)   (b)

	R^1	R^2	R^3	R^4	R^5	R^6	R^7	R^8	R^9
π	$\begin{pmatrix}+\\+\end{pmatrix}$	$\begin{pmatrix}0\\+\end{pmatrix}$	$\begin{pmatrix}-\\+\end{pmatrix}$	$\begin{pmatrix}-\\0\end{pmatrix}$	$\begin{pmatrix}0\\0\end{pmatrix}$	$\begin{pmatrix}0\\-\end{pmatrix}$	$\begin{pmatrix}+\\-\end{pmatrix}$	$\begin{pmatrix}+\\0\end{pmatrix}$	$\begin{pmatrix}0\\0\end{pmatrix}$

	R^{10}	R^{11}	R^{12}	R^{13}	R^{14}	R^{15}	R^{16}	R^{17}
π	$\begin{pmatrix}0\\+\end{pmatrix}$	$\begin{pmatrix}-\\+\end{pmatrix}$	$\begin{pmatrix}-\\0\end{pmatrix}$	$\begin{pmatrix}0\\0\end{pmatrix}$	$\begin{pmatrix}+\\0\end{pmatrix}$	$\begin{pmatrix}+\\-\end{pmatrix}$	$\begin{pmatrix}0\\-\end{pmatrix}$	$\begin{pmatrix}-\\-\end{pmatrix}$

(c)

Figure 7.5 (a) Partition of the phase space for the differential equation model of the cross-inhibition network (figure 7.2), using the nullclines $dx_a/dt = 0$ and $dx_b/dt = 0$. (b) Transition graph consisting of domains and transitions between domains. The small dots next to domains indicate self-transitions. (c) Sign of dx_a/dt and dx_b/dt in the regions of the phase space, as defined by the function π.

suggesting that the latter are stable. This is not the case for R^9, which corresponds to an unstable equilibrium point. The hysteresis property can also be inferred from the transition graph, bearing in mind that for a region R' to be reachable from another region R, there must be a path running from R to R'. It is then immediately seen from figure 7.5b that a perturbation from R^5 to R^1 may cause the system to attain the other equilibrium point at R^{13}. More generally, the transition graph can be shown to be a conservative approximation (Alur et al., 2000; Chutinan and Krogh, 2001) of the differential equation system, in the sense that every solution of the latter corresponds to some path in the former (although the converse does not necessarily hold). This means that the transition graph can be safely used to study the qualitative dynamics of the differential equation system.

The above reformulation of the study of qualitative properties of differential equation systems raises two important questions. First, for which range of parameter values is the transition graph invariant? Second, how can we actually compute the transition graph in the absence of precise numerical information on the parameters? These problems have been addressed in several areas of computer science and control theory, in particular in the context of work on *qualitative simulation* in artificial intelligence (de Jong, 2005; Kuipers, 1994) and on *discrete abstractions* in hybrid systems theory (Alur et al., 2000; Chutinan and Krogh, 2001). Examples of the application of these approaches to the analysis of genetic regulatory networks are the hybrid automaton models of Ghosh and Tomlin (2004) and the qualitative differential equation models of Heidtke and Schulze-Kremer (1998). Below we will discuss in more detail the qualitative simulation method developed in our group (Batt et al., 2005a; de Jong et al., 2004b), which has been specifically designed so as to favor scaling up to large and complex networks.

The qualitative simulation method uses simplified models of gene regulation proposed by Glass and Kauffman in the early seventies (Edwards et al., 2001b; Glass and Kauffman, 1973; Gouzé and Sari, 2002; Mestl et al., 1995). The major difference between these so-called *piecewise-linear* differential equations and the differential equations used to model the cross-inhibition network is that the sigmoid function h^- in figure 7.2b is replaced by a *step function* s^- that abruptly changes from 1 to 0 at a threshold value θ:

$$s^-(x, \theta) = \left\{ \begin{array}{l} 1, \text{ if } x < \theta, \\ 0, \text{ if } x > \theta. \end{array} \right. \tag{7.3}$$

By means of the threshold values of the variables, the phase space can be partitioned into hyperrectangular regions, in each of which the system behaves in a qualitatively-homogeneous manner.[1] It has been proven that the transition graph defined on this partition is invariant for certain inequality constraints on the parameters that can often be inferred from the experimental literature. Moreover, it is possible to compute the transition graph, by means of simple symbolic rules, from a piecewise-linear differential equation model of the network supplemented by inequality constraints. The qualitative simulation method has been implemented in the computer tool Genetic Network Analyzer (GNA) (de Jong et al., 2003a).

The method and the computer tool have been applied to the analysis of the complex genetic regulatory network controlling the initiation of sporulation in the Gram-positive soil bacterium *Bacillus subtilis* (de Jong et al., 2004a). Under conditions of nutrient deprivation, *B. subtilis* cells may not divide and form a dormant, environmentally-resistant spore instead. The decision to abandon growth and division and initiate sporulation involves a radical change in the pattern of gene expression in the cell. The switch of the genetic program is controlled by a complex regulatory network integrating various environmental, cell-cycle, and metabolic signals. A graphical representation of the network controlling the initiation of sporulation is shown in figure 7.6a, displaying key genes and their promoters, proteins encoded by the genes, and the regulatory action of the proteins.

The graphical representation of the network has been translated into a piecewise-linear differential equation model of the network supplemented by inequality constraints on the parameters. The resulting model consists of 11 variables and 48 parameters constrained by 70 parameter inequalities. The choice of the latter is largely determined by biological data. Using this model, the response of wild-type and mutant cells to nutrient deprivation has been simulated by means of GNA. This

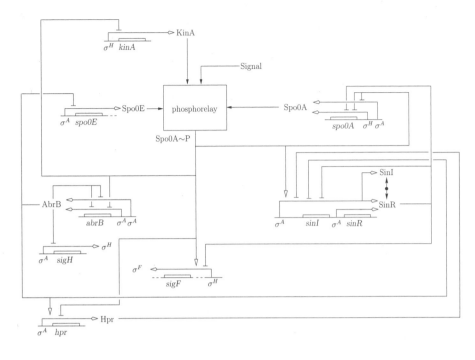

(a)

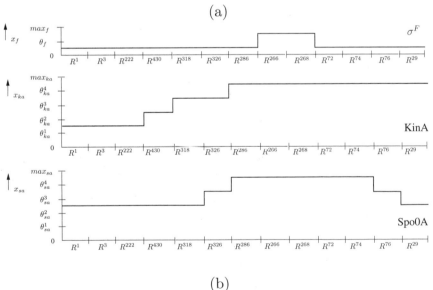

(b)

Figure 7.6 (a) Key genes, proteins, and regulatory interactions making up the network involved in *B. subtilis* sporulation (de Jong et al., 2004a). (b) Path in the transition graph produced by qualitative simulation of the response of a *B. subtilis* cell to nutrient deprivation. The figure shows how the threshold boundaries on the concentrations of σ^F, KinA, and Spo0A evolve as a consequence of the successive region transitions. The concentration of σ^F transiently crosses the threshold θ_f, above which this sigma factor directs the transcription of genes essential for later stages of the sporulation process.

has given rise to transition graphs consisting of up to several hundreds of regions, most of which can be simplified by eliminating regions that are instantaneously traversed and therefore of limited biological interest. An example of a path in such a transition graph is shown in figure 7.6b.

Analysis of the sporulation network by means of GNA has revealed that essential features of the initiation of sporulation in wild-type and mutant strains of *B. subtilis* are reproduced by the model (de Jong et al., 2004a). In particular, the choice between division and sporulation is seen to be determined by competing positive and negative feedback loops influencing the accumulation of the phosphorylated form of the transcription factor Spo0A. Above a certain threshold concentration, Spo0A~P activates various genes whose expression commits the bacterium to sporulation, such as genes encoding sigma factors that control the alternative developmental fates of the mother cell and the spore (figure 7.6b). Other examples of the application of GNA are the qualitative modeling and simulation of quorum sensing in the pathogenic bacterium *Pseudomonas aeruginosa* (Usseglio Viretta and Fussenegger, 2004) and the carbon starvation response in *E. coli* (Ropers et al., 2005).

In summary, the basic idea informing discrete abstractions of the dynamics of continuous systems is that they partition the phase space into regions in which the system behaves in a qualitatively homogeneous manner. The state of the system is henceforward described by the region in which it resides, and a change of state by a transition from one region to another. In comparison with the underlying continuous system, the use of discrete abstractions leads to a loss of quantitative precision. However, for many questions the abstract description is sufficiently informative and well-adapted to the available biological data. Moreover, transition graphs are easy to analyze and capture qualitative properties of the system that are invariant for moderate changes in parameter values. The transition graphs grow exponentially with the number of genes in the network, though, which limits the scalability of the approach. Tools for the formal verification of qualitative dynamical properties reduce this problem to some extent (Batt et al., 2005b; Bernot et al., 2004; Chabrier-Rivier et al., 2004), but cannot entirely avoid it.

7.4 Boolean Networks

Instead of discretizing the dynamics of a continuous model, one could also study qualitative properties of genetic regulatory networks by directly starting with a discrete model. The sigmoid shape of the Hill function in figure 7.2b suggests that, to a first approximation, a gene can be described as either active (on) or inactive (off). That is, if the gene is active (inactive), the protein it encodes is assumed present (absent) in the cell. The change in gene expression can be described by making the assumption that the change in activation state of a gene is determined in a combinatorial fashion by the activation state of other genes, in particular genes encoding regulatory proteins. The above intuitions have been formalized by

Boolean networks, which has become popular in the wake of a groundbreaking study by Kauffman (1969; see Kauffman (1993); Somogyi and Sniegoski (1996) for reviews).

Let the vector $\hat{\boldsymbol{x}} = (\hat{x}_1, \ldots, \hat{x}_n)' \in \{0,1\}^n$ of Boolean variables represent the state of a network of n genes. Each $\hat{x}_i$ has the value 0 (inactive) or 1 (active). The activation state $\hat{x}_i$ of a gene at a discrete time-point $t+1$ is determined by a Boolean function $\hat{b}_i : \{0,1\}^n \to \{0,1\}$, which defines $\hat{x}_i(t+1)$ in terms of $\hat{\boldsymbol{x}}(t)$. Most of the time, $\hat{b}_i$ will effectively depend on the state of only k_i of the n genes. The variable $\hat{x}_i$ is also referred to as the *output* of the gene and the k_i variables from which it is computed the *inputs*. For k_i inputs the total number of possible Boolean functions $\hat{b}_i$ mapping the inputs to the output is $2^{2^{k_i}}$. This means that for $k_i = 2$ there are 16 possible functions, the logical "AND" and the logical "OR" being two examples. In summary, the dynamics of a genetic regulatory network can be described by a Boolean network defined by the following equations:

$$\hat{x}_i(t+1) = \hat{b}_i(\hat{\boldsymbol{x}}(t)), \ i \in \{1, \ldots, n\}, \tag{7.4}$$

where $\hat{b}_i$ maps k_i inputs to an output value. The Boolean network corresponding to the cross-inhibition network is shown in figure 7.7a. The network is quite simple, consisting of two Boolean variables, $\hat{x}_a$ and $\hat{x}_b$, each connected to the other variable by means of a logical "NOT". For illustrative purposes, an example of a slightly more complex Boolean network—involving three variables, two inputs per variable, and various logical functions—is shown in figure 7.7b.

The dynamics of a Boolean network are conveniently represented by means of a transition graph. Let $\hat{\boldsymbol{x}}, \hat{\boldsymbol{x}}' \in \{0,1\}^n$ be two states of a Boolean network with n genes. There exists a *transition* from $\hat{\boldsymbol{x}}$ to $\hat{\boldsymbol{x}}'$, denoted by $\hat{\boldsymbol{x}} \to_s \hat{\boldsymbol{x}}'$, if and only if $\hat{x}'_i = \hat{b}_i(\hat{\boldsymbol{x}})$, for every $i \in \{1, \ldots, n\}$. Formally, $\to_s \subseteq \{0,1\}^n \times \{0,1\}^n$. The subscript s indicates that the transitions are *synchronous*, that is, the states of all genes are updated simultaneously. Notice that the transitions are *deterministic*, in the sense that every state of the system has a single successor. The transition graph can be formally defined as follows:

$$\text{BTG} = (\{0,1\}^n, \to_s). \tag{7.5}$$

A sequence of states $\hat{\sigma}$ in the transition graph is a *path* if and only if $\hat{\sigma} = (\hat{\boldsymbol{x}}^0)$ or $\hat{\sigma} = (\hat{\boldsymbol{x}}^0, \ldots, \hat{\boldsymbol{x}}^m)$, $m > 0$, and for all $i \in [0, \ldots, m-1]$, we have $\hat{\boldsymbol{x}}^i \to_s \hat{\boldsymbol{x}}^{i+1}$. Because the number of states in the state space of a Boolean network is finite, when extended every path will eventually reach an *attractor*, either a state having itself as a successor (*point attractor*) or a state cycle (*cyclic attractor*). The attractor states and the states leading to the attractor together constitute the *basin of attraction* of the attractor. For simple networks the attractors and their basins of attraction can be calculated by hand, but for larger systems the use of computer programs becomes inevitable, given that the size of the transition graphs scales exponentially with the number of genes (section 7.3). Examples of such computer programs are DDLab (Wuensche, 2003) and GINsim (Chaouiya et al., 2003).

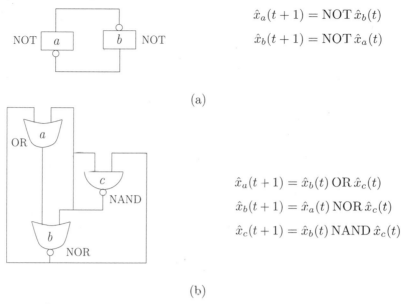

$$\hat{x}_a(t+1) = \text{NOT}\,\hat{x}_b(t)$$

$$\hat{x}_b(t+1) = \text{NOT}\,\hat{x}_a(t)$$

(a)

$$\hat{x}_a(t+1) = \hat{x}_b(t)\,\text{OR}\,\hat{x}_c(t)$$

$$\hat{x}_b(t+1) = \hat{x}_a(t)\,\text{NOR}\,\hat{x}_c(t)$$

$$\hat{x}_c(t+1) = \hat{x}_b(t)\,\text{NAND}\,\hat{x}_c(t)$$

(b)

Figure 7.7 (a) Boolean network model of the cross-inhibition network in figure 7.1, in the form of an electronic circuit and a system of equations. (b) Illustration of a more complex Boolean network model. $\hat{x}_a$, $\hat{x}_b$, $\hat{x}_c$ are Boolean variables; NOT, OR, NOR, NAND are Boolean functions.

The transition graph for the cross-inhibition network is shown in figure 7.8a. Because the network consists of two genes, we have a total of four states, denoted by 00, 01, 10, and 11. For instance, 01 means that gene a is off and gene b is on. The graph consists of three unconnected attractors: the point attractors 10 and 01, and a cyclic attractor consisting of 00 and 11. Notice that one of the proteins is present and the other absent in the point attractors, which allows these states to be related to the stable equilibrium points in the differential equation model (figure 7.3a). On the other hand, the cyclic attractor has no obvious counterpart in the differential equation model. Moreover, the hysteresis property of the cross-inhibition network is not preserved. When perturbing the activation state of one of the genes in a point attractor, that is, when randomly flipping the Boolean value of $\hat{x}_a$ or $\hat{x}_b$, the system makes a transition to 00 or 11. From there it can neither return to its original state nor reach the other point attractor. This illustrates that there are situations in which the idealizations underlying Boolean networks are not appropriate, in the sense that the models cannot account for experimentally-observed dynamical properties.

Several generalizations of the standard Boolean network formalism have been proposed, based on assumptions that are more realistic from a biological point of view. Instead of using synchronous transitions between states, one could resort to *asynchronous* transitions, in which the activation state of only a single gene is updated at a time. Formally, this amounts to replacing $\rightarrow_s$ by a new transition relation $\rightarrow_a \subseteq \{0,1\}^n \times \{0,1\}^n$, where $\hat{\boldsymbol{x}} \rightarrow_a \hat{\boldsymbol{x}}'$ if and only if $\hat{x}_i' = \hat{b}_i(\hat{\boldsymbol{x}})$, for

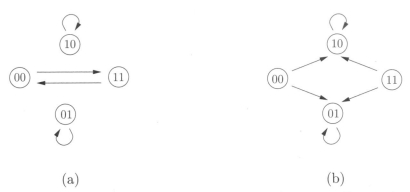

(a)   (b)

Figure 7.8 Transition graphs for the Boolean network corresponding to the cross-inhibition network (figure 7.7a). (a) Transition graph for synchronous transitions ($\rightarrow_s$). (b) Transition graph for asynchronous transitions ($\rightarrow_a$).

some $i \in \{1, \ldots, n\}$, and $\hat{x}'_j = \hat{x}_j$, for all $j \neq i$.[2] The use of $\rightarrow_a$ makes the Boolean network *nondeterministic*, in the sense that a state may have up to n successors. The asynchronous transition graph for the cross-inhibition network is shown in figure 7.8b. As can be immediately verified, both the bistability and the hysteresis property are now reproduced. Asynchronous Boolean networks underlie the logical method introduced by Thomas, who also proposes the use of multivalued instead of Boolean variables, in order to distinguish multiple levels of gene expression (Thomas, 1973; Thomas and d'Ari, 1990). Another generalization of the standard formalism are *probabilistic Boolean networks*, Boolean network which do not associate a single Boolean function with a gene, but rather a probability distribution on a set of Boolean functions, thus taking into account uncertainty in the state transitions (Shmulevich et al., 2002a,b).

Boolean network models and their generalizations have been able to give insights into the functioning of actual genetic regulatory networks, as demonstrated by studies of pattern formation in early *Drosophila* development (Albert and Othmer, 2003; Sánchez et al., 1997; Sánchez and Thieffry, 2001, 2003), flower morphogenesis in *Arabidopsis* (Mendoza et al., 1999), and mucus production in *Pseudomonas aeruginosa* (Bernot et al., 2004). The results confirm the basic assumption of qualitative approaches that many important dynamical properties of an organism do not depend on specific reaction mechanisms or precise numerical values for kinetic constants, but are to a large extent determined by the structure of interactions of the network (chapter 2).

More generally, standard Boolean networks have been a popular model for theoretical investigations of the relation between the structure and dynamics of genetic regulatory networks. The basic idea of this so-called *ensemble approach*, proposed by Kauffman (1993, 2004), is to consider the ensemble of Boolean networks sharing some structural properties, such as a particular number of inputs per gene or Boolean functions of a particular type. We can then randomly sample networks from the ensemble and provide statistics on their dynamical properties, such as the number of attractors and the size of their basins of attractions. Under the above assumptions, it follows that the dynamical properties typically found for the sampled networks must be attributed to the structural properties defining the ensemble, and can hopefully be explained by the latter. The biological relevance of the ensemble approach is based on the hypothesis of Kauffman that real genetic regulatory networks belong to an ensemble whose defining properties remain to be discovered (2004). The aim is to identify this ensemble by the iterative generation of ensembles and the comparison of their typical dynamical properties with experimental data. The simplicity of standard Boolean networks makes them excellent models for doing the extensive computations required by the ensemble approach.

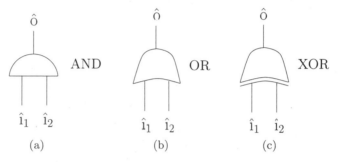

Figure 7.9 Examples of (a)–(b) canalizing and (c) non-canalizing Boolean functions with inputs $\hat{\imath}_1$ and $\hat{\imath}_2$, and output $\hat{o}$. In the case of the AND (OR) function, a value of 0(1) for one of the inputs forces the output to 0(1). In the case of the XOR function no such value for one of the inputs exists.

An interesting recent application of the ensemble approach is a study of the logical functions expected to play a role in gene regulation (Kauffman et al., 2003b). A gene is regulated by many transcription factors, which may combine to yield a complex regulatory logic, as demonstrated by the analysis of the control of expression of the *Endo16* gene in the sea urchin (Yuh et al., 1998). However, there is some evidence that one particular class of logical functions, so-called *canalizing functions*, are overrepresented in gene regulation (Harris et al., 2002). In terms of Boolean logic, a canalizing function has at least one input, such that for at least one value of this input and for any other value of the remaining inputs, the output value is fixed to either 0 or 1 (Kauffman, 1993) (figure 7.9). Kauffman et al. have generated random Boolean networks of thirty genes having a structure of interactions equal to that of the core of the yeast transcriptional regulatory network. The Boolean functions in the networks are chosen from either a distribution of all Boolean functions or a distribution of canalizing Boolean functions. The networks sampled from the two ensembles show different stability properties, that is, they tend to react differently to random perturbations of an initial state. In fact, networks with canalizing Boolean functions are on average more stable than networks with arbitrary Boolean functions, in the sense that in the former case the state after the perturbation remains closer to the initial state (Kauffman et al., 2003b, 2004). Since this stability or robustness is expected on biological grounds (chapter 2), the results could be taken as evidence that actual genetic regulatory networks belong to an ensemble of canalizing networks.

Contrary to the approach discussed in the previous section, Boolean networks do not have any intrinsic abstraction relation to an underlying continuous system. Standard Boolean networks describe the state of the system by a vector of Boolean values, indicating for each of the genes whether it is on or off. The state of the system evolves in discrete time, as a consequence of transitions that may change the activation state of one or more genes. The attractiveness of Boolean network models is based on the intuitiveness of the representation of gene regulation by means of Boolean functions and the simplicity of the algorithms used for computing the transition graphs. However, the classical formalisms make strong simplifying assumptions, in particular the use of binary values for gene activation and synchronous transitions. These assumptions are relaxed in the generalized formalisms mentioned, thus increasing the biological validity of the models, but at the price of losing some of the computational and mathematical simplicity of the standard approach.

7.5   Graphs

The previous section suggests another way to predict the behavior of genetic regulatory networks. If certain structural properties can be shown to imply specific dynamical properties, then the behavior of the system could be inferred, at least tentatively, by verifying whether the network possesses these structural properties. As shown in section 7.2, the cross-inhibition structure of the example network endows it with bistability and hysteresis properties for a large range of parameter values. Therefore, one might argue, identifying the cross-inhibition pattern in a network could provide us with a clue as to the dynamics of the system. This demands a study of the structural properties of genetic regulatory networks, for which graph models are well-suited.

A *graph* is defined as a tuple (V, E), with V a set of *vertices* and $E \subseteq V \times V$ a set of *edges* (Berge, 2001):

$$G = (V, E). \tag{7.6}$$

The edges represent the relation between vertices and may be directed or undirected. A *directed* edge is a pair $(i, j) \in E$ of vertices, where i denotes the head and j the tail of the edge. The order of the vertices is of no importance, if (i, j) is an *undirected* edge. A genetic regulatory network can now be seen as a directed graph in which the vertices represent genes and the edges regulatory interactions. The edges are directed from regulating to regulated genes, from genes encoding transcription factors to the targets of the transcription factors. In order to express the nature of the regulatory interactions, we can label the edges. By defining a directed edge as a tuple $(i, j, s) \in E$, with $s \in \{+, -\}$, it can be indicated whether i is activated or inhibited by j. As an illustration, the graph corresponding to the cross-inhibition network is shown in figure 7.10a. This simple graph is composed of two vertices, a and b, as well as two directed edges. Figure 7.10b shows a slightly more complex example of a graph model, added for illustrative purposes.

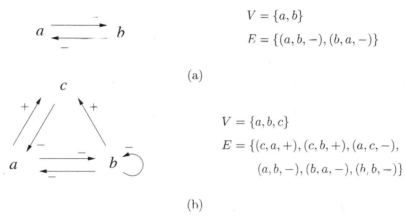

$$V = \{a, b\}$$

$$E = \{(a, b, -), (b, a, -)\}$$

(a)

$$V = \{a, b, c\}$$

$$E = \{(c, a, +), (c, b, +), (a, c, -),$$

$$(a, b, -), (b, a, -), (b, b, -)\}$$

(b)

Figure 7.10   Directed graphs representing (a) the cross-inhibition network in figure 7.1, and (b) a more complex network, added for comparison.

The representation of a genetic regulatory network as a graph allows the analysis of its structural properties by means of graph-theoretical techniques (Barabási and Oltvai, 2004; Newman, 2003). The global connectivity properties of the network can, for instance, be described by the average degree and the degree distribution of the vertices. The *degree k* of a vertex indicates the number of edges to which it is connected (if necessary, incoming and outgoing edges can be distinguished). $\langle k \rangle$ denotes the *average degree* and $P(k)$ the *degree distribution* of the graph. The properties give an indication of the complexity of the graph and allow different types of graphs, and therefore networks, to be distinguished (figure 7.11). In classical *random graphs* (figure 7.11a), also called Erdős-Rényi graphs, the probability that a given vertex has k edges follows a Poisson distribution $P(k)$. That is, the vertices typically have $\langle k \rangle$ edges and the vertices having significantly more or less edges than $\langle k \rangle$ are extremely rare, as shown in part (c) of the figure. By contrast, in *scale-free graphs* (figure 7.11b), the vertex degrees obey a power-law distribution $P(k) \sim k^{-\gamma}$, shown in part (d) of the figure. Scale-free graphs are inhomogeneous, in that most of the vertices have few edges, whereas some vertices, called *hubs*, have many edges and hold the graph together.

For values of the degree exponent γ between 2 and 3, scale-free graphs have a number of surprising properties. First, the average length of the path between two vertices of the graph is proportional to $\log \log |V|$, where $|V|$ denotes the number of vertices of the graph (Barabási and Oltvai, 2004; Newman, 2003). This is even

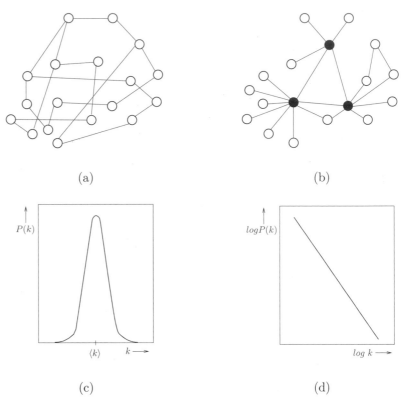

(a)   (b)

(c)   (d)

Figure 7.11   Schematic illustration of the architecture of (a) random and (b) scale-free undirected graphs (Bray, 2003). The degree distribution follows (c) a Poisson distribution in random graphs and (d) a power-law distribution in scale-free graphs. k denotes the degree of a vertex and $P(k)$ the degree distribution. The filled vertices in (b) are hubs.

shorter than the average path length in random graphs, which scales as $\log |V|$ and confers on them the *small-world property* (Watts and Strogatz, 1998). The small-world property implies that local perturbations can quickly spread out through the entire regulatory network. Second, the presence of hubs makes scale-free graphs robust against accidental failures (Albert et al., 2000; Jeong et al., 2000, 2001). Whereas randomly removing a certain number of vertices disintegrates a random graph, in a scale-free graph this mainly affects the numerous low-degree vertices, the absence of which does not decompose the graph. Third, unlike classical random graphs, scale-free graphs can possess a modular structure (Ravasz et al. (2002); chapter 3). Such graphs are constructed by iteratively combining small and tightly-clustered modules of vertices into a hierarchical structure.

There is now quite some evidence that genetic regulatory networks, and many other biological and non-biological networks, are scale-free (Dobrin et al., 2004; Featherstone and Broadie, 2002; Guelzim et al., 2002; Jeong et al., 2000, 2001; Lee et al., 2002; Maslov and Sneppen, 2002; Tong et al., 2004; Wagner and Fell, 2001). Some caution should be observed in interpreting the results though. Because current data on regulatory interactions are incomplete, a subnetwork of the actual network is analyzed, which may have a different degree distribution (Stumpf et al., 2005). Moreover, the particular graph representation chosen to model the network may bias the results, as shown by Arita for the *E. coli* metabolic network (Arita, 2004). In the case of genetic regulatory networks, graph models are usually restricted to direct transcription regulation interactions, thus ignoring indirect interactions that are mediated by metabolites binding to transcriptional regulators (Alm and Arkin, 2003).

Although the analysis of structural properties like the degree distribution may yield insights that seem intuitively important for understanding the dynamics of a network, it is not so easy to actually pin down their behavioral consequences. Some studies have begun to explore the topic, using a combination of graph theory and Boolean networks (Aldana and Cluzel, 2003; Fox and Hill, 2001; Oosawa and Savageau, 2002), but the relation between global structural properties and network dynamics is still largely an open question (Strogatz, 2001). Alternatively, one could follow an approach focusing on local structural properties, in particular specific patterns of interactions between the network components. In this vein, Thomas has conjectured that positive feedback loops in regulatory networks are a necessary condition for the occurrence of multiple equilibrium points (Thomas and d'Ari, 1990), a conjecture that has been proven since by a number of authors (Cinquin and Demongeot (2002); Gouzé (1998); Plahte et al. (1995); Snoussi (1998); Soulé (2003); see also Remy et al. (2003)).[3] In the remainder of this section, we will discuss another example of the latter approach, the identification and functional analysis of motifs.

Loosely speaking, *network motifs* are recurring patterns of interactions between a small number of network components (Milo et al. (2002); Shen-Orr et al. (2002); see Wolf and Arkin (2003), for a review). Their functional importance has been suggested by the evolutionary conservation of motifs within the yeast protein-protein interaction network (Wuchty et al., 2003) and the convergent evolution towards the same motifs in the transcriptional regulatory network of diverse species (Conant and Wagner, 2003).

Techniques for discovering motifs consist in the identification of small patterns in the graph that are overrepresented when compared to a randomized version of the same graph (Milo et al., 2002, 2004b). More precisely, all possible patterns of a fixed number of vertices occurring in the graph are enumerated in a first step. The statistical significance of a pattern is then inferred from the comparison of the original graph, corresponding to the biological network, with a set of randomized graphs, in which each vertex has the same number of incoming and outgoing edges as the corresponding vertex in the original graph. A pattern is a motif if it occurs significantly more often in the original graph than in the randomized graphs. Since randomized networks are supposed to be free of any type of natural selection, the overrepresentation of the motifs can be assumed to have an evolutionary origin, reflecting the importance of the function performed by the motif. This conclusion is sensitive, though, to the particular randomization procedure followed and requires careful statistical validation (Artzy-Randrup et al., 2004; Milo et al., 2004a).

Shen-Orr et al. have searched the transcriptional regulatory network of *E. coli* for motifs, using information from the database RegulonDB and the literature (Shen-Orr et al., 2002). In this network, consisting of 855 genes and 1,330 regulatory interactions, three overrepresented motifs have been identified: the *feedforward loop*, in which a transcription factor regulates a second transcription factor and both regulate together their target gene (figure 7.12a); the *single-input motif*, in which a group of genes is controlled by a single transcription factor; and the *dense-overlapping regulons*, in which genes and the transcription factors controlling their expression form a highly-overlapping structure. The feedforward loop is the motif occurring most frequently (40 times) in the *E. coli* network. This has been subsequently confirmed for an extended version of the same network, in which an even higher number of feedforward loops have been found (Ma et al., 2004a).

What could be the functional role of the feedforward loop? In a follow-up study the group of Alon has theoretically and experimentally demonstrated the information processing task carried out by this motif. Using a differential equation model of the feedforward motif, they show that its role might be to filter out fluctuations in input stimuli and allow a rapid response when the stimuli disappear (Mangan et al., 2003; Mangan and Alon, 2003). Consider the feedforward loop in figure 7.12b, where the transcription factors X and Y together activate the gene *z*. When X is active and above a threshold concentration, the input signal activating X is transmitted to the output Z through a direct path from X and an indirect path from X through Y. Hence, a transient signal is not transmitted to Z, since it does not allow the concentration of Y to reach a threshold level high enough to stimulate the expression of gene *z* (figure 7.12c). On the other hand, a persistent input signal enables the concentration of Y to rise and eventually allows Z to pass its threshold level. The functioning of the feedforward loop is asymmetric, since the inactivation of X leads to the rapid downregulation of *z*. The above predictions have been experimentally verified for the L-arabinose utilization system in *E. coli* using reporter genes (Mangan et al., 2003). In this feedforward loop, CRP corresponds to the general transcription factor X and AraC to the specific transcription factor Y, while *z* is the operon *araBAD*.

The discussion of the feedforward motif illustrates how a clear, well-defined function can be assigned to a pattern of interactions that is overrepresented in the network. Unfortunately, it is not always possible to make such a straightforward connection between structure and function. Usually, motifs do not occur in isolation, but rather overlap to generate complex *motif clusters* (Dobrin et al., 2004). This makes it difficult to draw definite conclusions on the function of an individual pattern of interactions occurring in a cluster. For instance, it is not obvious that the network in figure 7.10b, in which the cross-inhibition pattern is embedded in a more complex feedback structure, also possesses the bistability and hysteresis properties for a large range of parameter values. In order to establish this, the

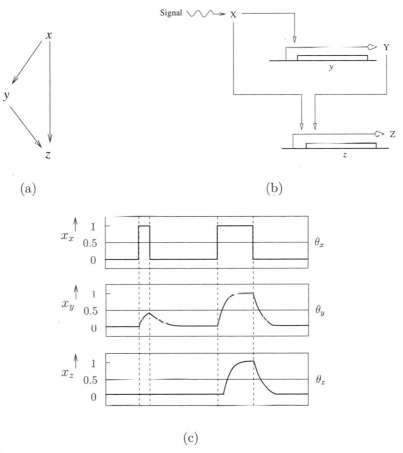

Figure 7.12   (a) Feedforward loop motif in graph representation. (b) Feedforward loop in genetic regulatory network, where it is assumed that both X and Y are necessary for expression of z. (c) Dynamic properties of the feedforward loop (Shen-Orr et al., 2002). x_x, x_y, and x_z denote the concentrations of X, Y, and Z respectively, and θ_x, θ_y, and θ_z, their threshold levels. The input signal activates X.

static graph analysis need to be complemented by a dynamic analysis of the type discussed in earlier sections of this chapter.

7.6   Discussion

In order to understand how the functions and development of living organisms are controlled by the networks of interactions between genes, proteins, and small molecules within and between cells, we need mathematical methods and computer tools. We have insisted on the demand for qualitative approaches for the modeling, analysis, and simulation of genetic regulatory networks, that is, approaches capable of inferring properties of the dynamics of genetic regulatory networks that are invariant for a range of reaction mechanisms and values of kinetic constants (section 7.2). The interest in these qualitative approaches derives from the fact that, for most networks of biological interest, we do not dispose of detailed information on the reaction mechanisms and precise numerical values for the kinetic constants. Moreover, it is reasonable to assume that many dynamical properties of living organisms are robust to at least some variations in mechanisms and numerical values. This does not mean that qualitative approaches always impose themselves: there are biological questions for which quantitative precision is required, and there do exist systems for which detailed, quantitative information is available. Quantitative and qualitative approaches should be seen as complementary rather than mutually exclusive.

In this chapter we have reviewed three approaches for the analysis of qualitative properties of the dynamics of genetic regulatory networks, based on increasingly-abstract modeling formalisms: discrete abstractions of differential equations, Boolean networks, and graphs. Whereas the first two approaches use models that explicitly describe the dynamics of the system, the third approach is based on the assumption that an analysis of the structure of the system provides useful insights into its dynamics. The structural and dynamic approaches focus on distinct, but complementary aspects of the networks, and in practice need to be applied in combination. As discussed in section 7.5, the assignment of a function to a network motif or module requires tools for studying the network dynamics. On the other hand, tools for analyzing the network structure are critical for dealing with the problem that the transition graphs generated by the qualitative approaches scale exponentially with the size of the network. Instead of studying the dynamics of very large networks directly, it seems more judicious to distinguish network modules, study the dynamics of these modules individually, and then analyze the interactions between the modules on a higher level of abstraction, using simplified models for each of the modules (chapter 3).

What are the main future directions for research on qualitative approaches towards the analysis of genetic regulatory networks? Of the many challenges that could be mentioned, two deserve special attention in our view. The first concerns the impact of qualitative modeling of genetic regulatory networks on experimental biology. The qualitative approaches have some features that make them particularly suitable for the systems studied at the forefront of experimental research, notably the ability to deal with incomplete and non-quantitative information. However, while many excellent qualitative models have been developed and described in the literature, examples of the experimental verification of novel predictions made by these models are still relatively rare.

A second challenge is the development of qualitative methods that allow the integrated analysis of genetic regulatory networks and other types of molecular interaction networks, such as metabolic and signal transduction networks. In this chapter we have focused on the interactions occurring in gene regulation, but some of the methods could be applied or extended to the modeling of other types of interactions, such as enzyme-catalyzed reactions or protein-protein interactions.

The search for motifs in a network composed of transcription regulation and protein-protein interactions in yeast is a good example (Yeger-Lotem et al., 2004). Alternatively, hybrid approaches could be followed, in which methods adapted to the specific problems of each type of network are combined. The use of Boolean network models to add the effects of gene regulation to flux balance models of *E. coli* metabolism is a case in point (Covert et al. (2001, 2004); see also chapter 5). Both directions are promising but have been little explored thus far.

Acknowledgments

We would like to thank Grégory Batt, Julien Gagneur, Jean-Luc Gouzé, and Vincent Lacroix for their careful reading of and helpful comments on earlier versions of this chapter. We acknowledge the financial support of the ARC initiative of INRIA (project GDyn), the ACI IMPBio initiative of the French Ministry for Research (project BacAttract), and the NEST Adventure program of the European Commission (project Hygeia).

Notes

1.   The resulting partition does not actually preserve the derivative sign pattern, as in figure 7.5b, but a finer-grained partition can be formulated for which this is the case (Batt et al., 2005b).

2.   Mixtures of synchronous and asynchronous transitions can also be proposed. Such transition relations allow some, but not necessarily all, genes to change their activity state simultaneously.

3.   The cross-inhibition network in figure 7.1 is an example of a network with a positive feedback loop, due to the fact that each protein positively influences the expression of its own gene, by inhibiting the expression of the gene encoding the inhibitor of its own gene.

8   Stochastic Modeling of Intracellular Kinetics

Johan Paulsson and Johan Elf

Cellular events are triggered by random collisions between molecules. If each type of event occurred numerous times per generation, this randomness could possibly average out and cells could behave deterministically. But many central cellular reactions by contrast occur so infrequently that substantial relative fluctuations arise spontaneously. By affecting the rates of other reactions, these fluctuations can propagate through networks and spread to any cellular process. The tendencies to correct fluctuations also range from strong to insignificant depending on the kinetic mechanisms, causing some systems to behave with high precision and others to accumulate extreme variability. Many aspects of life in the individual cell are therefore best understood probabilistically. This is further supported by a rapidly growing body of experimental work. Most macromolecules are found to be present in very low numbers per chemical species, and studies measuring single cell concentrations almost invariably report large variation from cell to cell. This chapter introduces some theoretical aspects of randomness in simple genetic and metabolic networks, including both general mathematical techniques and specific biological phenomena.

8.1   Chapter Overview

The text is organized as follows: section 8.2 discusses the assumptions behind stochastic modeling of chemical reactions. Section 8.3 presents a multivariate model for stochastic gene expression that can be solved exactly. Section 8.4 gives an interpretation of the fluctuation-dissipation theorem (FDT), tailored to biochemical processes. Section 8.5 uses simulations and FDT approximations for systems that operate near critical points, and section 8.6 shows how such fluctuations can be tamed by negative feedback. Section 8.7, finally, gives some examples of noise-induced transitions and constructive roles of noise. For Monte Carlo methods we refer to chapter 9 and chapter 16 for spatial and nonspatial descriptions respectively.

For more extensive theoretical treatments of chemical fluctuations we refer to the several textbooks available (Erdi and Tóth, 1989; Gardiner, 1985; Keizer, 1987; van Kampen, 1992).

8.2   Basic Models for Stochastic Kinetics

8.2.1   Differential Equations for Probabilities, Averages, and Macroscopic Concentrations

A starting point for stochastic descriptions of chemical reactions is to define a sufficiently complete set of state variables such that changes only depend on the current state (Lax, 1960). This could in principle include continuous variables, such as temperature, cell age, or volume, but to simplify the notation we here only account for discrete jumps corresponding to changes in the number of molecules of each species.

Consider i_{max} different chemical species homogeneously distributed[1] in a volume Ω. The state of the system is defined by state vector $\mathbf{n} = [n_1 \cdots n_{i\,\max}]^T$ where n_i is the number of molecules of species i. Let there be j_{max} types of reactions and let reaction j change component i from n_i to $n_i + \nu_{ij}$ with a rate r_j that depends only on the current state of the system, $\mathbf{n}$. The probability that reaction j occurs in a small time interval Δt is then $r_j(\mathbf{n})\Delta t$. The integers ν_{ij} form an $i_{max} \times j_{max}$ stoichiometric matrix ν where the j:th column ν_j corresponds to the change in the state vectors when a reaction of type j occurs.

The probability of arriving in state $\mathbf{n}$ during a short time interval Δt is the sum of the probabilities for leaving from other states $\mathbf{n} - \nu_j$ to state $\mathbf{n}$ in a single reaction, $\Delta t \sum_j r_j (\mathbf{n} - \nu_j) P (\mathbf{n} - \nu_j, t)$. Similarly, the probability of leaving state $\mathbf{n}$ in this time interval is $\Delta t \sum_j r_j (\mathbf{n}) P (\mathbf{n}, t)$. The probability $P(\mathbf{n}, t + \Delta t)$ to be in state $\mathbf{n}$ at time $t + \Delta t$ is thus:

$$P(\mathbf{n}, t + \Delta t) = P(\mathbf{n}, t) - \underbrace{\Delta t \sum_j r_j (\mathbf{n}) P (\mathbf{n})}_{\text{leaving}} + \underbrace{\Delta t \sum_j r_j (\mathbf{n} - \nu_j) P (\mathbf{n} - \nu_j)}_{\text{arriving}}$$

(8.1)

Rearranging equation 8.1, dividing by Δt and taking the continuity limit $\Delta t \to 0$ leads to a time-continuous state-discrete Markov process—the *master equation* for the system of chemical reactions (Singer, 1953; van Kampen, 1992):

$$\frac{d}{dt} P(\mathbf{n}, t) = \sum_{j=1}^{M} \left(r_j(\mathbf{n} - \nu_j) P(\mathbf{n} - \nu_j, t) - r_j(\mathbf{n}) P(\mathbf{n}, t) \right)$$

(8.2)

The motion of the averages $\langle n_i \rangle$ can be postulated directly from the rates and event sizes:

$$\frac{d \langle n_i \rangle}{dt} = \sum_j v_{ij} \langle r_j (\mathbf{n}) \rangle \tag{8.3}$$

The corresponding macroscopic concentrations x_i are the number of molecules per unit volume in an infinitely large system, that is, a system where the rates and the volume go to infinity in such a way that x_i converges. For many processes this could be at least hypothetically achieved by taking an infinite population of cells, removing walls and membranes, and keeping the remaining cell components well mixed, similar to some *in vitro* experiments. In practice, however, rate equations dx_i/dt are typically constructed directly from first principles and often used to approximate average cell behavior. For nonlinear systems it must then be remembered that there are qualitative differences between true averages and their macroscopic idealizations (Bharucha-Reid, 1960; Renyi, 1954), something we will return to in section 8.7.

8.2.2 Simulating Paths of the Master Equation

From the analysis above equation 8.1 we saw that the probability that some reaction will occur in a short time interval Δt is the sum of their individual probabilities, $\Delta t \sum_j r_j (\mathbf{n}) = \Delta t r_0$. Let $p(t)$ be the probability that the system has *not* left state $\mathbf{n}$ at time t, given that it was in state $\mathbf{n}$ at time $t=0$. The change in $p(t)$ between time t and $t + \Delta t$ is then $\Delta p(t) = p(t + \Delta t) - p(t) = -r_0 p(t) \Delta t$. Taking the limit of continuous time, $\Delta t \to 0$, and solving the resulting differential equation $dp(t)/dt = -r_0 p(t)$, gives $p(t) = \exp(-r_0 t)$. The probability that a reaction has occurred at time t is thus $F(t) = 1 - p(t) = 1 - \exp(-r_0 t)$, that is, the system resides an exponentially distributed time in state $\mathbf{n}$, with an average $\langle t \rangle = 1/r_0$. The probability that the first event is reaction j is in turn given by its relative contribution to the total rate, $\Pr (reaction_j \,| any\ reaction) = r_j (\mathbf{n})/r_0$.

This defines a simple algorithm for generating individual paths of the random process: pick the next reaction time from an exponential distribution and choose event type according to the fractional rates. Physical or chemical considerations are then only important when choosing what states and jumps to include to make the description Markovian. The algorithm itself is indistinguishable from the definition of a homogenous time-continuous state-discrete Markov processes (Doob, 1945). Daniel Gillespie—who effectively pioneered its use for generating sample paths of chemical reaction networks (Cao et al., 2004b; Gillespie, 1976, 1977)—refers to it as the stochastic simulation algorithm in chapter 16, but here we will call it the Gillespie-Doob algorithm.

8.2.3   Elementary Reactions

All kinetic modeling relies on condensing fast transitions between short-lived physical states into single reaction steps. The most common example is the approximation that a system is "well-stirred," where bimolecular reactions are described without accounting for spatial positions. This essentially assumes that the molecules involved in a chemical reaction have time to diffuse over the whole volume before they are likely to be involved in a reaction again (Gardiner and Steyn-Ross, 1984). Descriptions of unimolecular reactions may similarly assume rapid internal equilibrations, so that transitions to functionally different states again effectively behave as if they had no memory. Because these assumptions are so common and more or less similar from system to system, they are often considered "elemental." However, the key assumption that the time-scales are separated can be equally true for more complicated chemical reactions. For example, transcription involves an enormous number of small steps, yet for many purposes the whole process of making RNA molecules could possibly be approximated as Poissonian, that is, with exponentially distributed dwell times between births of new RNA molecules.

Complicated reactions that can be represented by a single step are said to be "elementary complex"—elementary because they effectively behave as elemental reactions on the time-scale studied, and complex because they could be broken down into several more elemental reactions (Keizer, 1987). For example consider a protein that rapidly equilibrates between two conformations

$$A_1 \xrightleftharpoons[\lambda_2]{\lambda_1} A_2$$

such that it is in conformation A_2 during $p = \lambda_1 (\lambda_1 + \lambda_2)^{-1}$ percent of the time. If the A_2 conformation participates in another reaction $A_2 \xrightarrow{\lambda_3} B$ which occurs on a much slower timescale than the conformational changes in A, then this reaction can be considered elementary complex with rate $\lambda_3 p n_A$ where n_A is the total number of A molecules. For the purpose of modeling changes in B it is then not necessary to include the two different conformations of A.

In some cases it is important to also account for the fact that molecules involved in intermediate states of complex reactions cannot participate in other reactions. For example, let a protein autorepress its own transcription such that active genes are repressed with rate $\lambda_2 n$ and inactive genes are derepressed with rate λ_1. Each gene then switches on and off as

$$on \xrightleftharpoons[\lambda_1]{\lambda_2 n} off$$

If these reactions equilibrate rapidly compared to the other reactions, it is again tempting to simply assume that genes are active for $p = \lambda_1 (\lambda_1 + \lambda_2 n)^{-1}$ percent of the time, ignoring the details of binding and unbinding. However, it may also be important to account for the fact that bound repressors are unavailable for other reactions, which can have large and qualitative effects on the dynamics (Berg et al., 2000b). That being said, it should also be emphasized that simplifications of this type are often necessary to make the models tractable, and they should not be avoided out of a superstitious fear of condensations. Only allowing uni- or bimolecular reactions can easily give an unwarranted impression of legitimacy: even in the simplest and best characterized cellular systems, it is simply not the case that we can make accurate quantitative assumptions with any confidence. When running the substantial risk of leaving out critical variables and reactions, the possibility of explicit simplifications that greatly facilitate modeling is one of the few blessings we can count. Increasing the state space will make a system less transparent and its dynamics less intelligible, which in turn makes it more difficult to identify gross inaccuracies in the assumptions. Furthermore, just like seemingly complicated reactions can behave as if they were elemental, seemingly simple reactions can hide a non-trivial behavior. For example, first-order unimolecular reactions are often automatically assumed to be elemental, though in many cases there are long-lived intermediate states such that the transitions are not memory-lacking (Xie, 2002; Yang et al., 2003).

8.2.4   Brief History of Stochastic Modeling of Chemical Reactions

Stochastic modeling of chemical reactions has come a full century from the first studies of Brownian motion (Bachelier, 1900; Einstein, 1905). Models of fluctuating concentrations in turn date back to the 1930s (Leontovich, 1935) and were soon followed by biological applications. Several theoretical studies in the 1940s emphasized the intracellular randomness associated with small numbers: The "$1/\sqrt{N}$-rule" of relative fluctuations at equilibrium (Schrödinger, 1944) influenced generations of biologists, and autocatalysis was shown to further amplify variation (Delbruck, 1940). The 1950s saw the first experimental analyses of heterogeneity in bacterial gene expression (Benzer, 1953), and after a decade of focusing more on stochastic enzyme kinetics (Bartholomay, 1962), theory for stochastic gene expression was developed in some detail (Berg, 1978a; Rigney, 1979a,c; Rigney and Schieve, 1977). Both theoretical and experimental efforts intensified in the 1980s and 90s, but it is only in the last five or ten years that the field has truly taken off. This is largely due to the possibility of systematic quantitative studies of protein fluctuations using green fluorescent protein (GFP) (Elowitz et al., 2002; Ozbudak et al., 2002), but also to a wider appreciation of the stochastic foundation of kinetic theory.

8.3   Stochastic Gene Expression

Gene expression is stochastic by nature: genes are activated and inactivated by random association and dissociation of repressors or transcription factors to DNA, transcription of a specific gene often occurs a few times per cell cycle, and many proteins are present in low numbers per cell. Detailed models of these phenomena have been published for almost three decades, and the presentation here (sections 8.3 and 8.4) is an abbreviated version of a recent technical review (Paulsson, 2005).

Most models have focused on the same principles: Changes in the number of proteins per cell n_3 depend on the number of mRNAs n_2, and changes in n_2 in turn depend on the number of active genes n_1. Fluctuations in n_1 will thus enslave n_2 which in turn enslaves n_3. The simplest model includes six chemical events: Each individual gene copy (n_1^{tot} totally) switches on and off with constant rates λ_1^+ and λ_1^-, transcription and translation follow Poisson processes with rates $\lambda_2 n_1$ and $\lambda_3 n_2$, and both transcripts and proteins decay exponentially with average lifetimes τ_2 and τ_3:

$$
\begin{array}{lll}
\text{Gene activation:} & n_1 \xrightarrow{\lambda_1^+\left(n_1^{tot}-n_1\right)} n_1 + 1 & \\[2mm]
\text{Gene inactivation:} & n_1 \xrightarrow{\lambda_1^- n_1} n_1 - 1 & \frac{d\langle n_1\rangle}{dt} = \lambda_1^+\left(n_1^{tot} - \langle n_1\rangle\right) - \lambda_1^-\langle n_1\rangle \\[2mm]
\text{Transcription:} & n_2 \xrightarrow{\lambda_2 n_1} n_2 + 1 & = \lambda_1^+ n_1^{tot} - \langle n_1\rangle/\tau_1 \\[2mm]
\text{mRNA degradation:} & n_2 \xrightarrow{n_2/\tau_2} n_2 - 1 & \frac{d\langle n_2\rangle}{dt} = \lambda_2\langle n_1\rangle - \langle n_2\rangle/\tau_2 \\[2mm]
\text{Translation:} & n_3 \xrightarrow{\lambda_3 n_2} n_3 + 1 & \frac{d\langle n_3\rangle}{dt} = \lambda_3\langle n_2\rangle - \langle n_3\rangle/\tau_3 \\[2mm]
\text{Proteolysis:} & n_3 \xrightarrow{n_3/\tau_3} n_3 - 1 &
\end{array}
$$

$$(8.4)$$

where $\tau_1 = \left(\lambda_1^+ + \lambda_1^-\right)^{-1}$. The average dynamics can be postulated immediately since all rates are linear in the state variables, but fluctuations around the average must either be simulated or analytically derived from the three-variable master equation for the probability $P(n_1,\, n_2, n_3) = P(\mathbf{n})$. Because all rates are linear in the state variables $\mathbf{n}$ we can get (see methods in Berg (1978a)) an exact expression for the normalized protein variance, which in the stationary state follows

$$
\underbrace{\frac{\sigma_3^2}{\langle n_3\rangle^2}}_{\substack{\text{Total} \\ \text{protein noise}}} = \underbrace{\frac{1}{\langle n_3\rangle}}_{\substack{\text{From individual} \\ \text{births and deaths} \\ \text{Poisson}}} + \underbrace{\frac{1}{\langle n_2\rangle}}_{\substack{\text{From spontaneous} \\ \text{mRNA noise} \\ \text{Poisson}}}\underbrace{\frac{\tau_2}{\tau_3 + \tau_2}}_{\substack{\text{One-step} \\ \text{time-averaging}}}
$$

$$
+ \underbrace{\frac{1 - P_{on}}{\langle n_1\rangle}}_{\text{Binomal}}\underbrace{\frac{\tau_2}{\tau_2 + \tau_3}\frac{\tau_1}{\tau_1 + \tau_3}\frac{\tau_1 + \tau_3 + \tau_1\tau_3/\tau_2}{\tau_1 + \tau_2}}_{\substack{\text{From forced mRNA noise,} \\ \text{in gene activation-inactivation} \\ \text{Two-step} \\ \text{time-averaging}}} \qquad (8.5)
$$

In addition to spontaneous Poisson fluctuations (first term), proteins are also randomized by mRNA fluctuations (second and third terms), that follow:

$$\frac{\sigma_2^2}{\langle n_2 \rangle^2} = \frac{1}{\langle n_2 \rangle} + \frac{1 - P_{on}}{\langle n_1 \rangle} \frac{\tau_1}{\tau_2 + \tau_1} \tag{8.6}$$

The first term of equation 8.6 again reflects Poisson fluctuations, now coming from the random births and deaths of individual transcripts. The first factor of the second term can in turn be interpreted as the normalized stationary variance in the (binomially distributed) number of active genes:

$$\frac{\sigma_1^2}{\langle n_1 \rangle^2} = \frac{1}{n_1^{\max}} \frac{\lambda_1^-}{\lambda_1^+} = \frac{1 - P_{on}}{\langle n_1 \rangle} \tag{8.7}$$

where $P_{on} = \lambda_1^+ / (\lambda_1^+ + \lambda_1^-)$ is the stationary probability that a given gene is on. At any given average, stationary fluctuations are smaller than Poissonian because the total number of genes is fixed. The second factor of the second term of equation 8.6 comes from time-averaging and can be explained by solving the second average equation in equation 8.4 for fixed $\langle n_1 \rangle$

$$\underbrace{\langle n_2 \rangle_{t_1 + t_2} - \langle n_2 \rangle_\infty}_{\substack{\text{Deviation from stationary} \\ \text{average at time } t = t_1 + t_2}} = \underbrace{\left(\langle n_2 \rangle_{t_1} - \langle n_2 \rangle_\infty \right)}_{\substack{\text{Deviation from stationary} \\ \text{average at time } t = t_1}} e^{-t_2/\tau_2} \tag{8.8}$$

This means that n_2 exponentially forgets its initial conditions with rate $1/\tau_2$, that is, events that occurred more than τ_2 time units ago are almost forgotten. The same principles apply to n_1 and the ratio τ_2/τ_1 thus determines how much the number of active genes changes within the kinetic memory of the mRNA concentration. If τ_2/τ_1 is large, the time-averaging factor in equation 8.6 is close to zero, reducing mRNA fluctuations just like throwing many dice reduces relative fluctuations in the total outcome. These principles also apply to proteins in equation 8.5. The second term comes from time-averaged spontaneous mRNA fluctuations, and the third term comes from low-copy gene fluctuations that are first time-averaged by mRNAs and then by proteins.

Many processes including fluctuations in the protein synthesis machinery (Elf and Ehrenberg, 2005a), feedback regulation of transcription and translation (Becskei and Serrano, 2000; Swain, 2004; Tomioka et al., 2004; Elf and Ehrenberg, 2005b), localization of transcription factors, as well as controlled transport, maturation, folding, and degradation of mRNA and protein could similarly affect the rate of gene expression and fluctuations in protein concentrations. To make the results accessible in the current space, these effects are ignored above by implicitly absorbing all other processes into effective rate constants.

For readers interested in the original literature, we recommend the pioneering and excellent papers by David Rigney, Otto Berg and colleagues (Berg, 1978a; Rigney, 1979b,a,c; Rigney and Schieve, 1977) as well as the numerous other studies that have appeared since (Kepler and Elston, 2001; Paulsson, 2004; Paulsson and Ehrenberg,

2001; Peccoud and Ycart, 1995; Raser and O'Shea, 2004; Sasai and Wolynes, 2003; Swain et al., 2002; Tapaswi et al., 1987; Thattai and van Oudenaarden, 2001).

8.4   Fluctuation-Dissipation Approximations

Most models of stochastic reaction networks include nonlinear rates and can therefore rarely be solved exactly. They can still be simulated using the Gillespie-Doob algorithm described in section 8.2 and chapter 16, but each numerical simulation only shows the behavior of a single trajectory for a single combination of parameters. Simulations are therefore easier to evaluate if they are complemented by more generic approximations. To exemplify straightforward interpretations of generic approximations, we here discuss a nonequilibrium version of the fluctuation-dissipation theorem (Keizer, 1987; Lax, 1960) (FDT). This states that the matrix of covariances σ (with notation $\sigma_{ii} = \sigma_i^2$) follows:

$$\frac{d\sigma}{dt} = \mathbf{A}\sigma + \sigma\mathbf{A^T} + \langle\mathbf{B}\rangle \tag{8.9}$$

where "drift" matrix $\mathbf{A}$ reflects the dynamics for relaxation to steady state and "diffusion" $\mathbf{B}$ reflects the randomness of the individual events. This equation is used under different names in many areas of study and it can be derived in many ways (Elf and Ehrenberg, 2003; Gardiner, 1985; Keizer, 1987; Lax, 1960; van Kampen, 1992). However, it is always assumed, explicitly or implicitly, that the responses in the reaction rates can be linearized in the parts of state space that are reached by fluctuations.

To define $\mathbf{A}$, let $J_i^{tot} = \sum_j v_{ij} r_j(\mathbf{n})$ be the total flux of component i at state $\mathbf{n}$ where reaction number j occurs with rate r_j, producing v_{ij} molecules of species i. The averages then exactly follow

$$\frac{d\langle n\rangle}{dt} = \langle J^{tot}\rangle = \langle J^+\rangle - \langle J^-\rangle \tag{8.10}$$

where J_i^+ and J_i^- are the total production and elimination fluxes of species n_i. The covariance matrix in turn exactly follows

$$\frac{d\sigma}{dt} = \langle J^{tot}\alpha^T\rangle + \langle\alpha J^{tot^T}\rangle + \langle B\rangle \tag{8.11}$$

where $\alpha_i = n_i - \langle n_i\rangle$ is the displacement from the average and $B_{ik} = \sum_j \nu_{ij}\nu_{kj}r_j(\mathbf{n})$ (see below). The FDT formulation in equation 8.9 follows from equation 8.11 by approximating the flux $\mathbf{J}$ as linear in $\mathbf{n}$, that is, by Taylor expanding the rates around the average. That is not a trivial procedure though. When $\mathbf{J}$ depends nonlinearly on $\mathbf{n}$, the equations for the average dynamics cannot even be solved exactly for the steady state because $\langle J(n)\rangle \neq J(\langle n\rangle)$. One approach (van Kampen, 1961) solves this problem by approximating fluctuations close to the macroscopic limit where numbers are large and fluctuations are small, that is, starting with rate equations. A related but slightly different approach is to simply make the direct mean-field approximation $\langle J(n) \rangle \approx J(\langle n \rangle)$ that the average rate is the rate at the average number. The difference between the two approaches is clear when approximating the average rate of bimolecular homodimerization, $r^- = \lambda n(n-1)$:

$$\langle r^- \rangle = \lambda \langle n(n-1) \rangle = \lambda \left(\langle n \rangle^2 + \sigma^2 - \langle n \rangle \right) \approx \begin{cases} \overbrace{\lambda \left(\langle n \rangle^2 - \langle n \rangle \right)}^{\text{mean-field approx.}} \\ \underbrace{\lambda \langle n \rangle^2}_{\text{macroscopic approx.}} \end{cases} \tag{8.12}$$

Both ignore the variance, but only the latter additionally assume high numbers. Using either method in equation 8.12, the dynamic matrix $\mathbf{A}$ can then be calculated as the Jacobian matrix of the average dynamics:

$$A_{ik} \approx \frac{\partial J_i^+ (\langle \mathbf{n} \rangle)}{\partial \langle n_k \rangle} - \frac{\partial J_i^- (\langle \mathbf{n} \rangle)}{\partial \langle n_k \rangle} \tag{8.13}$$

This is a measure for how the fluxes are affected by changes in $\mathbf{n}$ and thus summarizes, in an approximate and local way, how fluctuations are amplified or corrected. Neither the mean-field or macroscopic method necessarily provides a good approximation, though, and when applied to specific nonlinear schemes they should always be checked numerically.

Matrix $\mathbf{B}$ in the exact equation 8.11 can be similarly approximated using $\langle B(\mathbf{n}) \rangle \approx B(\langle \mathbf{n} \rangle)$:

$$\langle B_{ik} \rangle \approx \sum_j \nu_{ij} \nu_{kj} r_j (\langle \mathbf{n} \rangle) \tag{8.14}$$

This is a measure of the size and frequency of the random events that introduce fluctuations in the first place. The FDT thus captures how the overall variability of the system as measured by σ depends on fluctuations introduced in the diffusion matrix $\mathbf{B}$ and the dissipation of fluctuations introduced in the Jacobian matrix $\mathbf{A}$.

8.4.1   Interpreting the FDT in Terms of Physical Observables

The approximations above may or may not be accurate, but greatly facilitate first order approximations. Calculating stationary variances is now technically straightforward, and the conceptual challenge lies in interpreting the results in terms of general physical principles. To facilitate interpretations, the stationary FDT can be reinterpreted in terms of more straightforward physical properties, following Paulsson (2004, 2005) and starting with:

$$\mathbf{M}\eta + \eta \mathbf{M}^T = \mathbf{D} \tag{8.15}$$

where:

$$\eta_{ik} = \frac{\sigma_{ik}}{\langle n_i \rangle \langle n_k \rangle} \quad , \quad M_{ik} = -\frac{\langle n_k \rangle}{\langle n_i \rangle} A_{ik} \quad \text{and} \quad D_{ik} = \frac{B_{ik}}{\langle n_i \rangle \langle n_k \rangle} \tag{8.16}$$

The normalized Jacobian matrix $\mathbf{M}$ can be further rewritten by using the rules for differentiation of logarithms:

$$\frac{\partial \ln f}{\partial \ln x} = \frac{x}{f} \frac{\partial f}{\partial x} \quad \text{and} \quad \frac{\partial \ln (f/g)}{\partial \ln x} = \frac{\partial \ln f}{\partial \ln x} - \frac{\partial \ln g}{\partial \ln x} \tag{8.17}$$

Applying these to equation 8.13 and using the steady state condition $\langle J_i^+ \rangle = \langle J_i^- \rangle \equiv \langle J_i \rangle$ gives:

$$\begin{aligned} A_{ik} &= \frac{\partial \langle J_i^+ \rangle}{\partial \langle n_k \rangle} - \frac{\partial \langle J_i^- \rangle}{\partial \langle n_k \rangle} = \frac{\langle J_i \rangle}{\langle n_k \rangle} \left(\frac{\partial \ln \langle J_i^+ \rangle}{\partial \ln \langle n_k \rangle} - \frac{\partial \ln \langle J_i^- \rangle}{\partial \ln \langle n_k \rangle} \right) \\ &= -\frac{\langle J_i \rangle}{\langle n_k \rangle} \frac{\partial \ln \langle J_i^- \rangle / \langle J_i^+ \rangle}{\partial \ln \langle n_k \rangle} \end{aligned} \tag{8.18}$$

To reduce notation we here use the true averages $\langle J_i^+ \rangle$ rather than $J_i^+ (\langle \mathbf{n} \rangle)$, but interpret the matrices in the macroscopic limit where they are interchangeable. Matrix $\mathbf{M}$ now becomes

$$M_{ik} = -\frac{\langle J_i \rangle}{\langle n_i \rangle} \frac{\partial \ln \langle J_i^- \rangle / \langle J_i^+ \rangle}{\partial \ln \langle n_k \rangle} \tag{8.19}$$

At steady state, the average degradation (or synthesis) rate per molecule is approximately equal to the inverse of the average lifetime τ_i:

$$\frac{\langle J_i^- \rangle}{\langle n_i \rangle} = \frac{\langle J_i^+ \rangle}{\langle n_i \rangle} = \frac{\langle J_i \rangle}{\langle n_i \rangle} \approx \frac{1}{\tau_i} \tag{8.20}$$

This is only exact for exponential first-order decay and approximate for all nonlinear degradation mechanisms. However, it is not an *additional* approximation. As shown above, one version of the stationary FDT approximation evaluates all parameters at the macroscopic steady state, that is, in the hypothetical limit where each molecule is immersed in a constant environment of other concentrations. Within this approximation even nonlinear degradation mechanisms perfectly mimic first-order exponential decay at steady state. Matrix $\mathbf{M}$ thus follows:

$$M_{ik} = -\frac{H_{ik}}{\tau_i} \quad \text{where} \quad H_{ik} = \frac{\partial \ln \langle J_i^- \rangle / \langle J_i^+ \rangle}{\partial \ln \langle n_k \rangle} \tag{8.21}$$

The $\mathbf{H}$ parameters are logarithmic susceptibilities or elasticities and measure how the birth-to-death ratio is affected by concentration changes: If $H_{ik}=2$, a 1% increase in n_k will approximately cause a 2% increase in the degradation rate relative to the synthesis rate of n_i.

A full reinterpretation of matrix $\mathbf{D}$ for arbitrary chemical events will be published separately. Here we restrict the analysis to nonlinear versions of cases where each chemical event adds or removes a single molecule of a single species[2] :

$$D_{ii} = \frac{\langle J_i^+ \rangle + \langle J_i^- \rangle}{\langle n_i \rangle^2} = 2 \frac{\langle J_i \rangle}{\langle n_i \rangle^2} \quad \text{and} \quad D_{ik} = 0 \text{ for } i \neq k \qquad (8.22)$$

Using equation 8.20 we then get

$$D_{ii} = \frac{1}{\tau_i} \frac{2}{\langle n_i \rangle} \qquad (8.23)$$

Within the approximation, the randomness *introduced* by the ith component is thus inversely proportional to its average copy number, $\langle n_i \rangle$. That does not mean that the actual fluctuations in the ith are inversely proportional to $\langle n_i \rangle$—the final effect of the probabilistic events is filtered through the dynamic responses.

8.4.2   Examples of Elasticities

A univariate example illustrates the basic principle:

$$\frac{d\langle n_1 \rangle}{dt} = \lambda^+ \langle n_1 \rangle^\alpha - \lambda^- \langle n_1 \rangle^\beta \quad \Rightarrow \quad H_{11} = \frac{\partial \ln \left(\lambda^- \langle n_1 \rangle^\beta \big/ \lambda^+ \langle n_1 \rangle^\alpha \right)}{\partial \ln \langle n_1 \rangle} = \beta - \alpha$$
$$(8.24)$$

The univariate elasticity thus equals the difference in effective kinetic order of the degradation and synthesis fluxes. If the stochastic process is compared to a random walk in a valley, the elasticity estimates the normalized steepness of the walls. For an unbiased and unbounded random walk in one variable, $H_{11} = 0$ and the dynamics is neutrally stable. The multivariate cases are equally simple. Excluding genes from the gene expression model above (and shifting the indices so that n_2 becomes n_1 and n_3 becomes n_2), the mRNA-protein part gives:

$$\begin{array}{l} \frac{d\langle n_1 \rangle}{dt} = \lambda_1^+ - \lambda_1^- \langle n_1 \rangle \\ \frac{d\langle n_2 \rangle}{dt} = \lambda_2^+ \langle n_1 \rangle - \lambda_2^- \langle n_2 \rangle \end{array} \quad \Rightarrow \quad \mathbf{H} = \begin{bmatrix} 1 & 0 \\ -1 & 1 \end{bmatrix} \qquad (8.25)$$

where $\tau_i = 1/\lambda_i^-$. Both the uni- and multivariate examples above are particularly simple because both synthesis and degradation follow power-laws, but many more complicated mechanisms are also easy to evaluate using the definitions in equations 8.18–8.21 or simply eyeballing lower and upper bounds.

8.4.3   A Generalized Pseudo-Bivariate Example

Rewriting the FDT in terms of physical observables greatly facilitates interpretation and makes it possible to collectively address families of dynamic processes without losing interpretability. For example consider the following extension of the mRNA-protein model:

$$
\begin{aligned}
n_1 &\xrightarrow{J_1^+(n_1)} n_1 + 1 \\
n_1 &\xrightarrow{J_1^-(n_1)} n_1 - 1 \\
n_2 &\xrightarrow{J_2^+(n_1,n_2)} n_2 + 1 \\
n_2 &\xrightarrow{J_2^-(n_1,n_2)} n_2 - 1
\end{aligned}
\quad \text{so that} \quad
\begin{aligned}
\frac{d\langle n_1\rangle}{dt} &= \left\langle J_1^+(n_1)\right\rangle - \left\langle J_1^-(n_1)\right\rangle \\
\frac{d\langle n_2\rangle}{dt} &= \left\langle J_2^+(n_1,n_2)\right\rangle - \left\langle J_2^-(n_1,n_2)\right\rangle
\end{aligned}
\tag{8.26}
$$

Because the second variable does not affect the first, this is a pseudo-bivariate stochastic system. According to the FDT approach above any process with a stable fixed point then has:

$$
M = -\begin{bmatrix} H_{11}/\tau_1 & 0 \\ H_{21}/\tau_2 & H_{22}/\tau_2 \end{bmatrix}
\quad \text{and} \quad
D = \begin{bmatrix} 2/(\langle n_1\rangle \tau_1) & 0 \\ 0 & 2/(\langle n_2\rangle \tau_2) \end{bmatrix}
\tag{8.27}
$$

Solving equation 8.15 gives

$$
\eta_{22} = \frac{\sigma_2^2}{\langle n_2\rangle^2} \approx \overbrace{\underbrace{\frac{1}{\langle n_2\rangle}}_{\substack{\text{low-copy} \\ \text{fluctuations}}} \times \underbrace{\frac{1}{H_{22}}}_{\substack{\text{effective} \\ \text{stability}}}}^{\text{spontaneous or intrinsic } x_2 \text{ noise}} +
$$

$$
\overbrace{\underbrace{\frac{\sigma_1^2}{\langle n_1\rangle^2}}_{\substack{\text{environmental} \\ \text{fluctuations}}} \times \underbrace{\frac{H_{21}^2}{H_{22}^2}}_{\substack{\text{static} \\ \text{susceptibility}}} \times \underbrace{\frac{H_{22}/\tau_2}{H_{22}/\tau_2 + H_{11}/\tau_1}}_{\substack{\text{one-step} \\ \text{time-averaging}}}}^{\text{forced or extrinsic } x_2 \text{ noise}}
\tag{8.28}
$$

where $\eta_{11} = \sigma_1^2 \langle n_1\rangle^{-2} \approx \langle n_1\rangle^{-1} H_{11}^{-1}$. The intrinsic noise term comes from the spontaneous randomness of X_2 itself, introduced by element D_{22}. Its first factor represents population smallness—each birth and death event has a larger relative effect in a smaller population. The second factor of the first term represents the dynamic response to perturbations and can be interpreted in several different ways. Normalized deviations $\Delta\tilde{x}$ from steady state in the corresponding deterministic system follow

$$
\frac{\partial \Delta\tilde{x}}{\partial t} = -\mathbf{M}\Delta\tilde{x} = -\begin{bmatrix} \frac{H_{11}}{\tau_1} & 0 \\ \frac{H_{21}}{\tau_2} & \frac{H_{22}}{\tau_2} \end{bmatrix} \Delta\tilde{x}
\tag{8.29}
$$

Parameter H_{22}/τ_2 is thus the adjustment rate constant to steady state following a perturbation. The rate can be changed in two ways: by changing the nonlinearity as measured by H_{22} or by changing the average lifetime τ_2. However, as seen in equation 8.27, the latter would also affect the rate of spontaneous randomization by the same factor. This makes intuitive sense: one time-constant can always be chosen arbitrarily to define the time scale. The elasticity H_{22} is thus the rate of adjustment normalized by the rate of perturbation. This can also be seen in the probability $P(birth)$ that the next event is a birth rather than a death

$$P\left(birth\right) = \frac{J_2^+}{J_2^+ + J_2^-} = \frac{1}{1 + J_2^-/J_2^+} \tag{8.30}$$

To reduce fluctuations, $P(birth)$ should be as low as possible above the average $\langle n_2 \rangle$ and as high as possible below $\langle n_2 \rangle$, that is, J_2^-/J_2^+ should increase sharply with n_2. Here the dimensionless elasticities come into play. If n_2 changes by 1%, the ratio J_2^-/J_2^+ will change approximately H_{22} percent. The elasticity thus estimates the preference for returning to steady state rather than deviating further. Returning to the analogy of a random walker in a valley, $1/\langle n_1 \rangle$ is a normalized measure of the frenzy of the walker and H_{22} is a scale-free measure of the steepness of the walls, forcing the walker back to the bottom.

The extrinsic noise term in equation 8.28 comes from environmental variable n_1, that is, from element D_{11} in the diffusion matrix. Factor $\sigma_1^2/\langle n_1 \rangle^2$ thus measures the extent of environmental fluctuations and the second two factors determine the response of the system. The response can in turn be separated into a temporal component and a steady state component. The steady state component can be separated by first considering an environment with very slow dynamics, that is, an environmental variable whose average lifetime approaches infinity, $\tau_1 \rightarrow \infty$. Consider the following deterministic equation

$$\frac{dx_2}{dt} = \frac{\lambda_2 x_1}{x_2^2} - \beta_2 x_2 \tag{8.31}$$

where $H_{21}/H_{22} = -1/3$. If x_1 changed permanently from one value to another, the steady state of x_2 would eventually adjust to

$$x_2 = \sqrt[3]{\lambda_2/\beta_2 x_1} \tag{8.32}$$

An eightfold increase in x_1 then gives a two-fold change in x_2. More generally assume the following relation between the steady states:

$$x_2 \approx c x_1^{-H_{21}/H_{22}} \tag{8.33}$$

where c is a proportionality constant. A 1% change in x_1 then eventually produces a $-H_{21}/H_{22}$ percent change in x_2. Here elasticity H_{21} measures the capacity of x_1 to affect the birth and death balance of x_2, and $1/H_{22}$ measures the sensitivity of the x_2 steady state to changes in its own rates. The static susceptibility $(H_{21}/H_{22})^2$ in equation 8.28 is squared because the equation is formulated for the normalized variance rather than the normalized standard deviation.

In most applications, fluctuations in n_1 are not infinitely slow and the temporal responses then also affect fluctuations in n_2. As noted above, H_{11}/τ_1 and H_{22}/τ_2

are the normalized rate constants for adjustments to steady state in concentrations x_1 and x_2 respectively ($-H_{11}/\tau_1$ and $-H_{22}/\tau_2$ are the eigenvalues of M). The ratio

$$\frac{H_{22}/\tau_2}{H_{11}/\tau_1} \tag{8.34}$$

is thus a measure for how rapidly x_2 changes relative to x_1. Comparing with equation 8.28 illustrates a qualitative difference between spontaneous and forced fluctuations. If the fluctuations come from the randomness of the environment, rapid adjustments simply make the system more responsive to underlying fluctuations. In other words: The current state of a more slowly adjusting system depends on a longer history of ups and downs in the environment that then partially cancel out.

Increasing parameter H_{22} thus has several effects: increasing the tendency to return to a preferred average, reducing the susceptibility to permanent changes in the environment, and increasing the temporal responsiveness. The susceptibility decreases quadratically with H_{22} while the temporal responsiveness at most increases in proportion to H_{22}, so the net effect should be lower noise. This is not fully general though. In feedback systems with lags or delays, the temporal response factor can increase more than quadratically in H_{22}, so that a higher H_{22} can increase total noise and cause oscillatory responses to perturbations (for experimental observations of oscillations in feedback systems, see Lahav et al., 2004).

8.5   Fluctuations near Critical Points

From the FDT analysis above we saw that the size of the stationary fluctuations depend on two opposing forces: the turnover of molecules in random events that contributes to diffusion in state space and the rate of relaxation back towards the average. These principles are illustrated in figure 8.1 At thermodynamic equilibrium these forces are fundamentally coupled such that for each substance the variance in the number of molecules is smaller than or equal to the average (as in a binomial distribution). However, in systems away from equilibrium the flux can be large although the rate of relaxation to steady state is small and vice versa. That means that relative fluctuations can be arbitrarily large even if the number of molecules is high, or arbitrarily small even if the number of molecules is low.

For univariate linearized systems where one molecule is synthesized or consumed per reaction, the stationary FDT boils down to

$$\sigma^2 = \frac{B}{-2A} = -\frac{J^+}{A} = -\frac{J^-}{A}, \tag{8.35}$$

where $J^\pm$ are the total stationary fluxes of either production or elimination and A is the rate of relaxation back to steady state.

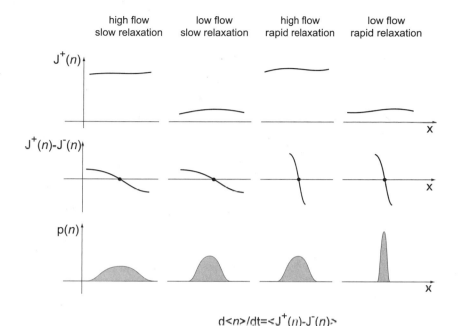

Figure 8.1   Flow, relaxation, and fluctuations. The figure illustrates in a univariate example the four different combinations of high/low flux and fast/slow relaxation to steady state. Near-critical fluctuations arise when the rate of relaxation is low at the same time as the flow is relatively high.

In normalized variables the same relation takes the form

$$\frac{\sigma^2}{\langle n \rangle^2} = \frac{1}{\langle n \rangle H},$$

(8.36)

Poissonian-sized fluctuations with $\sigma^2 = \langle n \rangle$ is thus only obtained when the turnover of the pool $J/\langle n \rangle$ is equal to the rate of relaxation, $-A$, or, equivalently, when the elasticity $H = 1$ in equation 8.36. Here we consider some simple kinetic systems that operate near dynamically unstable points ($H \approx 0$ in univariate systems) and therefore display large fluctuations.

8.5.1   Autocatalysis

Consider the autocatalytic system with three reactions

$$n \xrightarrow{\lambda_1} n+1 \qquad n \xrightarrow{\lambda_2 n} n+1 \qquad n \xrightarrow{\mu n} n-1$$

(8.37)

Because the rates depend linearly on state, the time evolution of the average value is given by the exact

$$\frac{d \langle n \rangle}{dt} = \lambda_1 + (\lambda_2 - \mu) \langle n \rangle$$

(8.38)

For $\mu > \lambda_2$ the system has a stationary state with average $\langle n \rangle = \lambda_1/(\mu - \lambda_2)$. The rate of relaxation to steady state is $\mu - \lambda_2$ and the average flow though the system is $\mu \langle n \rangle = \mu \lambda_1/(\mu - \lambda_2)$. The steady state elasticity is

$$H = \frac{\partial \ln \left(\mu \langle n \rangle / (\lambda_1 + \lambda_2 \langle n \rangle) \right)}{\partial \ln \langle n \rangle} = \frac{\lambda_1}{\lambda_1 + \lambda_2 \langle n \rangle} = \frac{\mu - \lambda_2}{\mu} \tag{8.39}$$

The FDT in turn gives

$$\eta_{11} = \frac{\sigma^2}{\langle n \rangle^2} = \frac{1}{\langle n \rangle} H_{11}^{-1} = \frac{1}{\langle n \rangle} \frac{\mu}{\mu - \lambda_2} = \frac{\mu}{\lambda_1}. \tag{8.40}$$

This expression is exact, as can be shown by calculating the full stationary distribution (Gardiner, 1985) or by noting that the reaction rates are linear in n.

8.5.2   Covalent Modification-Demodification Cycles

Assume that a substrate is converted from unmodified state X_1 to modified state X_2 by one enzyme, and back again by another enzyme. With a constant total number n_{max} of modified and unmodified molecules, the state of the system is described by the number n of X_1 molecules. The reaction scheme is

$$X_1 \xrightarrow{J^-(n)} X_2 \qquad X_2 \xrightarrow{J^+(n_{max} - n)} X_1 \tag{8.41}$$

Here we assume Michaelis-Menten type reactions, where $J^-(n) = k_1 n/(n + K)$ and $J^+(n_{max} - n) = k_2(n_{max} - n)/((n_{max} - n) + K)$. The Michaelis-Menten approximation relies on condensations of several elemental reactions, and was originally derived for macroscopically large systems. The stochastic behavior (Bartholomay, 1962) can be different, so this should only be considered a first approximation. However, it still accounts for the major dynamic effect of first order degradation at low n and saturation at high n.

The modification-demodification cycles in equation 8.41 can display so-called zero-order ultrasensitivity if $n_{max} \gg K$, such that both enzymes can be saturated for $k_1 = k_2 = k$ (Goldbeter and Koshland, 1981). Ultrasensitivity refers to the fact that a small fractional change in the rate constant k for modification rates makes a large difference in the fractional level of modification. A slight increase in an enzyme level can thus push almost all molecules into one state or the other. As a first approximation, the FDT method gives $\sigma^2/\langle n \rangle = H^{-1} = d \ln \langle n \rangle / d \ln k$, that is, the ultra-sensitive response to changes in k is again tightly connected to large random fluctuations. For more details and exact theoretical expressions see Berg et al. (2000a). Large fluctuations in modification-demodification cycles were also recently experimentally observed by Korobkova et al. (2004).

8.5.3   Multisubstrate Reactions

Consider the anabolic reaction scheme with two species and the following transitions:

$$n_1 \xrightarrow{k} n_1 + 1 \qquad n_2 \xrightarrow{k} n_2 + 1$$
$$n_1 \xrightarrow{\mu n_1} n_1 - 1 \qquad n_2 \xrightarrow{\mu n_2} n_2 - 1 \qquad (8.42)$$
$$[n_1, n_2] \xrightarrow{n_1 n_2 k_2} [n_1 - 1, n_2 - 1]$$

It is possible to treat more general cases analytically (Elf et al., 2003), but here we only consider situations where the consumption flux of the bimolecular reaction is much higher than the first order consumption events, that is $\mu n_1 \approx \mu n_2 \ll k_2 n_1 n_2$. The averages follow

$$d \langle n_1 \rangle / dt = k - k_2 \langle n_1 n_2 \rangle - \mu \langle n_1 \rangle$$
$$d \langle n_2 \rangle / dt = k - k_2 \langle n_1 n_2 \rangle - \mu \langle n_2 \rangle \qquad . \qquad (8.43)$$

where $\langle n_1 n_2 \rangle \approx \langle n_1 \rangle \langle n_2 \rangle$ in the mean-field or macroscopic approximation. When $\mu > 0$ this system has one attracting steady state where $\langle n_1 \rangle = \langle n_2 \rangle = \sqrt{k/k_2 + \mu^2/k_2^2} - \mu/2k_2 \approx \sqrt{k/k_2}$. For the linearized dynamics around steady state, the Jacobian matrix has a slow eigenvalue μ which determines the rate of relaxation to the steady state. In the limit $\mu \to 0$ the single attracting steady state bifurcates into a curve of steady state points satisfying $k_2 n_1 n_2 = k$. Thus in the limit $\mu \to 0$ the fluctuations become macroscopically large, as the system can diffuse freely on the curve of stationary states (Elf and Ehrenberg, 2003). When $0 < \mu \ll \sqrt{kk_2}$ the proximity to the critical point makes the fluctuations large in any system of finite size.

The FDT approximation can also be applied to the system, but in this case it is advantageous to first make a linear transformation of the variables, such that the new stochastic variables correspond to fluctuations in the two perpendicular eigendirections of the linearized system. In the slow eigen-direction, [1 -1] corresponding to the eigenvalue, μ, fluctuations are large and slow compared to the fast and small fluctuations in the perpendicular direction [1 1]. By combined application of FDT on the two separated timescales an accurate analytical solution can be obtained (Elf and Ehrenberg, 2003). Here we will only focus on the large and slow fluctuations in the variable $w = n_1 - n_2$. In this variable the state transitions and their rates are

$$w \xrightarrow{k} w + 1 \qquad w \xrightarrow{k} w - 1$$
$$w \xrightarrow{\mu n_1} w - 1 \qquad w \xrightarrow{\mu n_2} w + 1 \qquad (8.44)$$

The macroscopic rate equation for w is $d \langle w \rangle / dt = k - k + \mu \langle n_2 \rangle - \mu \langle n_1 \rangle = -\mu \langle w \rangle$. That is, deviations in the difference decay exponentially with rate $-A_{11} = \mu$. In steady state $\langle n_2 \rangle = \langle n_1 \rangle \equiv \langle n \rangle$ and the fluxes are $J^+ = J^- = J = k + \mu \langle n \rangle$. The FDT approximation for the variance in w then gives

$$\sigma_w^2 = -\frac{k + k + \mu n_1 + \mu n_2}{-2\mu} = \frac{k}{\mu} + \langle n \rangle \qquad (8.45)$$

The fluctuations in the original variables n_1 and n_2 are $\sigma_n^2 \geq \sigma_w^2/4$, were equality holds if the fluctuations in n_1 and n_2 are perfectly anti-correlated. The relative fluctuations in the individual pools are thus $\sigma^2 / \langle n \rangle^2 > k/4\mu$. This result indicates very large fluctuation as shown in figure 8.2. The variance in w can also be calculated exactly (Elf, 2004) using moment generating functions with the result that $\sigma_w^2 = k/\mu + \langle n \rangle$. The FDT result equation 8.45 is therefore exact because we could change variables so that the rates are linear in the new variables.

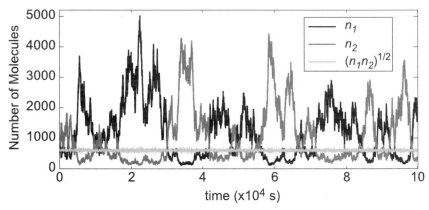

Figure 8.2 Near-critical fluctuations. The system is simulated using the Gillespie-Doob algorithm. The large fluctuations in n_1 and n_2 are anti-correlated such that $k_2 n_1 n_2 \approx k$. Parameters: $k = 600s^{-1}$, $k_2 = 0.001s^{-1}$, and $\mu = 2 \cdot 10^{-4}s^{-1}$.

The large fluctuations can in this case be explained by a multivariate version of the argument behind the zero-order behavior above: an increase in n_1 is compensated by a decrease in n_2 such that the total flow is unchanged. The same phenomenon is predicted to occur in the pools of aminoacylated tRNA used as substrates in protein synthesis. Here many different concentration combinations give the same total rate of protein synthesis, such that the fluctuations in the individual ternary complex pool can be very large (Elf and Ehrenberg, 2005a).

8.6 Negative Feedback of Replication Control

In some systems—including replication control, cell division, and central metabolic pathways—fluctuations pose a threat that cells must carefully eliminate. The most studied mechanism for noise suppression is perhaps negative feedback control, where a random fluctuation to a higher concentration leads to lower synthesis rate and forces the system back towards the average. Negative feedback is particularly effective in systems that otherwise would display near-critical dynamics (Elf et al., 2003; Paulsson and Ehrenberg, 2001), but its effects can also be significant in other types of systems (Becskei and Serrano, 2000; Lahav et al., 2004; Swain, 2004). Here we will focus on plasmid replication control as a model system but keep the arguments general.

Plasmids are unrivaled model systems for noise suppression for several reasons:

1. The average plasmid loss rate—the risk that a plasmid-containing mother cell gives rise to a plasmid-free daughter cell—increases drastically with random fluctuations.

2. Plasmids self-replicate and would thus generate enormous fluctuations without negative regulation, causing both high plasmid losses and slowed growth.

3. Most plasmid species have copy numbers of about 1–100 per cell, such low numbers that spontaneous fluctuations could be substantial.

4. Numerous plasmid replication control systems only include two or three gene products and have been as well characterized as λ phage. They also tend to be more independent of background processes than almost any other cellular network.

For many plasmid species, an increase in the number of plasmid copies (n_1) increases the average synthesis flux (J_2^+) of a replication inhibitor (n_2) and thereby decreases the plasmid replication flux (J_1^+). The average dynamics can often be modeled by:

$$\frac{d \langle n_1 \rangle}{dt} = \langle J_1^+ (n_1, n_2) \rangle - \langle J_1^- (n_1) \rangle \ \text{ and } \ \frac{d \langle n_2 \rangle}{dt} = \langle J_2^+ (n_1, n_2) \rangle - \langle J_2^- (n_1, n_2) \rangle \tag{8.46}$$

Both J_1^+ and J_1^- are assumed proportional to n_1 because every plasmid copy can self-replicate (one molecule at a time) and because plasmid segregation in growing populations can be qualitatively approximated by first-order degradation (again eliminating one molecule at a time). Further assuming that J_1^+ and J_2^+ monotonically decrease and increase with n_2 and n_1 respectively, the equations have a stable steady state (Kurosawa et al., 2002) around which we can approximate stationary variances using the fluctuation-dissipation theorem. To reduce the algebraic complexity we will here assume that the inhibitors are present in such high numbers that they do not contribute their own plasmid-independent fluctuations through noisy signaling $(D_{22} = 0)$. This is not always a reasonable approximation and it will certainly hide some interesting principles. However, a full analysis will be published elsewhere, showing the effect of both noisy signaling and fluctuations in environmental variables.

Without any further assumptions or restrictions, the reinterpretation of the FDT in section 8.4 then leads to

$$\eta_{11} = \frac{\sigma_1^2}{\langle n_1 \rangle^2} \approx \frac{1}{\langle n_1 \rangle} \times \left(\frac{H_{22}}{|H_{12}H_{21}|} + \frac{\tau_2}{\tau_1} \frac{1}{H_{22}} \right)$$

$$\approx \frac{1}{\langle n_1 \rangle} \times \underbrace{\frac{1}{G}}_{\substack{\text{correcting} \\ \text{fluctuations}}} \times \underbrace{\frac{T_1 + T_2}{T_1}}_{\substack{\text{inverse effect} \\ \text{of time-averaging}}} \tag{8.47}$$

where τ_1 is the generation time of the host cell and τ_2 is the average lifetime of the inhibitor (including dilution effects). The compounded parameters are defined by $G = -H_{12}H_{21}/H_{22}$, $T_1 = \tau_1/G$, and $T_2 = \tau_2/H_{22}$. A high negative H_{21} means that a relative increase in plasmid concentration gives a high relative increase in the inhibitor synthesis rate. However, plasmids affect inhibitors by encoding their genes, so there are few exceptions to $J_2^+ = kn_1 f(n_2)$ for which $H_{21} = -1$. This is not only true for constitutively expressed genes: if inhibitors would feed back on their own synthesis, that would instead affect H_{22}. A low H_{22} means that the ratio J_2^-/J_2^+ is insensitive to changes in n_2, so that the steady state average of n_2 conversely is sensitive to J_2^-/J_2^+. If inhibitors were made at constant rates per plasmid copy and decayed exponentially, corresponding to $J_2^+ = kn_1$ and $J_2^- = n_2/\tau_2$, then $H_{22} = 1$. If inhibitors instead were degraded by enzymes that operate close to saturation, H_{22} could be arbitrarily close to zero (zero-order ultrasensitivity (Elf et al., 2003)), and if inhibitors autorepressed their own synthesis, H_{22} could be arbitrarily increased. A high H_{12} means that an increase in the inhibitor concentration sharply turns off replication. Plasmids use numerous strategies to increase H_{12}, including multistep proofreading control, inhibitor multimerization, and cooperative binding of inhibitors where H_{12} is the Hill coefficient far from saturation.

The compounded parameter $G = -H_{12}H_{21}/H_{22}$ is the total sensitivity gain over the feedback loop. If $G = 3$, then a 1% change in the plasmid concentration would eventually lead to a 3% change in the plasmid birth-to-death balance. The time constants $T_1 = \tau_1/G$ and $T_2 = \tau_2/H_{22}$ determine how rapidly the plasmid changes and how rapidly the inhibitor adjusts to the plasmid.

This is best illustrated by some examples. Several plasmids have been described by

$$\frac{d\langle n_1 \rangle}{dt} = \lambda_1 \left\langle n_1 \frac{K}{K + n_2^h} \right\rangle - \mu_1 \langle n_1 \rangle \text{ and } \frac{d\langle n_2 \rangle}{dt} = \lambda_2 \langle n_1 \rangle - \mu_2 \langle n_2 \rangle \tag{8.48}$$

In the first equation, λ_1 is the frequency with which each plasmid copy attempts to initiate replication, h is the Hill coefficient of inhibition, and μ_1 is the dilution rate due to cell growth. In the second equation, λ_2 is the per plasmid inhibitor synthesis rate, and μ_2 is the sum of the dilution rate due to cell growth and the inherent inhibitor degradation rate. The parameters in equation 8.48 relate to the parameters in equation 8.47 as:

$$\begin{array}{ll} \tau_1 = \mu_1^{-1} \\ \tau_2 = \mu_2^{-1} \end{array} \quad \text{and} \quad \begin{array}{l} H_{22} = -H_{21} = 1 \\ H_{12} = h\frac{\langle n_1 \rangle}{K + \langle n_1 \rangle} = h\left(1 - \mu_1/\lambda_1\right) \end{array} \tag{8.49}$$

If the inhibitors instead were degraded by a Michaelis-Menten mechanism, as

$$\frac{d\langle n_2 \rangle}{dt} = \lambda_2 \langle n_1 \rangle - \mu_1 \langle n_2 \rangle - \mu_2 \left\langle \frac{n_2}{K + n_2} \right\rangle \tag{8.50}$$

then the effective H_{22} can be substantially lower. If the enzymatic degradation term dominates over the first-order term, and the enzymes operate close to saturation, the inhibitor displays zero-order ultrasensitivity (see section 8.5) (Elf et al., 2003) and H_{22} approaches zero. This could reduce plasmid fluctuations greatly: a small relative increase in the plasmid concentration will produce an enormous relative increase in the quasi–steady state of the inhibitor, which leads to an enormous relative decrease in the plasmid replication frequency. However, as seen in equation 8.47 increasing H_{22}^{-1} too much will eventually increase fluctuations. This is because a higher H_{22}^{-1} will also slow down the dynamic response of the inhibitor, i.e., it will take more inhibitor lifetimes τ_2 before the inhibitor adjusts to its new quasi–steady state after a change in plasmid concentration. Parameter G only represents the effective gain over the loop if the inhibitor response is fast. If the inhibitor has a finite response time, it will instead lag behind and depend on the history of plasmid concentrations. The current inhibitor value is thus an effective average over a history of plasmid concentrations—just like protein fluctuations average over gene or mRNA fluctuations in equation 8.5—and therefore tends to display smaller relative deviations from steady state. In other words, the inhibitor underestimates the deviation from steady state and the response is weaker than it otherwise would have been. The inhibitor time-averaging thus increases plasmid fluctuations. This can be thought of in terms of corrections: tighter feedback loops (higher G) correct spontaneous fluctuations more rapidly, but if inhibitors lag behind the corrections are slowed on average.

The analysis above illustrates a few principles that are common to negative feedback systems—the importance of zero-order effects, sensitivity amplification and time-lags. However, numerous other principles are not accounted for in the analysis above. The fact that inhibitors are made and degraded by inherently stochastic mechanisms will produce a signaling noise that can enslave the plasmid. In many systems, negative feedback can thereby increase fluctuations by introducing its own randomness. For example, in gene expression autorepression of transcription may increase the tendency to correct fluctuations, but would also introduce random association and dissociation of the repressor to DNA, which can have enormous randomizing effects (Paulsson, 2004; Tomioka et al., 2004).

8.7   Noise Induced Transitions

8.7.1   Stochastic Focusing and Noise-Suppression-by-Noise

Plasmids with the replication control systems described in section 8.6 by equation 8.48 seem to be limited by $G < h$, that is, the total gain over the feedback loop cannot be higher than the Hill coefficient of inhibition. Some of these plasmids have further been suggested to use so-called hyperbolic mechanisms (Hill coefficient $h = 1$) and thus seem doomed to inefficient noise suppression. However, the FDT approach is a so-called mean-field approximation and depends on linearized responses. It can be qualitatively misleading in some nonlinear systems.

When inhibitors fluctuate rapidly ($T_1 \ll T_2$ in equation 8.47), the bivariate master equation behind equation 8.48 can be replaced by a univariate master equation for n_1 and a conditional master equation for the probability $P(n_2|n_1)$ of n_2 given n_1. The simplest inhibitor dynamics that give rise to equation 8.48 are Poissonian synthesis with rate $\lambda_2 n_1$ combined with exponential decay with rate $\mu_2 n_2$. This would generate a Poissonian conditional probability $P(n_2|n_1)$ with conditional average $\langle n_2 \rangle = \lambda_2 n_1 \mu_1^{-1}$. If we assume that the probability that a replication attempt is successful is given by the hyperbolic function

$$q\left(n_2\right) = \frac{K}{K + n_2} \tag{8.51}$$

then the true average q is

$$\langle q\,|n_1 \rangle = \sum_{n_1=0}^{\infty} q\left(n_2\right) P\left(n_1\,|n_2\right) \neq q\left(\langle n_2 \rangle\right) \tag{8.52}$$

This reflects the fact that $\langle q|n_1 \rangle$ receives a disproportional contribution from the left tail of the distribution where n_2 is low. However, not only the actual value is affected, but also the normalized sensitivity to changes in n_1. Even with simple Poisson fluctuations, the inhibition function above with Hill coefficient of $h = 1$ can locally behave as if $h = 2$ or higher. That is, it is possible to have

$$G = -\frac{\partial \ln\left(\langle q\,|n_1 \rangle\right)}{\partial \ln\left(n_1\right)} \gg 1 \tag{8.53}$$

The low-copy noise can thus make for a more sharply changing function. This can be understood as follows: The probability that the next inhibitor event is a birth rather than death is $\lambda_2 n_1 \left(\lambda_2 n_1 + \mu_2 n_2\right)^{-1}$. The probability that n_2 randomly walks away from its average $\langle n_2 \rangle = \lambda_2 n_1 \mu_1^{-1}$ to the lower values where the nonlinear $\langle q|n_1 \rangle$ receives a disproportional contribution thus depends on the birth intensity $\lambda_2 n_1$ at every step. The concentration n_1 thus affects multiple transitions, and the final effect is indeed similar to other schemes for multistep sensitivity amplification (Ehrenberg and Blomberg, 1980; Freter and Savageau, 1980), like kinetic proofreading (Hopfield, 1974; Ninio, 1975). This principle was called *stochastic focusing* in biochemistry (Paulsson et al., 2000; Paulsson and Ehrenberg, 2001). It is a type of noise-induced transition (Horsthemke and Lefever, 1984) and reminiscent of the nonlinear effect in the bimolecular rate $r = \lambda n^2$ where $\langle r \rangle = \lambda \left(\langle n \rangle^2 + \sigma_n^2 \right)$ as pointed out early in the literature on stochastic chemistry (Renyi, 1954). If x represents the protein from the gene expression model in section 8.3 we can tune the rates of transcription and translation in such ways that the average $\langle n \rangle^2$ goes down but the variance σ_n^2 goes up so much that the sum $\langle n \rangle^2 + \sigma_n^2$ actually increases. The average rate of the bimolecular reaction could then go up even if the average concentration went down. This again means that the effective nonlinearities can be modulated almost arbitrarily by modulating the fluctuations around an average.

So far we only looked at how the underlying fluctuations affect the *average* value of the nonlinear function, but they also affect *fluctuations* in the same. The effect much depends on the relative timescale of fluctuations. If the inhibitor fluctuations above are fast compared to the lifetime of the plasmid ($\tau_2 \ll \tau_1$), then only the conditional average of q will have an effect on the plasmid dynamics. This is because only persistent fluctuations enslave dependent processes. However, if the inhibitor fluctuations are not fast enough, then inhibitor fluctuations can have disastrous consequences for the plasmid regulation, drastically widening the distributions. This phenomenon is analyzed in more detail in (Paulsson and Ehrenberg, 2000).

There is also another timescale that is important in this context. The hyperbolic function above comes from a condensation of uni- and bimolecular reactions. For many plasmids, the initiation frequency λ_2 is the rate with which they enter an intermediate state I from which they then decide to continue with replication or abortion, according to

$$
\begin{array}{c}
\mathrm{I} \xrightarrow{\ k_r\ } \text{replicate} \\
\ \ {\scriptstyle k_a n_2} \downarrow \\
\text{abort}
\end{array}
\tag{8.54}
$$

The probability for replication is then

$$
q = \frac{k_r}{k_r + k_a n_2} = \frac{K}{K + n_2}
\tag{8.55}
$$

The averaging in equation 8.52 thus implies that the inhibitor number n_2 remains constant for the duration of the event. If the inhibitors fluctuate infinitely fast, the abortion rate would simply be $k_a \langle n_2 \rangle$ and the stochastic focusing effect would disappear (this effect was discussed in more mathematical detail in (Paulsson et al., 2000; Paulsson and Ehrenberg, 2000). For plasmids to exploit stochastic focusing for noise suppression, there are thus two restrictions on the timescales: Inhibitor fluctuations must be much slower than the duration of the individual event to affect the average inhibition (including diffusion), and much faster than plasmid fluctuations to avoid randomizing the same. Both conditions seem satisfied for the best characterized plasmids. Plasmids change on a timescale on the order of hours, inhibitors change on the order of a few minutes, and the chemical decision in equation 8.54 takes about 10–20 seconds. The inhibitors are small RNAs and probably capable of diffusing many times across the cell during this time.

For applications to other systems, it may also be worth mentioning that the hyperbolic control function also can arise from

$$\text{on} \underset{k_d}{\overset{k_a n_a}{\rightleftarrows}} \text{off} \tag{8.56}$$

where n_a is the number of available repressor molecules. There is a subtle difference though. Immediately after the repressor has fallen off, there is at least one repressor molecule in the system. This affects the distribution of repressor molecules slightly. If the repressors are made in a Poisson process and decay exponentially, this subtlety is in fact enough to entirely abolish the stochastic focusing effect due to detailed balance constraints. However, if the inhibitors display other and perhaps more realistic types of fluctuations, for example if they are produced in bursts, then stochastic focusing can have large effects on mechanisms like equation 8.56. For a more detailed analysis see (Berg et al., 2000b).

Signal noise can thus in principle be used to make control more regular and deterministic—even in the simplest monostable negative feedback systems without sensitive bifurcations. It does require a separation of timescales, which rules out some candidate mechanisms, but is still a very real possibility in some of the best characterized negative feedback systems, like replication control of plasmids R1 or ColE1. It can also be used to create bistability in mechanisms that otherwise would be doomed to monostability. For example, macroscopic analyses of some mutually repressive systems have shown that hyperbolic repression functions (equation 8.55) are not sensitive enough to generate bistability (Cherry and Adler, 2000; Gardner et al., 2000). That conclusion no longer holds true when spontaneous fluctuations in concentrations are taken into account; stochastic focusing can make the hyperbolic functions sensitive enough to support bistability.

8.7.2   Noise-Induced Escape from Macroscopic Attractors

Some biochemical systems can exhibit distinctly different, self-perpetuating states depending on previous stimuli (Angeli et al., 2004; Ferrell Jr., 2002; Monod and Jacob, 1961; Ozbudak et al., 2004)—including irreversible developmental switches in the cell cycle (Tyson et al., 2001), the maturation of oocytes (Xiong and Ferrell Jr., 2003), the ubiquitous phosphorylation switches in signal transduction pathways (Bhalla et al., 2002), and the lysis-lysogeny decision system of phage lambda (Ptashne, 1992). The attractors in these multistable systems are by definition locally but not globally stable. A series of random fluctuations—originating in the random births and deaths of individual molecules—can thus force the system to escape one basin of attraction and allow it to be captured by another (Erdi and Tóth, 1989; Horsthemke and Lefever, 1984; Kramers, 1940). Depending on the size of the fluctuations and the strength of the local stability, the escape rates can be arbitrarily low. For example, the probability for phage lambda to spontaneously switch from a lysogenic to a lytic stage is on the order of 10^{-8} per generation (Aurell and Sneppen, 2002). Seen across large populations, however, this can still be a large enough number to have dramatic consequences for the population as a whole. In addition to spontaneous escapes from a given attractor, there is also a probabilistic initial choice of attractor when an infecting phage commits to either lysis or lysogeny, something that has been extensively studied using Monte Carlo sampling (Arkin et al., 1998).

The escape between distant attractors can not be analyzed by local linearized models, including FDT, since the escape characteristics mostly depends on what happens in between the attractors. Analytical approximations can sometimes be useful to characterize the escape rates (Aurell and Sneppen, 2002), but often numerical methods are the only practically useful way to study global dynamics of these systems. Straightforward Monte Carlo sampling typically converges too slowly for such problems, though, and the full master equation typically has too many states for direct integration. An attractive alternative is to approximate the master equation by a Fokker-Planck equation (FPE) (Risken, 1984), which is a partial differential equation for the time-dependent probability density function. The FPE approximation is good when the probability distribution function varies smoothly over state space. Since the FPE is non-local it can be used to analyze escape from macroscopic attractors (Qian et al., 2002), and it is also suitable for numerical integration using the extensively refined methods developed for partial differential equations (Ferm et al., 2004).

An example of FPE integration for a noise induced escape from a macroscopic attractor is illustrated in figure 8.3. In this example a trajectory that escapes the macroscopic point-attractor ends up in a limit cycle attractor in a model of a circadian oscillator (Vilar et al., 2002). Many organisms have evolved internal clocks to keep track of time. These are often based on biochemical oscillators that then must be resistant to some environmental and internal cues (Barkai and Leibler, 2000; Mihalcescu et al., 2004). One possible mechanism for generating a circadian oscillator is to use a transcriptional activator protein that promotes both its own expression and the expression of a repressor protein which in turn sequesters the activator. Vilar et al. present a quantitative model for such a system including the activator and repressor proteins, their respective mRNAs, the activity state of their promoters, and the activator-repressor complex. The model is reported to display regular oscillations in activator activity, even for relative large internal fluctuations in the levels of some of the chemical species (Vilar et al., 2002). In fact, for some parameters, internal fluctuations can drive the oscillation even if the corresponding macroscopic system has a single stable nonoscillating attractor. Rather than destroying the regular oscillations, the random fluctuations thus make them possible. This is illustrated in figure 8.3.

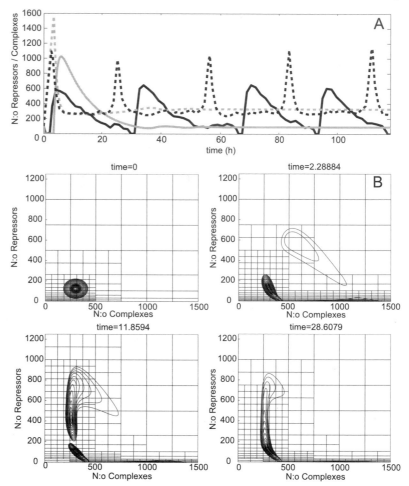

Figure 8.3 Noise induced oscillations. In A we see an example of the time evolution of two components of the circadian clock system with repressor (solid) and activator-repressor complex (dashed) in a macroscopic (gray) or a stochastic (black) model. The macroscopic model settles in a steady state whereas the stochastic model oscillates. In B we see the time evolution of the whole probability density function (PDF) as modeled by the Fokker-Planck approximation of the master equation. Initially the PDF was localized close to the macroscopic attracting stationary state. Throughout the time evolution the FPE was adaptively discretized as indicated by the grids (Ferm et al., 2004). The equations and parameters are those given for the bivariate RC-system by (Vilar et al., 2002) except for $\delta R=0.1h^{-1}$.

Acknowledgements

Thanks to Paul Sjöberg, Lars Ferm, and Per Lötstedt at Uppsala University for the simulation in figure 8.3. J.E. was funded by grants from the Swedish research council to Måns Ehrenberg and by Knut och Alice Wallenbergs stiftelse.

Notes

1. By homogeneous we mean that each molecule has an equal probability to be anywhere in the volume on the timescale of the chemical reactions that change the state (see chapter 9).

2. One component always turns into another in chemical reactions, but in condensed descriptions some species are approximated as constant sources or sinks of matter and not included as state variables.

9   Kinetics in Spatially Extended Systems

Karsten Kruse and Johan Elf

From cells to tissues and organisms, biological systems display spatially inhomogeneous structures. They result from processes in which the time for the transport of proteins across the whole system is long compared to typical reaction times. In this chapter, theoretical approaches for describing the dynamics of such systems are presented. In the first part, continuum descriptions in terms of partial differential equations are discussed. Such a description is appropriate if one is interested in the dynamics on scales that are large compared to molecular length scales as, for example, interaction distances of single molecules. In this context, a key concept is that of currents, which account for the transport of particles. Several techniques for deriving expressions for currents are discussed. On smaller scales, the discrete nature of the molecules cannot be neglected and a stochastic description is required. In particular, this is the case when a molecule has only a few potential reaction partners within the diffusion range. A stochastic description in terms of the reaction-diffusion master equation is presented in the second part of this chapter. It is a generalization of techniques presented in chapter 8 to account for inhomogeneous particle distributions. As will be shown, in the limit of many reactants within the diffusion range, the reaction-diffusion master equation is well approximated by a continuum description. The different approaches are illustrated by application to the Min-system of the bacterium *Escherichia coli* as well as other subcellular systems.

9.1   Continuum Descriptions

Continuum theories describe the dynamics of spatially extended systems on scales that are large compared to molecular scales (Landau and Lifshitz, 1995). In such a description, the discrete nature of the single molecules forming the system is neglected. Instead, the state of the system is given in terms of continuous functions of space and time, the fields[1]. In the simplest case, the fields represent densities, for example of proteins. In addition, they can represent additional features of the molecules involved, for example the mean orientation of elongated molecules like cytoskeletal filaments. The fields are linked to microscopic representations of the system state in terms of individual molecules by local averages. Local averages are performed over volume elements that are small compared to the length scales of the structures one is interested in, but large enough to contain a sufficient number of particles such that spatial fluctuations within a volume element are negligible.

Example: Min-oscillations. Division of the bacterium *Escherichia coli* usually occurs in the cell center leading to two daughter cells of equal size. Selection of the center as the division site is in part achieved by the Min-system which consists of three proteins, MinC, MinD, and MinE (de Boer et al., 1989; Bi and Lutkenhaus, 1993). While MinC inhibits assembly of the division apparatus on the cytoplasmic membrane, MinD and MinE regulate the spatial distribution of MinC. Fluorescence microscopy of Min-proteins tagged with green fluorescent protein (GFP) has revealed that the distributions of the Min-proteins change periodically with time (Raskin and de Boer, 1999b; Hu and Lutkenhaus, 1999; Raskin and de Boer, 1999a; Hale et al., 2001). During one half of the period, most proteins are localized in one cell half, while during the other half of the period they predominantly reside in the opposite cell half. The oscillation periods vary from cell to cell and range from 40 seconds to 120 seconds. As a consequence of the oscillations, MinC suppresses formation of the division apparatus close to the cell poles, but not in the center. The oscillations are generated by MinD and MinE alone, while MinC oscillates because it co-localizes with MinD. Over the last few years, several continuum descriptions of the Min-protein dynamics have been developed (Meinhardt and de Boer, 2001; Howard et al., 2001; Kruse, 2002; Huang et al., 2003; Drew et al., 2005; Meacci and Kruse, 2005). In these descriptions, the fields are given by the surface densities of MinD and MinE on the cytoplasmic membrane and the volume densities of MinD and MinE in the cytoplasm. As the distribution of MinC is directly related to the distribution of MinD, it is not incorporated.

In a continuum description, the dynamics of the fields is commonly given by partial differential equations[2]. The dynamic equations depend on a number of phenomenological parameters. While the values of these parameters are determined by details of the molecular interactions, the form of the dynamic equations is largely independent of these details. Rather, it is imposed by the symmetries displayed by the system. For example, the equations must transform correctly if the system is rotated[3]; see sect. 9.1.3 for further discussion of the role of symmetries. Hence, on large scales the system's behavior is independent of most properties of the microscopic molecular interactions.

As an example, consider fluid water. Water molecules are characterized by their charge distribution, and their interactions involve dipole-dipole interactions and hydrogen bonds. For most practical purposes, however, the flow of water can be described by the Navier-Stokes equation (Landau and Lifshitz, 1995). Neglecting the very weak compressibility of water, this equation contains only two parameters, the water density and the shear viscosity. The same equation also describes the flow of all other simple fluids which may consist of molecules very different from water. The implications of these differences for the dynamics on large scales are fully captured by distinct numerical values of the phenomenological parameters. Therefore, appropriate continuum descriptions of spatially extended systems can be obtained from much less information than is needed for microscopic descriptions. On a more technical level, continuum descriptions in addition permit the use of the powerful methods of differential calculus for analysis. Taken together, these points make continuum descriptions an extremely helpful tool to investigate mechanisms underlying the formation of spatiotemporal structures.

In the following, general principles that guide the formulation of continuum descriptions will be presented. Before continuing with the general discussion, however, first the very important class of reaction-diffusion systems is introduced.

9.1.1 Reaction-Diffusion Systems

In his groundbreaking paper on the chemical basis of morphogenesis, Turing introduced the idea that the diffusion of particles together with chemical reactions can lead to the formation of spatiotemporal patterns (Turing, 1952). Having biological systems in mind, Turing suggested that these patterns might be at the origin of structures in living systems such as the regular arrangement of the tentacles of hydra. In fact, the formation of compartments in *Drosophila* and calcium dynamics in cell aggregates as well as within cells have been successfully described using a reaction-diffusion approach (Cross and Hohenberg, 1993; Koch and Meinhardt, 1994; Falcke, 2004). The application of reaction-diffusion systems to describe intracellular protein dynamics is a more recent development.

In a reaction-diffusion system each field represents the density of one particle species. The different species can, for example, represent different kinds of molecules or different states of one kind of molecule. The reaction terms correspondingly describe reactions involving the different molecules or transitions between the different states. In their most general form, the dynamics of two interacting species is described as

$$\frac{\partial}{\partial t}c_1(\mathbf{r}, t) = D_1\nabla^2 c_1(\mathbf{r}, t) + u_1(c_1, c_2) \tag{9.1}$$

$$\frac{\partial}{\partial t}c_2(\mathbf{r}, t) = D_2\nabla^2 c_2(\mathbf{r}, t) + u_2(c_1, c_2) \tag{9.2}$$

Here, $c_i(\mathbf{r}, t)$, $i = 1, 2$ denotes the densities of the two species at a point $\mathbf{r} = (x, y, z)$ in space[4] and at time t. The operator $\partial/\partial t$ denotes the partial derivative with respect to time, that is a derivative with respect to time while the space coordinates are kept constant. The first terms on the right hand sides describe particle diffusion. The parameters D_i are the respective diffusion constants and ∇^2 is the Laplace-operator. In three spatial dimensions, $\nabla^2 = \partial^2/\partial x^2 + \partial^2/\partial y^2 + \partial^2/\partial z^2$. Here $\partial^2/\partial x^2$ is the second partial derivative with respect to x and so on. The form of the diffusion term will be derived below. The functions u_i depend on the densities and account for the reactions in the system.

Example: Min-oscillations. Several models for the dynamics of the Min-proteins fall into the class of reaction-diffusion systems (Meinhardt and de Boer, 2001; Howard et al., 2001; Huang et al., 2003; Drew et al., 2005). There, cytosolic MinD and MinE diffuse, while diffusion of membrane-bound proteins is usually neglected. The reaction terms account for the exchange of proteins between the cytoplasm and the membrane. It was shown in vitro that the ATPase MinD has a high affinity for the inner bacterial membrane if ATP is present (Hu et al., 2002). Furthermore, for concentrations of MinD exceeding a critical value, filamentous MinD aggregates are formed on the membrane. MinE associates with the membrane only in the presence of MinD. There, it stimulates hydrolysis of the ATP bound to MinD, which eventually drives the proteins off the membrane. These results are compatible with the behavior of MinD and MinE in vivo. Several different reaction schemes have been developed that incorporate these findings. As an example, consider the model proposed by Huang et al. (2003). There, binding of MinD to the membrane is assumed to be cooperative, leading to the aggregation of membrane-bound MinD. The binding of MinE to the membrane is described by a second order process involving the concentration of membrane-bound MinD. On the membrane, MinE is assumed to exist only in complexes with MinD. Finally, the release of MinDE complexes is described as a first order process. Explicitly, the dynamic equations are

$$\partial_t c_D = D_D \partial_x^2 c_D - [\omega_D + \mu_{dD}(c_d + c_{de})]c_D + \omega_{de} c_{de} \tag{9.3}$$

$$\partial_t c_E = D_E \partial_x^2 c_E + \omega_{de} c_{de} - \omega_E c_d c_E \tag{9.4}$$

$$\partial_t c_d = -\omega_E c_d c_E + [\omega_D + \mu_{dD}(c_d + c_{de})]c_D \tag{9.5}$$

$$\partial_t c_{de} = -\omega_{de} c_{de} + \omega_E c_d c_E \tag{9.6}$$

For simplicity, the dynamic equations are given here in one spatial dimension and in the limiting case of immediate rebinding of ATP to MinD after it is released from the membrane. The distributions of cytosolic MinD and MinE are denoted by c_D and c_E, while c_d and c_{de} denote the densities of membrane-bound MinD and MinDE complexes, respectively. The remaining parameters denote rate constants for the different reactions. Note that in agreement with experimental results, the above equations conserve the numbers of MinD and MinE proteins.

The behavior of a reaction-diffusion system is determined by the values of the diffusion constants and of the various rates. An analysis of the dynamic equations 9.1 and 9.2 usually starts with the identification of spatially homogeneous stationary states, $c_i(\mathbf{r}, t) = c_i^{(0)}$ for all $\mathbf{r}$ and t (Cross and Hohenberg, 1993). For such a state, the time and space derivatives appearing in the dynamic equations vanish such that $u_i(c_1^{(0)}, c_2^{(0)}) = 0$ for $i = 1, 2$. Then, the stability of these states with respect to perturbations is analyzed. The basic idea is the same as for ordinary differential equations (chapter 6), but in the present context, the perturbation can depend on the space coordinate.

For the stability analysis, the perturbed distribution is written as $c_i = c_i^{(0)} + \delta c_i$. If the state is stable, the perturbation decays with time. In the opposite case, it grows and a pattern is formed. Inserting the above expression into the dynamic equations yields the time-evolution of the perturbation δc_i. If the initial perturbation is small, only terms linear in δc_i have to be retained; non-linear terms are much smaller and can therefore be neglected. This leads to

$$\frac{\partial}{\partial t}\left(\begin{array}{c} \delta c_1 \\ \delta c_2 \end{array}\right) = \mathsf{L}\left(\begin{array}{c} \delta c_1 \\ \delta c_2 \end{array}\right) \tag{9.7}$$

Here, the linear operator L is given by

$$\mathsf{L} = \left(\begin{array}{cc} D_1\nabla^2 + u_{11} & u_{12} \\ u_{21} & D_2\nabla^2 + u_{22} \end{array}\right) \tag{9.8}$$

The constants u_{ij} with $i, j = 1, 2$ are defined as $u_{ij} = \partial u_i/\partial c_j$, where the derivatives are evaluated at $c_i = c_i^{(0)}$.

To proceed further, the densities are decomposed into eigenmodes of the linear operator L. The effect of L on an eigenmode ϕ is to multiply the mode with a constant λ_ϕ, the corresponding eigenvalue. Explicitly, $\mathsf{L}\phi = \lambda_\phi\phi$. For an eigenmode, the dynamic equation

$$\frac{\partial}{\partial t}\phi = \mathsf{L}\phi = \lambda_\phi\phi \tag{9.9}$$

is readily solved

$$\phi(\mathbf{r}, t) = \exp(\lambda_\phi t)\phi(\mathbf{r}, 0) \tag{9.10}$$

where $\phi(\mathbf{r}, 0)$ is the initial perturbation at time $t = 0$. If the eigenvalue λ_ϕ has a negative real part, the mode will decay exponentially in time. In the opposite case, it will grow. The homogeneous state is, in this case, unstable under the corresponding perturbation, and a pattern will form. Parameter values for which the maximum of the eigenvalues' real parts becomes positive indicate a bifurcation or instability. In the vicinity of an instability, the pattern is well described by the unstable eigenmode. If the corresponding eigenvalue is real, the instability leads to a spatially nonhomogeneous pattern that is stationary in time. In the opposite case, the pattern will oscillate in time (Cross and Hohenberg, 1993).

For systems of a finite size, the determination of the eigenmodes and eigenvalues of the differential operator L is usually difficult. Therefore, an analysis is often carried out first assuming an infinite system size. Then, the perturbations can be decomposed into Fourier modes $c_\mathbf{k}$ with

$$\delta c(\mathbf{r}, t) = \int_{-\infty}^{\infty} \frac{d\mathbf{k}}{2\pi}\, c_\mathbf{k}(t)e^{i\mathbf{k}\cdot\mathbf{r}} \tag{9.11}$$

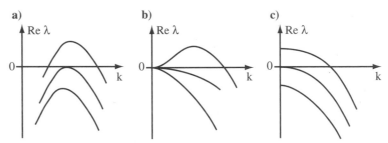

Figure 9.1 Schematic representation of the linear growth rate $\mathrm{Re}\lambda_k$ as a function of the wave number k. Graphs in one diagram can be obtained by changing a suitable phenomenological parameter in the dynamic equations. For the case displayed in **a)**, the instability occurs at a finite wave number, in **b)** and **c)** at $k = 0$, respectively implying a periodic state and a homogeneous state right after the instability.

Inserting this expression into equation 9.7, the dynamics of each mode is governed by a 2×2 matrix L_k with

$$\mathsf{L_k} = \begin{pmatrix} -D_1 k^2 + u_{11} & u_{12} \\ u_{21} & -D_2 k^2 + u_{22} \end{pmatrix} \tag{9.12}$$

The eigenvalues of this matrix are

$$\lambda_k = \frac{1}{2}\left[\mathrm{tr}\mathsf{L_k} \pm \sqrt{(\mathrm{tr}\mathsf{L_k})^2 - 4\det\mathsf{L_k}}\right] \tag{9.13}$$

where $\mathrm{tr}\mathsf{L_k} = -(D_1 + D_2)k^2 + u_{11} + u_{22}$ is the trace and $\det\mathsf{L_k} = (D_1 k^2 - u_{11})(D_2 k^2 - u_{22}) - u_{12}u_{21}$ is the determinant of $\mathsf{L_k}$. Stability requires the real parts of $\lambda_\mathbf{k}$ to be negative for all values of $\mathbf{k}$.

As a function of the mode number $\mathbf{k}$, the real parts of the eigenvalues will show one of the functional behaviors indicated in figure 9.1. At an instability, the wave number of the critical mode k_c as well as the imaginary part of the critical mode's eigenvalue can be zero or non-zero. This leads to essentially three classes of patterns that are formed at an instability: stationary patterns with a characteristic spatial wavelength and oscillatory patterns with or without a characteristic spatial wavelength (Cross and Hohenberg, 1993). Beyond the instability, the non-linear terms can no longer be neglected. Often one then has to fall back upon numerical solutions of the dynamic equations.

Summarizing, a linear stability analysis can give a rough understanding of the system behavior as a function of the system parameter without the need of explicitly solving the dynamic equations. If there are more than two densities involved, an analytic calculation of the eigenvalues is in general not possible. However, it is no problem to obtain them in this case numerically.

Example: Min-oscillations. The homogeneous stationary state of the dynamic equations 9.3 to 9.6 is determined by the roots of a polynomial. For each value of the parameters it is unique and has to be determined numerically. The eigenvalue with the largest real part as a function of the wave-number k is shown in figure 9.2.

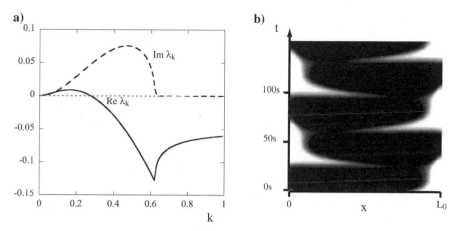

Figure 9.2 Analysis of the dynamic equations 9.3 to 9.6. **a)** Real (solid line) and imaginary (dashed line) part of the eigenvalue with the largest real part as a function of the wave number k. In an interval of k-values, the real part is positive. At the same time the imaginary part is non-zero in this range, indicating an oscillatory instability. The gray dotted line indicates $\mathrm{Re}\,\lambda_k = 0$. **b)** Space-time plot of the total MinD-distribution obtained from a numerical solution of the dynamic equations. The density is color coded with brighter gray levels indicating a higher density. MinD periodically shifts from one end to the opposite end.

The real parts are positive in an interval in which their imaginary part does not vanish. The instability is thus oscillatory. Correspondingly, a numerical solution of the dynamic equations yields a distribution that changes periodically in time.

9.1.2   Densities and Currents

The reaction-diffusion equations 9.1 and 9.2 are a special case of a continuity equation. The structure of a continuity equation reflects that at a given point in space, the number of molecules in a small volume element can change because of two possible events. First, particles can be transported into or out of this volume element and second, particles can be created or destroyed within the volume element. This applies to all conserved quantities[5] (de Groot and Mazur, 1984).

Transport across the surface of a volume is described by currents. They give the number of particles traversing a surface element per unit time. A current $\mathbf{j}$ is a vector with components j_i, where i indicates the three directions x, y, and z. Each component j_i is the current through a surface element perpendicular to the direction i. The net change of particle number in a small volume element is then obtained from the divergence of the corresponding current

$$\boldsymbol{\nabla} \cdot \mathbf{j} = \frac{\partial}{\partial x} j_x + \frac{\partial}{\partial y} j_y + \frac{\partial}{\partial z} j_z \tag{9.14}$$

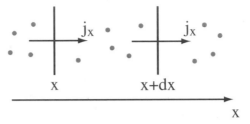

Figure 9.3 Change in particle number due to particle transport in one dimension. The current at x is $j_x(x)$, the current at $x + dx$ is $j_x(x + dx)$. The current is counted positive if directed to the right. The net change in particle number n per unit time is then the difference between the currents across the left and the right boundary of the interval, $dn/dt = j_x(x) - j_x(x + dx)$. Replacing the particle number by the density c through $n = c\, dx$ and taking the limit $dx \to 0$ one obtains $\partial c/\partial t = -\partial j/\partial x$. For example, in the case of diffusion we have $\mathbf{j} = -D\partial c/\partial x$, such that the diffusion equation is $\partial c/\partial t = D\partial^2 c/\partial x^2$. In higher dimensions the contributions of all directions have to be summed leading to expression 9.14.

See figure 9.3. Here, $\boldsymbol{\nabla}$ is the gradient operator with $\boldsymbol{\nabla} = (\partial/\partial x, \partial/\partial y, \partial/\partial z)$ in three dimensions. Creation or destruction of particles is captured by source and sink terms s.

Hence, the continuity equation for the evolution of a particle density has the form

$$\frac{\partial}{\partial t}c + \boldsymbol{\nabla} \cdot \mathbf{j} = s \tag{9.15}$$

In the reaction-diffusion equations 9.1 and 9.2, the current is a consequence of diffusion. It is given by $-D_i\boldsymbol{\nabla}c_i$, whereas the source and sink terms are given by the reaction terms u_i with $i = 1, 2$.

While the source and sink terms are usually given by kinetic equations for the reactions taking place in the system, there is no generally applicable procedure or framework for deriving expressions for the currents. As will be discussed in section 9.2, currents can be derived from microscopic descriptions by applying a mean-field approximation and then coarse-graining. Other approaches to the currents are phenomenological and do not require a microscopic model. At the system boundary, additional specifications have to be made. In situations where the system is confined by an impenetrable wall, currents across the system boundary have to vanish. In other situations, there might be a constant influx into the system, fixing the value of the current to a constant value. This would be the case for proteins that are generated at a constant rate in a source that is located at the system boundary. In more complicated situations, the current at the boundary depends on the present state of the system. An example is provided by cell walls containing receptor molecules to which proteins can bind. Here, the binding rate will depend on the occupancy of the receptors and probably on the presence of other molecules.

Before presenting the phenomenological approach, let us mention again that a continuum description will in general also contain fields of nonconserved quantities. An example is the orientation of cytoskeletal filaments. Obviously, there are no

currents associated with such quantities, and their time evolution is consequently
not given by a continuity equation. However, the same strategies that are used to
obtain expressions for the currents can also be applied to the rate of change of these
fields. Therefore, they will not be discussed explicitly in the following.

9.1.3   Phenomenological Currents

While there is no generally applicable framework for deriving phenomenological
expressions for the currents, there are some universal constraints on possible
expressions. They follow from the symmetries displayed by the system (Nicolis and
Prigogine, 1977; de Groot and Mazur, 1984; Chaikin and Lubensky, 1995). First
of all, if a cause is invariant under the action of a spatial symmetry operation like
rotation or reflection, then the same must be true for any effects due to this cause.
This condition is expressed by the Curie principle which furthermore states that if
an effect is not invariant under a certain symmetry operation then neither can be the
cause. One consequence of this principle is, for example, that the directed motion
of molecular motors is only possible because actin filaments and microtubules are
polar, that is because they have two different ends.

A second symmetry that imposes constraints on phenomenological theories is
the invariance under time-reversal of the microscopic equations of motion. That
is, even though the expressions in a macroscopic description may not be derived
from a microscopic description, the universal property of time-reversal invariance
of the microscopic equations of motion imposes constraints on these expressions.
This remarkable point was made by Onsager, who showed that for certain systems,
different phenomenological parameters are intimately related (Onsager, 1931a,b).

At thermodynamic equilibrium, a system is described by a set of macroscopic
state variables like temperature, volume, and particle number (Chaikin and Lubensky, 1995). The free energy F is a function of these variables. The equilibrium state
is the one that minimizes the free energy F, while respecting the constraints imposed on the system. If a constraint is released, the system will evolve towards a new
equilibrium state that is determined by a minimum in the free energy respecting
the new constraints.

The dynamics of systems out of thermodynamic equilibrium, but sufficiently
close to it, can still be obtained within a generally applicable framework. More
precisely, it applies to situations, where each of the volume elements introduced
above can be assumed to be in thermodynamic equilibrium. In this case, a free
energy F of the total system can be defined by summing the free energies associated
with the individual volume elements. Changes in the system's free energy can
then be expressed in terms of products of currents and associated causes, often
called generalized thermodynamic forces. The currents are naturally functions of
the generalized forces. In particular, if there are no forces there will be no currents.
An expansion of the currents in terms of the forces thus starts with the linear terms.
Keeping only these, one obtains phenomenological dynamic equations (de Groot
and Mazur, 1984; Chaikin and Lubensky, 1995). The phenomenological parameters

appearing in the expansion are called linear response coefficients and depend in general on the state of the system.

Example: Diffusion. Consider the case of a single species of noninteracting molecules at constant temperature and constant pressure. Then the free energy per unit volume f only depends on the particle density c, $f \equiv f(c)$. Consequently,

$$
\begin{aligned}
\frac{d}{dt}F &= \int d\mathbf{r}\, \frac{\partial}{\partial t} f(c) = \int d\mathbf{r}\, \left(\frac{\partial}{\partial c}f\right)\frac{\partial}{\partial t}c \\
&= -\int d\mathbf{r}\, \mu \nabla \cdot \mathbf{j} = \int d\mathbf{r}\, \mathbf{j} \cdot \nabla\mu
\end{aligned}
$$

In this calculation, the continuity equation 9.15 with vanishing source terms has been used and the chemical potential $\mu = \partial f/\partial c$ has been introduced. The chemical potential gives the change in free energy upon addition of a particle to the system. Expressing the current $\mathbf{j}$ in terms of the generalized force $\nabla\mu$ one finds in linear order

$$
\mathbf{j} = -\Lambda\nabla\mu \tag{9.16}
$$

where Λ is the phenomenological coefficient describing the response of the system, that is the current, to a gradient in the chemical potential. As the free energy, the chemical potential is a function of the particle density. Defining $D = \Lambda\partial\mu/\partial c$, the diffusion current can then be cast in the familiar form

$$
\mathbf{j} = -D\nabla c \tag{9.17}
$$

where D is the diffusion constant. The minus sign in equation 9.16 has been introduced to obtain $D > 0$. For an ideal gas $\mu(c) = k_B T \ln c$, where k_B is the Boltzmann constant and T temperature. Since $\Lambda/c = \xi^{-1}$ is the mobility of a particle, one finds $D\xi = k_B T$, which is the well-known Einstein relation (de Groot and Mazur, 1984).

Example: Molecular motors. As an example of nondiffusive transport consider the motion of molecular motors (Alberts et al., 2002). These proteins use the energy derived from ATP-hydrolysis to move along cytoskeletal filaments. Examples are kinesins that transport vesicles along microtubules or myosins that together with actin form the contractile machinery in muscles. The motion of motors along a filament is directional, where the direction of motion is determined by the orientation of the filament. For these motors, the current is given by $\mathbf{j} = \mathbf{v}c_{\mathrm{mot}}$, where c_{mot} is the motor density and $\mathbf{v}$ is the average motor velocity. The motion of motors is driven the hydrolysis of ATP, but also by external forces. Hydrolysis of ATP occurs in the presence of a difference $\Delta\mu$ in the chemical potentials of ATP and its hydrolysis products ADP and P_i. The associated rate is denoted r. The change in free energy is $\mathbf{f} \cdot \mathbf{v} + r\,\Delta\mu$ and an expansion of $\mathbf{v}$ and r in terms of $\mathbf{f}$ and $\Delta\mu$ yields in linear order (Jülicher et al., 1997)

$$\mathbf{v} \;=\; \lambda_{11}\mathbf{f} + \boldsymbol{\lambda}_{12}\Delta\mu \tag{9.18}$$

$$r \;=\; \boldsymbol{\lambda}_{21}\cdot\mathbf{f} + \lambda_{22}\Delta\mu \tag{9.19}$$

Note, that $\boldsymbol{\lambda}_{12}$ couples a vector quantity and a scalar quantity and must therefore be itself a vector illustrating the Curie principle. As mentioned above, the cross coefficients $\boldsymbol{\lambda}_{21}$ and $\boldsymbol{\lambda}_{12}$ are not independent of each other. In fact, the Onsager relations impose in the present case $\boldsymbol{\lambda}_{21} = \boldsymbol{\lambda}_{12}$.

There are situations in which the general framework indicated above is not sufficient for obtaining appropriate expressions for the currents. First of all, interactions with the environment can lead to anomalous diffusion. This would be, for example, the case when proteins get trapped in small regions of space with corresponding dwell times that are algebraically distributed, that is, the probability of having a large dwell time t is proportional to $t^{-(1+\gamma)}$ with $0 < \gamma < 1$. Another example is provided by particles moving on DNA that folds back on itself. As the particles detach from the DNA, diffuse through three-dimensional space and reattach at a different location on the DNA, the effective motion along the one dimensional DNA can be anomalous (Berg et al., 1981; Brockmann and Geisel, 2003). Such processes can be described in the frame of continuous time random walks (Montroll and Shlesinger, 1984). Secondly, systems can be far from thermodynamic equilibrium. Even though it is in general hard to measure the distance to thermodynamic equilibrium, presumably most people would tend to say that living cells are far away from it. Still, one might argue that the theory presented so far should in many situations describe the dominant effects. There are, however, situations in which linear terms are absent or dominated by non-linear terms such that more general expressions are needed.

Example: Attractive interactions. In the case of an attractive interaction between proteins, macroscopic currents are induced by gradients in the protein density. In contrast to the case of diffusion, however, the current will be directed towards higher concentrations. One might therefore be tempted to describe the aggregation process by a diffusion equation with negative diffusion constant. This equation is unphysical for several reasons. First of all, it can lead to negative densities. Secondly, infinite particle densities can be generated in finite times. Finally, it generates structures on arbitrarily fine length scales. All these problems can be avoided by modifying the diffusion current such that

$$j = c(c_{\max} - c)(k_1\partial_x c + k_2\partial_x^3 c) \tag{9.20}$$

with $k_1, k_2 > 0$. For simplicity, it has been assumed in this expression that the motion of particles is confined to the x-axis. The pre-factor c prevents the appearance of negative densities, while the pre-factor $c_{\max} - c$ prevents the density from growing beyond any limit by introducing a maximal density $c_{\max}$. Finally, the third order derivative avoids the formation of structures on arbitrarily small length scales.

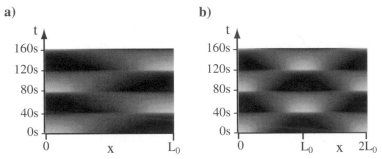

Figure 9.4 Space-time plot of the total MinD distribution on the membrane obtained from a numerical solution, the dynamic equations 9.21 and 9.22. Results are shown for a system size of $L_0 = 2\mu m$ **(a)** and $2L_0$ **(b)**, revealing the period doubling characteristic for the Min-oscillations in longer bacteria. The density is color coded with brighter gray levels indicating a higher density. Modified from (Meacci and Kruse, 2005).

Example: Min-oscillations. For the Min-oscillations a mechanism has been proposed that is based on an attractive interaction of membrane-bound MinD (Kruse, 2002; Meacci and Kruse, 2005). This attraction is assumed to be the dominant process for the formation of MinD aggregates. Assuming homogeneous cytosolic distributions C_D and C_E of MinD and MinE, respectively, the dynamics of the Min-system reduces to the evolution of the densities of MinD and MinDE complexes on the membrane. Explicitly,

$$\partial_t c_d = \omega_D C_D(c_{\max} - c_d - c_{de}) - \omega_E C_E c_d - \partial_x j \qquad (9.21)$$

$$\partial_t c_{de} = -\omega_{de} c_{de} + \omega_E C_E c_d \qquad (9.22)$$

Here, the aggregation current j is chosen to be of the form of equation 9.20. The reaction terms describing attachment of MinD and MinE to the membrane and detachment of MinDE complexes from the membrane are linear in the membrane densities (C_D and C_E are constants). A linear stability analysis of the homogeneous state can be performed analytically in this case. It reveals a critical value $k_{1,c}$ for the parameter k_1 such that the homogeneous state is unstable for $k_1 > k_{1,c}$. Furthermore, if the condition $\omega_{de}^2 < \omega_D \omega_E C_D C_E$ is met, then the instability is oscillatory and oscillatory solutions reminiscent of the Min-protein oscillations are obtained; see figure 9.4.

9.2 Stochastic Treatment of Nonhomogeneous Chemical Reactions

As stated above, continuum descriptions are appropriate only if the molecular densities are large enough, so that each molecule has many potential reaction partners within the diffusion range. If they are too few, then the discrete nature of the molecules becomes apparent and a stochastic description is required. In the derivation of the master equation in chapter 8 it was assumed that the spatial distribution of molecules equilibrates on a shorter time scale than the characteristic time scales for changes in the state variables. This was necessary in order to be able to take the transition rates $r_j(\mathbf{n})$ as constant for a state $\mathbf{n}$, where the numbers of molecules are counted for the whole reaction volume. However, if the molecules do not have time to diffuse through the reaction volume between their reactions, the rates will not only depend on the total number of different molecules, but also on when and where other reactions occurred. The Markovian property of the random process is then lost for descriptions that only include the total numbers of molecules as state variables.

The condition for homogeneity by diffusion is that

$$T_i \gg L^2/D_i \quad \text{for all i=1} \ldots \text{N} \tag{9.23}$$

where T_i is the average time between two reactions involving species i, D_i is its diffusion constant and L is the linear size of the system (Arnold, 1980; Gardiner and Steyn-Ross, 1984). When equation 9.23 is satisfied, each molecule has an equal probability to have its next reaction anywhere in the reaction volume. Thus the local deviations from the spatially averaged concentration induced by localized chemical reactions are spread throughout the system, such that the reaction rates are homogeneous. When equation 9.23 is not satisfied, the homogeneous master equation, used in chapter 8, is an approximation where it is assumed that $\langle r_j(\mathbf{n}) \rangle_\Omega \approx r_j(\langle \mathbf{n} \rangle_\Omega)$, that is the spatially averaged rates that we need in the homogeneous case equal the rates evaluated for the total number of molecules. This approximation may be good or bad depending on how sensitive the transition rates are to local perturbations in concentrations and how the molecules are distributed spatially.

9.2.1 The Reaction-Diffusion Master Equation

One way to model spatial heterogeneity is to introduce local concentrations. Here we do this by dividing the total volume Ω into C artificial cubic subvolumes of volume $\Delta = \Omega/C$ and by keeping track of how many molecules there are in each subvolume. The side length of the subvolumes ℓ is chosen so that equation 9.23 is satisfied if L replaced by ℓ. With this choice of subvolume size the mean reaction free path, the Kuramoto length (Kuramoto, 1974), is longer than a subvolume, and the spatial distribution of molecules within a subvolume can be considered homogeneous on the time scale of the chemical reactions.

At the same time, the subvolumes must be much larger than the mean free path. This is necessary to describe the movement in a subvolume as a diffusion process. The mean collision free path is very short in cells due to the high concentration of non-reactive molecules, for example solvents. This typically makes more detailed descriptions, including the velocity of the molecules, unnecessary, since the velocity distribution equilibrates on the time scale of non reactive collision with the solvent molecules.

Finally, the length of the subvolumes, ℓ, must also be significantly larger than the reaction radii[6] (Berg, 1978b; Ovchinnikov et al., 1989) of all interactions, which for biomolecules can be a more demanding requirement than the mean free path. This is required for well-defined association and dissociation rate constants within each subvolume (Elf and Ehrenberg, 2004). If the reaction subvolumes are made smaller than the reaction radii, molecules have a hard time finding each other, but when they do they never let go.

The extended state description is $\{\mathbf{n}\} = \{\mathbf{n}_1 \cdots \mathbf{n}_\kappa \cdots \mathbf{n}_C\}$, where $\mathbf{n}_\kappa = \{n_{1\kappa} \cdots n_{i\kappa} \cdots n_{N\kappa}\}$ and $n_{i\kappa}$ is the number of i molecules in subvolume κ. The state of the system is changed by chemical reactions within the subvolumes and diffusion events between the subvolumes. The chemical reactions have different rates in different subvolumes since they depend on the local concentrations of reactants $\Delta^{-1}\mathbf{n}_\kappa$. The probability that a reaction j will occur in subvolume κ during the infinitesimal time between t and $t + dt$ is $dt \cdot r(\Delta^{-1}\mathbf{n}_\kappa)$. If this reaction occurs, the local state is changed from $\mathbf{n}_\kappa$ to $\mathbf{n}_\kappa + \nu_j$.

Diffusion is modeled as a memory-lacking random walk in discrete space, as implemented by a set of first order diffusion events:

$$\{\cdots n_{i\gamma} \cdots n_{i\kappa} \cdots\} \xrightarrow{n_{i\kappa} d_i^{\kappa\gamma}} \{\cdots n_{i\gamma} + 1 \cdots n_{i\kappa} - 1 \cdots\} \qquad \begin{array}{l} i = 1, 2, ..., N \\ \kappa = 1, 2, ..., C \\ \gamma = 1, 2, ..., C \end{array} \qquad (9.24)$$

Here, an i-molecule diffuses from subvolume κ to subvolume γ. The first order diffusion rate constant for species i is taken to be $d_i^{\lambda\mu} = d_i^{\mu\lambda} = D_i/\ell^2$ for neighboring subvolumes and otherwise zero. This implies that the probability that an i-molecule diffuses from subvolume κ to its neighbor λ during the infinitesimal time between t and $t + dt$ is $dt \cdot d_i^{\kappa\lambda} n_{i\kappa}$.

Given the extended state description and a set of state transition rates for the local reaction and diffusion events, we can write down our *reaction-diffusion master equation* (RDME) as in chapter 8 (Kuramoto, 1974; Gardiner et al., 1976; Nicolis and Prigogine, 1977; Baras and Mansour, 1997):

$$\frac{dP(\{\mathbf{n}\},t)}{dt} =$$
$$\sum_\kappa \sum_{j=1}^M \left(r_j \left((\mathbf{n}_\kappa - \nu_j)\Delta^{-1} \right) P \left(\{\cdots \mathbf{n}_\kappa - \nu_j \cdots\}, t \right) - r_j(\mathbf{n}_\kappa \Delta^{-1}) P\left(\{\mathbf{n}\}, t\right) \right) +$$
$$\sum_\kappa \sum_{\gamma \neq \kappa} \sum_i \left(d_i^{\gamma\kappa} (n_{i\gamma} + 1) P \left(\{\cdots n_{i\gamma} + 1, n_{i\kappa} - 1 \cdots\}, t \right) - d_i^{\kappa\gamma} n_{i\kappa} P\left(\{\mathbf{n}\}, t\right) \right)$$
$$(9.25)$$

The upper row contains the state transition rates that are due to reactions ($j = 1, \ldots, M$). For each subvolume these rates are calculated for the local concentration of molecules. The lower row contains the terms for diffusion between neighboring subvolumes. Here, $\{\cdots n_{i\gamma} + 1, \cdots, n_{i\kappa} - 1 \cdots\}$ is the state where there are $n_{i\gamma}+1$ i molecules in subvolume γ and $n_{i\kappa}$-1 i molecules in subvolume κ compared to state $\{\mathbf{n}\}$. Figure 9.5 illustrates the principles from the perspective of the single molecule.

Let us mention that in cases where the microscopic transport of particles is not only due to diffusion, generalizations of the RDME can be used.

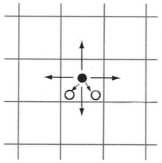

Figure 9.5 Example of how chemical reactions are modeled by the Reaction-Diffusion Master Equation illustrated in two spatial dimensions. The probability that the black molecule jumps to one of the neighboring subvolumes in the next infinitesimal time dt is $dt \times 4 \times D/\ell^2$, where D is the diffusion constant of the black molecule, ℓ is the length of the subvolume, and 4 is the number of neighbors. The probability that the black molecule instead binds one of the two white molecules is $dt \times \Delta \times k_a \times 1/\Delta \times 2/\Delta$, where k_a is the association rate constant, Δ is the volume of the subvolume, and $1/\Delta$ and $2/\Delta$ are the respective concentrations of black and white molecules in the subvolume.

Macroscopic Currents from the Reaction-Diffusion Master Equation
In the limit that there are macroscopically large numbers of reaction partners within the diffusion range of each molecule, the RDME converges to the macroscopic reaction-diffusion equation introduced in section 9.1.1 (Arnold and Theodosopulu, 1980).

To get an intuitive idea in one spatial dimension about how to reach this result, consider the average change in the number, $n_{i\kappa}$, of i molecules in subvolume κ between t and $t + dt$, under the assumption the system is in state $\{\mathbf{n}\}$ at time t.

$$\langle dn_{i\kappa} \rangle = \sum_j v_{ij} \left[dt \cdot r_j \left(\mathbf{n}_\kappa \Delta^{-1} \right) \right] + 1 \left[\frac{dt \cdot D_i n_{i(\kappa-1)}}{\ell^2} \right]$$
$$- 2 \left[\frac{dt \cdot D_i n_{i\kappa}}{\ell^2} \right] + 1 \left[\frac{dt \cdot D_i n_{i(\kappa+1)}}{\ell^2} \right] \qquad (9.26)$$

The stoichiometries of change in $n_{i\kappa}$ for the different events are here weighted by different events' probabilities given in brackets. Assuming that there are many molecules in each subvolume such that the molecule copy number distribution in each subvolume is well represented by the average concentration $c_{i\kappa} = \langle n_{i\kappa} \rangle \Delta^{-1}$ and $\langle r_j \left(\mathbf{n}_\kappa \Delta^{-1} \right) \rangle \approx r_j \left(\mathbf{c}_\kappa \right)$, we can approximate $\langle n_{i\kappa}(t + dt) \rangle = \langle n_{i\kappa}(t) \rangle + \langle dn_{i\kappa}(t) \rangle$ by

$$c_{i\kappa}(t + dt) = c_{i\kappa}(t) + dt \left[\Delta^{-1} \sum_j v_{ij} r_j \left(\mathbf{c}_\kappa \right) + D_i \frac{c_{i(\kappa-1)} - 2c_{i\kappa} + c_{i(\kappa+1)}}{\ell^2} \right] \qquad (9.27)$$

Further, if we recognize that the concentrations vary smoothly between the sub-volumes due to the constraint that neighbors should be in diffusion equilibrium equation 9.27 can be rewritten as

$$\frac{\partial c(x,t)}{\partial t} = \Delta^{-1} \sum_j v_{ij} r_j \left(c(x,t)\right) + D_i \frac{\partial^2 c(x,t)}{\partial x^2} \tag{9.28}$$

where $c_i(x,t)$ is the average concentration of species i at position x at time t. This approach can also be used in connection with a generalized RDME mentioned above. Then, the currents obtained in the high density limit differ in general from the diffusion current (see (Bollenbach et al., 2005) for an example).

In the limit of fast diffusion the RDME converges to the "ordinary" chemical master equation addressed in chapter 8 (Kuramoto, 1974; Gardiner and Steyn-Ross, 1984).

9.2.2 Sampling the Markov Process of the Reaction-Diffusion Master Equation

The RDME is too complicated for analytical approaches, especially if the system displays "exotic" properties, such as bi-stability, ultra-sensitivity, oscillations, spatial pattern formation, etc. As an alternative, it is possible to sample the Markov process one event at a time using an appropriate Monte Carlo method. For one-dimensional systems such simulations were pioneered in 1979 (Malek-Mansour and Houard, 1979) using the SSA (Gillespie, 1976), see chapter 8 and chapter 16. For three-dimensional systems, where the number of possible events is astronomical, a number of algorithmic improvements are required. See for instance Fricke and Wendt (1992); Hanusse and Blanché (1981).

The most recent improvements were made with the *next subvolume method* (Elf and Ehrenberg, 2004), which is an adaptation of Gillespie's SSA (Gillespie, 1976) and the *next reaction method* (Gibson and Bruck, 2000a) to the special structure of the RDME. The starting point is to sum the rates of all reaction and diffusion events in each subvolume. Let us call these sums r_κ for the different subvolumes κ. The time to the next event in each subvolume is then exponentially distributed with an average of $1/r_\kappa$. Given an initial distribution of molecules, the time for the first event in each subvolume is sampled from the respective exponential distribution, and the subvolumes are ordered in a priority queue according to when the events are scheduled to appear. The first event in the whole system occurs in the subvolume on top of the priority queue. The event that actually occurs in this subvolume is sampled in proportion to the rates of the different events that can occur in this subvolume. If it is a reaction event, it will change the state of the subvolume, which means that some of the rates for the events in the subvolume must be recalculated. Further, a new time for the next event in this subvolume must be sampled, and the corresponding element in the priority queue is sorted accordingly. If it is a diffusion event, two subvolumes are involved in the event and two elements in the priority queue must be reordered according to their new event times. The next subvolume method can be used to simulate systems with millions of subvolumes and molecules. It is implemented in the software tools McsoRD (Hattne et al., 2005) and SmartCell (Ander et al., 2004).

An alternative to the RDME approach to spatially dependent stochastic kinetics is to use particle-based simulation methods. These are usually discretized in time instead of space. The Brownian motion of the molecules is sampled at fixed time intervals, assuming that molecule displacements in space during the time interval follow a Gaussian distribution. Depending on the positions of the molecules in space, it is decided if nearby molecules have reacted or not during the last time interval. The available software tools (MCell (Stiles et al., 1998) and SmolDyn (Andrews and Bray, 2004)) make this decision in different ways. As a point of reference to the RDME treatment, one can consider the case where the time step is chosen as the mean time to the next diffusion event in the RDME description, that is $\Delta t = \ell^2/2D$ in one dimension. In this case the root mean square (RMS) displacement during the time step equals the length of one subvolume. If, in addition, the reaction probability during this time step is calculated from the local concentration within a radius equal to the RMS, the particle based and RDME based methods are very similar. Another algorithm that should be mentioned in this context is the Green's Function Reaction Diffusion algorithm (van Zon and ten Wolde, 2005). The GFRD can be used for very detailed reaction-diffusion simulations, since it neither is discretized in time nor in space.

9.2.3 Examples

Annihilation Kinetics $A + B \xrightarrow{k} \emptyset$

A simple example illustrates how spatial fluctuations can change the kinetics for very simple reaction schemes. Consider the reaction $A + B \xrightarrow{k} \emptyset$, with initial concentrations $n_1(0)/\Omega = n_2(0)/\Omega = 10\mu M$, $k=10^8 M^{-1}s^{-1}$, and $D=10^{-8} cm^2 s^{-1}$. The molecules are randomly (that is uniformly) distributed in a volume of $\Omega=10^{-12}$ liters divided into 10^6 cubic subvolumes of 10^{-18} liters.

Figure 9.6 shows the decay in the numbers of A and B molecules when the subvolumes are distributed in one, two, or three spatial dimensions. These decay rates should be compared to the corresponding (mean-field) deterministic reaction-diffusion description, where the geometry of the reaction volume does not matter if the initial distribution of molecules is uniform. The RDME treatment shows that the molecules disappear much slower than what would be deduced from the mean-field model. This is due to inevitable concentration imbalances in the systems, where, for instance, regions with more A than B molecules will consume all B molecules. Once regions dominated by one of the species have been established, further reactions can only occur at interfaces between A and B regions, which makes diffusion of molecules to the interfaces limiting for the rate of annihilation. For this simple reaction scheme, the simulated data can be compared to analytical work on the corresponding RDME using renormalization group methods (Lee and

Cardy, 1995), that demonstrate that $\langle n_1 \rangle = \langle n_2 \rangle \propto t^{-d/4}$, where d is the dimension of the system.

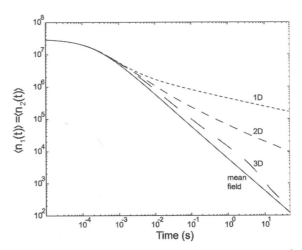

Figure 9.6   Geometry effect on the rate of annihilation in the $A + B \xrightarrow{k} \emptyset$ reaction. The simulation was run in MesoRD.

Noise Induced Domain Separation in Bistable Systems

Noise induced transitions were discussed in chapter 8. Here we consider additional noise induced properties that can arise in spatially extended systems. In particular, we will use a simple bistable system built on the double negative feedback principle (figure 9.7) to illustrate how internal fluctuations and slow diffusion can change the escape properties. The system can be either in a state where the E_A enzymes make a lot of A molecules that can bind and inhibit the E_B enzyme (dark grey ellipse) or in the state where E_B enzymes make a lot of B molecules that can bind and inhibit the E_A enzymes (light grey ellipse).

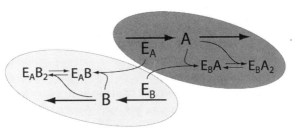

Figure 9.7   Double negative feedback schemes. E_A makes A and E_B makes B. A inhibits E_B and B inhibits E_A.

In a macroscopic analysis of the system (homogeneous or inhomogeneous) the system goes to one of the two attractors and stays there. However, when we consider the fluctuations, there is always a chance that the system escapes from one attractor to the other in a noise induced transition. Such escape problems have been studied for a long time in homogeneous systems (Erdi and Tóth, 1989; Horsthemke and Lefever, 1984). When the homogeneous system gets larger and larger, the average escape time from an attractor gets longer and longer. The escape time increases approximately exponentially with the volume of the system, as an escape requires that an increasing number of unlikely events occur in sequence. The exponential dependence of the escape time on the volume is for the double feedback system illustrated by the solid line in figure 9.8a.

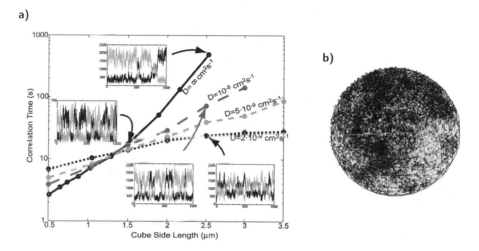

Figure 9.8   Reduction of escape time and domain separation. **a)** The correlation time for the number of A molecules in the double negative feedback system is plotted as a function of the linear extension of the cube shaped system. (The correlation time is the time, τ, at which the normalized autocorrelation function $\langle n_A(t) n_A(t+\tau) \rangle / \langle n_A \rangle^2 - 1$ has decreased to e^{-1} of its value at $\tau=0$. The correlation time is one half of the average time of escape from one of the attractors in a symmetric bi-stable system.) Inserts: Examples of time evolution of the total number of free A and B molecules are given for the points indicated by arrows. (The figure is reproduced from (Elf and Ehrenberg, 2004)) **b)** The black and white circles corresponds to free A and B molecules, diffusing together with all the other reactants in a sphere with radius 4μm. The sphere is divided into 268,096 subvolumes each of size $(0.1\mu\text{m})^3$. The rate of diffusion of all components is $d=2 \cdot 10^{-9}$ cm^2s^{-1}. The simulations were done with the next subvolume method using MesoRD.

When slow diffusion of the reactants is considered, such that the molecules do not have time to diffuse through the whole volume between two reactions, the escape from an attractor is faster than in the homogeneous case (Elf and Ehrenberg, 2004). In figure 9.8a this is seen in the reduced correlation times for lower diffusion constants with the same size of reaction volume.

At very low diffusion rates the correlation time even levels off and becomes independent of further increase of the system size. Systems at the plateau display domain separation, where different parts of the volume are in different attractors.

An example of domain separation of the double negative feedback system is seen in Fig. 9.8b, where all reactants are freely diffusing in a sphere of radius 4μm.

Not all bistable systems will display domain separation when the reactants are freely diffusing in three dimensions. However, if diffusion is geometrically obstructed such that the number of likely reaction partners is low, the spatial aspects of the stochastic bistable systems become important (Bhalla, 2004; Elf and Ehrenberg, 2004).

Min-oscillations

To give an example of how the Min-system behaves in a stochastic setting, the model by Huang et al. (2003) was simulated using MesoRD. For the wild type-shaped cell the stochastic and deterministic simulations are in good agreement, as shown in figure 9.9.

Figure 9.9   Stochastic simulation of the MinD oscillations. The figure shows MinDE complexes (black) and MinD (gray) bound to the bacterial membrane with 5 seconds intervals. The 4μm cell is modeled as a cylinder with two spherical caps and is divided into a membrane and an intracellular compartment. The three dimensional volume is discretized in subvolumes with side length 0.05μm. The concentrations of molecules and other parameters are close to those in the corresponding deterministic model (Huang et al., 2003).

9.3   Summary

We have presented in this chapter approaches for describing the dynamics of biological systems when spatial inhomogeneities cannot be neglected and the transport of particles have to be taken into account. In the simplest case, the transport of the molecules constituting the system is diffusive. If each molecule has many potential reaction partners, a mean-field description in terms of reaction-diffusion equations is then possible. An analysis of the dynamic equations can in this case make use of the powerful tools of differential calculus and often starts with a linear stability analysis of stationary homogeneous states of the system. This analysis is a systematic and not very time-consuming way to get a first impression of how the system behaves for different values of the parameters. Further analysis then commonly involves a numerical integration of the dynamic equations. If the change in local concentrations cannot be approximated as an average over a large number of random events in each volume element, a stochastic description must be used, for example, in terms of the reaction-diffusion master equation. This is a generalization of the chemical master equation presented in chapter 8 and is built upon a division of space into subvolumes. For its analysis, Monte Carlo methods have been developed. An extensive analysis of the system for many parameter values is often not possible because of the large simulation time needed to get good statistics. Instead, first the continuum limit can be used to identify possible interesting parameter values for which an extensive stochastic analysis is then performed.

In general, transport can rely on different mechanisms than diffusion. The evolution of particle densities is then still given by the continuity equation (9.15), but the currents will differ from the diffusion current. Macroscopic expressions for the current are constrained by the symmetries of the system. If the system is close to thermodynamic equilibrium, the currents can be expressed as linear combinations of the generalized thermodynamic forces. In general, however, there is no systematic procedure to arrive at macroscopic expressions, even though symmetries can guide their development. If a microscopic model is available, then macroscopic expressions can be obtained by a procedure similar to the one presented in section 9.2.1.

The techniques presented in this chapter have proven extremely useful to describe spatiotemporal structures in tissues and organisms. They are now also used to describe the dynamics of subcellular structures, as was illustrated by the example of the Min-system in *E. coli*. In general, they can be advantageously used whenever a rather limited number of different molecules is sufficient to characterize the state of a system. This does not imply that the number of different molecules constituting the system must be small. For example, the cytoskeleton contains numerous different proteins. Its state can, however, often be sufficiently well characterized by the distribution of the cytoskeletal filaments, while the effects of the associated proteins can be lumped into a number of parameters characterizing the interactions between the filaments. Thereby, the methods presented here complement those used for the analysis of biochemical networks presented in chapter 4 and chapter 5. It can be expected that they will continue to prove very valuable in discovering general principles underlying the formation of spatiotemporal structures in biological systems.

Acknowledgments

J.E. was supported by grants from the Swedish Research Council to Måns Ehrenberg, Uppsala University, and by Knut och Alice Wallenbergs stiftelse.

Notes

1. Therefore, continuum theories are often referred to as field theories.

2.   Another possibility are integro-differential equations. These contain integrals over the fields. Integrals over space reflect non-local interactions, for example by neurons that connect to distant neurons, while integrals over time reflect a memory in the system.

3.   Note that the solutions to the dynamic equations do not necessarily display the same symmetries as the equations. For example, the dynamic equations for the Min-system in three spatial dimensions do not change when the system is rotated around the long axis of the bacterium. In contrast, the fields do not have to be invariant under this transformation.

4.   Symbols printed in boldface denote vectors.

5.   In addition to particle numbers, other conserved quantities are momentum and energy. Momentum conservation needs to be taken into account if forces act on the system or are created within the system. The source terms in the continuity equation reflect in this case external forces. For systems operating at constant temperature, the continuity equation for energy does not lead to additional independent equations.

6.   The diffusion limited association rate constant for two spherical reactants freely diffusing in three dimensions is given by $k_a = \frac{(4\pi D\rho k)}{(k+4\pi D\rho)}$ (Noyes, 1961), D is the sum of the molecules' diffusion constants, ρ is the reaction radius, and k is the association rate constant at the reaction boundary. When $k \gg 4\pi D\rho$, the reaction is strictly diffusion controlled and $k_a = 4\pi DR$ (von Smoluchowski, 1917) . The dissociation rate constant k_d is similarly diffusion controlled $k_d = \frac{(4\pi D\rho\lambda)}{(k+4\pi D\rho)}$ (Berg, 1978b), whereas the equilibrium constant $K_d = \frac{k_d}{k_a} = \frac{\lambda}{k}$ is independent of the rate of diffusion. Here, λ is the microscopic dissociation rate constant.

III MODELS AND REALITY

10 Biological Data Acquisition for System Level Modeling—An Exercise in the Art of Compromise

Zoltan Szallasi

Most of the actual modeling of biological systems will be performed by researchers with strong foundations in the quantitative sciences. One of the most significant adjustments these experts have to make when entering the field of modeling of cellular systems is understanding and accepting the limitations of biological data. The system to be modeled, in most cases the living cell, is extremely complex, has rather limited observability and may be governed by principles that are beyond our current understanding. Most relevant to this book is the fact that we are trying to produce predictions for an entire system while only a subset of the variables can be measured, often with rather limited accuracy. It will probably take several years of intensive research to estimate the constraining effect of measurement techniques on system level modeling. This chapter reviews the various biological data acquisition techniques and compares their capabilities to the data requirements of various modeling techniques and to the estimated complexity of intracellular regulatory networks.

10.1 Chapter Overview

Most chapters in this volume are dedicated to the theoretical foundations and practical realization of complex system modeling. To a significant extent these considerations are independent from the fact that our intention is to model biological systems. The general rules or limitations of ODE-based models (see chapter 6) are basically the same when modeling the cell cycle or weather patterns. In fact, it is an exciting and open question, whether biological networks display any specific systemic property that would sharply set them apart from other complex networks (Milo et al., 2004b, 2002). It is currently rather certain, however, that the overall nature of biological networks, such as robustness, or the limitations of accuracy while measuring parameters will restrict the models in such a way that will have an impact on both the theoretical and practical aspects of biological modeling. In terms of theory, the remarkable robustness of biological networks may restrict complex networks of ODEs in such a way that the predictive power based on them may enable the meaningful modeling of large subcellular networks (Stelling et al., 2004b). In terms of practical limitations, insufficient accuracy or coverage of biological parameter estimations may prevent models from producing any meaningful predictions. Therefore, it is important for the prospective modeler to get familiar with several relevant aspects of biological data acquisition including:

A) The overall size and complexity of intracellular networks: This includes estimating the overall size of the genome in terms of active biochemical units, the number of relevant biochemical derivatives per gene and the average connectivity of the network. It must be also emphasized that deciphering the active part of the genome, especially for higher organisms, is far from being complete, and this dynamic research field yielded several major surprises during the last couple of years leading to a significant reevaluation of our understanding of how the genome is organized in functional terms.

B) The general principles of biological measurements – their technical and conceptual limitations: The various modeling approaches rely on rather different data requirements. Therefore, clear estimates on the accuracy, coverage, and sensitivity of a given data acquisition technology will determine its suitability for a given computational task. Graph theoretic approaches or flux balance analysis, for example, usually involve a significant part of the entire intracellular networks without a need for estimating kinetic parameters or concentrations. Ordinary differential equation–based dynamic modeling, on the other hand, is usually performed on rather limited subnetworks, with its success highly dependent on accurately estimating kinetic parameters and concentrations. The very crux of data acquisition in systems biology is the current trade-off between coverage and accuracy.

C) Concentration measurement versus kinetic parameter measurements: Although the detection technology is the same for both types of measurements, estimating kinetic parameters relies on time-dependent measurements, and they are also highly dependent on the experimental environment, whether, for example, the measurement was performed in free solution or directly in the intracellular environment. Consequently, determining kinetic parameters that reflect the intracellular reality will take more specialized approaches than those applied for concentration measurements.

D) The actual target of the measurements: Averaging biological measurements across cell populations, as most currently applied methods require, will mask important aspects of regulatory interactions in the individual cells. Therefore, the prospective modeler should be aware of situations when single cell measurements provide more relevant data.

10.2   The Estimated Size and Complexity of Intracellular Regulatory Networks

As we will see in chapter 15, modeling large dynamic networks leads to formidable computational challenges. Therefore, it is desirable to start with the least complex network, or with the smallest subset of a large network that provides correct predictions or helps answer a set of specific questions. However, it is not known how large a segment of the entire network has to be modeled in order to be able to predict a certain cellular behavior. This is also a context–dependent problem: a dynamic cell cycle model of limited complexity may provide a good description of the behavior of normal cells but may fail to provide a meaningful description of the neoplastic cell cycle, which may involve several regulatory interactions that can be ignored in the normal cell. For example, the BCR-ABL fusion protein is never present in normal cells and gets created during chromosomal translocation, a hallmark mechanism of cancer cells. This abnormal protein, which is not part of the mechanistic description of the normal cell cycle, has a significant regulatory input on some parts of the cell cycle machinery in leukemia (Gesbert et al., 2000). A modular view of biology has been proposed to alleviate some of the computational problems associated with system level modeling (Hartwell et al., 1999; Stelling et al., 2004b). However, the existence of modular structure in biological networks is far from being resolved (see chapter 3). Therefore, it is worth providing some quantitative estimates on the size and complexity of the entire intracellular regulatory network.

10.2.1   The Inventory of Biochemical Entities in an Intracellular Regulatory Network

A reasonable starting point for estimating the size of intracellular networks is the number of active genes in a given cell. This can be fairly well estimated for prokaryotes. The genome of these relatively simple unicellular organisms contains from 500 up to 6,000–7,000 genes, which are rather tightly packed and free of the complexities observed in higher organisms, such as introns or splice variants (Brown, 2002). The vast majority of transcribed genes code for proteins, therefore the number of potentially relevant network variables can be relatively safely deduced from the number of genes for these organisms.

Estimating the total size of the intracellular network of higher organisms is a significantly more difficult task, especially in the light of several recent unexpected discoveries. The first pass analysis of the recently finished genomes focused on the identification of protein coding regions and other widely studied nonprotein coding genes such as ribosomal RNA or micro RNAs. This yielded an estimate for the total number of genes between 6,000 for yeast and about 20,000–25,000 genes for humans (Brown, 2002) although this latter number is still changing considerably. First it was being adjusted from a much higher number to a lower level (International Human Genome Sequencing Consortium, 2004), then new experimental evidence suggested that several thousand genes were missed by earlier analysis (Saha et al., 2002). Despite these uncertainties, it is generally accepted that for most organisms the number of protein coding genes will not exceed 30,000–35,000 (Johnson et al., 2005). However, three surprising lines of evidence suggest that protein coding genes may not be the full story and we might be considerably underestimating both the total number of genes and the active part of the genome. First, a significant portion of all genes seems to be transcribed in the antisense direction, in addition to the sense transcription, both in prokaryotes and eukaryotes. In human cells, for example, up to 20% of genes may be transcribed in the antisense direction as well (Lehner et al., 2002; Yelin et al., 2003). Second, splice variants will considerably increase the overall diversity of the transcriptome of most eukaryotes, especially higher organisms. It is estimated that at least half of the human genes are alternatively spliced and a single gene may have a large number of potential splice variants (Modrek et al., 2001). Third, a recent set of papers employed the so-called "tiling" microarray technology, in which large regions of entire genomes are expression profiled using oligonucleotide microarray probes that cover the genome at regular, closely placed intervals. These probes are designed in an unbiased fashion and cover intronic and intergenic regions of the genome in addition to the usually examined exonic regions. Surprisingly, a large number of nonexonic probes showed significant expression levels, suggesting the existence of a large number of thus far unidentified RNA species (Johnson et al., 2005). It remains to be seen whether these regions of the genome code for proteins or regulatory RNA. Nevertheless, the recent, revolutionary impact of short regulatory RNA for biological research should serve as ample warning that we should be prepared for further surprises (Bartel, 2004).

Actively transcribed distinct RNA sequences comprise only the first layer of complexity of intracellular biological networks. RNA strands associated with protein coding genes are transported to the ribosomes where they serve as templates for protein production. This, however, is only the starting point for a series of posttranslational modifications that are necessary for proteins to exert their respective effects. It should be noted that a whole series of regulatory events exists between the transcription of a certain gene and the various active protein derivatives of the same gene, and these regulatory events often receive multiple conditional inputs from an array of other elements in the network. Therefore, a given protein-coding gene may have several biochemical derivatives, which may require separate introduction into a given model (Hoffmann et al., 2002; Schoeberl et al., 2002). A demonstrative example is shown in Figure 10.1 for the several steps involved from the production of mRNA of a transcription factor until the production of mRNA of a downstream-regulated gene.

The protein product of the gene "relA" is part of the NF-κB transcription factor complex, either with another identical RELA molecule as a homodimer, or with one of several other proteins as a heterodimer (Karin et al., 2002) (small letters usually designate RNA whereas capital letters are used for the protein product of the same gene). We start at the state when the mRNA of relA is already produced. In addition to the transcriptional regulation, the level of mRNA of this gene can

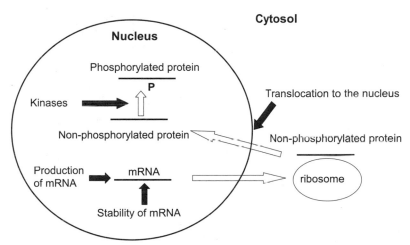

Figure 10.1 Independently regulated derivatives of the relA gene. (For details see text.) The black arrows indicate independent regulatory inputs.

be regulated by the stabilization or destabilization of mRNA. The level of protein production will be proportional to the net amount of relA mRNA and not only to the transcriptional activation of this gene alone. mRNA is the first regulated derivative of the relA gene. All proteins are produced in the cytoplasm in a non-modified form, and the RELA protein has to be first translocated to the nucleus to exert its transcriptional activity. The non-phosphorylated cytoplasmic and non-phosphorylated nuclear RELA protein, therefore, can be considered as two further derivatives of the relA gene, since the IκB proteins will regulate the localization of the NFκB complex in a conditional manner (Karin et al., 2002). The activity of the nuclear RELA protein is further regulated by phosphorylation at various serine residues (Duran et al., 2003). Therefore, the nuclear phosphorylated form of RELA can be considered as an additional derivative, since both the function and the regulation by stabilization differ for the phosphorylated and non-phosphorylated form. As shown in Figure 10.1, the gene relA has at least four independently regulated derivatives: its mRNA, the non-phosphorylated cytoplasmic, non-phosphorylated nuclear, and the phosphorylated nuclear forms. When building a dynamic model, these derivatives have to be entered into the model as separate entities (Hoffmann et al., 2002).

Therefore, the second step in determining the overall size of intracellular regulatory networks is estimating the number of relevant posttranslational modifications per protein. Various estimates put the number of distinct posttranslational forms of a given protein in yeast to about 3 and in humans to 3–6 (Banks et al., 2000; Papin et al., 2005). For prokaryotes, this number seems to be less than two. These are obviously rough and preliminary estimates, and high quality, manually curated protein databases will certainly provide more reliable numbers in the future (O'Donovan et al., 2001; Peri et al., 2003).

These numbers will be further increased by the fact that the same protein may have to be accounted for according to various, relevant localizations, as certainly seems to be the case for membrane associated receptors or nuclear proteins (Schoeberl et al., 2002; Smith et al., 2002)

The above-described rather staggering complexity seems to dwarf the more moderately sized collection of small molecules in a cell, which is commonly referred to as the metabolome. The total number of small molecules in any given organism, including humans, will probably not exceed 2,000–2,500 (Kell, 2004).

Taken together, based on the collection of genes and their derivatives, it seems that the number of independently regulated biochemical species will be between a few thousand for the simplest organisms and several hundreds of thousands for more complex organisms, such as humans. These numbers will probably elicit a wide variety of responses in the newcomers to the field, varying from total hopelessness to cautious optimism. On the one hand, even a relatively small network of ordinary differential equations can get out of hand rapidly (see chapter 6). On the other hand, the various constrained models, such as flux balance analysis (see chapter 5), provide meaningful predictions about biological systems based on networks of approximately a thousand metabolites (Edwards et al., 2001a). Control theoreticians also like to point out that a Boeing 777 contains about 150,000 subsystem modules, significantly more than the number of "relevant parts" that seem to be in a simple bacterium (Csete and Doyle, 2002). We are, probably, far from certain whether this is a fair comparison. The effect of stochasticity in biology (see chapter 8), and the implications of human, control theory–based design (see chapter 12) need to be accounted for and understood before a modeler gets carried away by such an optimistic comparison.

It may also be informative to take a look at the total number and the dynamic range of macromolecules per cell. From the size and dry weight content of cells and the average size of proteins or RNA, one may easily arrive at the following estimates (see for example www.dur.ac.uk/biological.sciences/Staff/Croy/GENNET1.HTM). A large cell, such as a hepatocyte (liver cell) is estimated to contain about $8*10^9$ protein molecules. This number is distributed amongst 10,000 different types of proteins with a dynamic range of about 5 orders of magnitude.

10.2.2   The Inventory of Regulatory Interactions in Intracellular Networks

In addition to the "number of parts," (nodes in graph representations, see chapter 7), the number of regulatory interactions (edges in a graph), is also an important characteristic of intracellular regulatory networks. These interactions can be derived by a wide variety of large-scale measurement techniques that detect various types of regulatory interactions. In principle, any method can be (and has been) used, which can produce some measurable phenotype on a large enough number of genes. Some types of interactions are directly probed by an actual measurement and readily interpretable in biochemical terms, such as protein binding. Others are derived indirectly by computational analysis of other data sets. For example, regulatory interactions have been postulated for genes that are coregulated in a large number of perturbed microarray based gene expression profiles (Ihmels et al., 2002; Tirosh and Barkai, 2005), or between genes that show synthetic lethality in double genetic knock-outs (Tong et al., 2004). In these latter cases it is often difficult to directly identify the actual biochemical mechanism(s) behind the postulated regulatory interactions.

Thanks to the accessibility of appropriate data, large protein interaction networks have been extensively studied, and cross-validating the various data sets has led to reliable estimates on the number of these interactions. Experimental data were produced both by small-scale protein interaction assays deposited in databases such as the Database of Interacting Proteins (DIP), and Human Protein Reference Database (Peri et al., 2003) and by high throughput technologies such as yeast two-hybrid measurements, high throughput mass spectrometry, etcetera. (see for example (Lee et al., 2004)). For yeast, most recent estimates suggest on the order of 30,000–35,000 interactions for the entire genome, yielding roughly 6–7 interactions per protein.

Human data sets are currently more sparse and biased (remember we do not even know the total number of human protein coding genes, let alone their identity). It may still be informative that the existing, supervised data sets currently contain 3–4 interactions per protein (Peri et al., 2003).

In summary, intracellular regulatory networks for the various organisms can be probably visualized by graphs with a total node number between a few thousand and perhaps a few hundreds of thousands, and with an average connectivity of less than 10.

10.3 Classifying Measurement Techniques from a Computational Modeling Perspective

The purpose of the following classification is to provide guidelines for prospective modelers when looking for or intending to produce appropriate data sets for a given modeling problem. An appropriate classification will reflect both the needs of a given modeling approach and the capabilities of the various measurement techniques (see Figure 10.2).

10.3.1 The Target of the Measurement

Section 10.2 is essentially a "list of parts" of intracellular networks. The most numerous group of relevant quantifiable variables comprises, of course, the genes and their derivatives. They can be measured at the DNA level, RNA level, and the various levels of protein modifications along with their localization. Metabolic network analysis requires the quantification of metabolites. The recently popularized suffixes -omics or -ome describe the measurement or cataloging of an entire collection of one type of biochemical molecules. The "genome" refers to the entire genetic information stored in the DNA strands of a given organism, "transcriptome" comprises all the genetic information that gets transcribed into RNA, etcetera. From a practical point of view, it is worth noting that all members of a given "-ome" can be usually measured by the same type of technology, whereas measurement technologies usually transfer poorly between the various "-omes." For example, a given oligonucleotide-based microarray platform can be, at least in principle, used for the measurement for any RNA species. On the other hand this technology is not suitable for the measurement of, for example, phosphoproteins. Merging data sets across various "-omes" and across various technologies is not an obvious task (Luscombe et al., 2004) and usually requires well thought-out, specialized methods such as Bayesian approaches (see chapter 4) (Lee et al., 2004).

10.3.2　Concentration versus Interaction Measurements

Concentration measurements can be performed accurately in a "context independent" manner. After destroying the cell, a necessary preparative step in many cases, the number of molecules can still be counted accurately by any of the well-established measurement technologies as outlined below. Measuring molecular interactions, however, is dependent on the "cellular context." An interaction detected in free solution may never occur inside the cell. Therefore, measurement techniques that reflect the reality of the inside of the cell had to be developed.

10.3.3　The Information Content of Measurements

In order to perform a given computational modeling task, a certain amount of experimental information is needed. For example, one can estimate the amount of data that is necessary to reverse engineer a given regulatory network (see chapter 11) (Andrec et al., 2005; Sontag et al., 2004). The accuracy and sensitivity of measurement techniques along with the strategy of selecting appropriate conditions or time points of the samples to be quantified all have a profound impact on the useful information content of an experiment (Szallasi, 1999). Therefore, one might be able to estimate whether a given experimental technique is suitable to produce appropriate data for a modeling approach. Some of the experimental techniques are mainly able to identify biochemical molecules without the power of providing anything more than semi-quantitative concentration estimates. Mass spectrometry–based proteomics without isotope labeling is an example that is discussed later on. However, these measurements, even without a quantitative dimension, can be used for topological understanding and modeling of protein networks (Lee et al., 2004). Other measurements, such as gene expression microarrays, provide information about the direction and magnitude of gene expression changes. Whereas the direction of changes seems to be rather reliable, the ratio of changes seems to be compressed in an intensity dependent manner (Yuen et al., 2002). Finally, several low throughput techniques, such as Western blot analysis (see below), provide accurate measurements with an error rate well within 10–15%.

Level of precision	Example	Modeling approach
Absent/present	Mass spectrometry analysis of protein interaction clusters	Graph models
Reliable detection of expression changes with limited precision of the actual expression ratios	Microarray measurements of gene expression profiles	Probabilistic/ Qualitative models
Concentration measurements with an experimental error of 10–20%	Western blot analysis of protein concentrations	Detailed, ODE-based dynamic models

Figure 10.2 Different biological data acquisition technologies produce results with rather different measurement accuracy. The precision of the method will determine its potential utility for a given modeling approach.

10.4   Low Throughput, Accurate Measurements of Gene Derivative Concentrations

The history of biochemical measurements has been a long struggle for increased specificity and sensitivity. In order to describe a biological system at the quantitative level, one would like to measure the various biochemical derivatives, such as the various posttranslational modifications of the relevant genes at various localizations within the cell, preferably with a reasonable time resolution. Needless to say, this is hard to achieve, and current methods involve various levels of trade-offs between the number of biochemical species to be quantified and accuracy.

Measuring the concentration of a single, well-defined biochemical species in a solution is well within the capabilities of modern molecular biology. Most high precision methods are based on the combination of size separation and the application of a specific, high affinity reporter system. Size separation methods are usually based either on gel or capillary electrophoresis, where the macromolecules of various sizes are driven through a molecular sieve by electrostatic potential. The sieve is formed of polymers such as polyacrylamide or agarose, that are cross-linked to produce the appropriate pore size that is best suited to separate the molecular weight range of the molecules of interest. Gel electrophoresis produces only a rather limited size resolution, therefore high specificity reporters are needed for accurate identification and quantification. As outlined later, for certain types of biopolymers, in particular nucleic acid chains, highly specific reporters can be produced with relative ease, whereas for others, such as polypeptide chains, the availability of specific reporters (antibodies) still depends on processes that cannot be easily controlled.

DNA or RNA fragments are first size separated by agarose gels during Southern and Northern blot analysis, respectively. The specificity of reporters is based on the Watson-Crick pairing. Under appropriate experimental conditions, a long enough nucleotide sequence with limited sequence homology to other parts of the genome will hybridize only to its target sequence. The probes are labeled by the incorporation of radioactive or fluorescently labeled nucleotides, which, in turn, will produce readily measurable signals that could be used for quantifying the DNA or RNA fragment in question. The experimental error is usually below or around 10–20% in the hands of an experienced user, and detection limits, thanks to new technology such as quantum dots, are in the sub-femtomolar range (Liang et al., 2005). This should allow the quantification of RNA molecules that are expressed at the level of 1 molecule per cell. This is, in fact, a necessary level of sensitivity because a significant number of transcripts are expressed at this level (Holland, 2002).

Quantitative RT-PCR also exploits the specificity of Watson–Crick pairing. In this case, two specific, most often fluorescently labeled PCR primers are used that will initiate the amplification of only the target nucleotide sequence. This process is kinetically measured and can be reliably used to estimate the starting concentration of the target sequence. It has a similar accuracy to Northern blots, and it was used to measure the concentration of RNA species below the concentration of one molecule per cell (Holland, 2002).

Protein concentrations are routinely quantified by Western blot analysis. In this, a protein mix is first size separated by polyacrylamide gel electrophoresis (Laemmli, 1970), then the specific reporter system is applied. For proteins, no convenient method similar to the Watson-Crick pairing exists to produce highly specific probes. It took decades and a great deal of ingenuity to work out effective methods to produce antibodies for analytical purposes (Harlow and Lane, 1988). Quite remarkably, today's antibodies provide highly specific probes not only for individual proteins but for the various posttranslational modifications, such as specific phosphorylation states, of a given protein as well (Czernik et al., 1991). Several ordinary differential equation–based modeling studies took advantage of the specificity and accuracy of data produced by such antibodies in Western blot analysis (Hoffmann et al., 2002; Schoeberl et al., 2002). The accuracy and sensitivity is similar to Northern blots. However, it should be noted that the production of specific antibodies against a large number of diverse proteins is still a significantly more labor-intensive experimental project than producing probes against nucleic acid sequences, which, to some extent, could be reduced to a computational problem.

From a modeling perspective, it is important to note that the above–described low throughput methods usually produce reliable, specific and rather accurate measurements. Hence their popularity for parameter fitting in dynamic models (Hoffmann et al., 2002; Schoeberl et al., 2002).

It should be also remembered, however, that in most cases the high specificity and accuracy of the above described methods relies heavily on the actual analytical conditions applied, which should be carefully adjusted for each probe separately. Antibody and nucleotide probe concentrations, salt concentrations of the hybridization and washing buffers, and hybridization temperatures should all be carefully optimized for maximum specificity. These parameters will of course vary significantly from target to target, and this fact already forecasts the difficulties encountered with high throughput methods, such as microarrays, that are based on multiplexing the above-described methods.

In principle, given high specificity probes, one can eliminate the size-separation step. In this, the so-called dot blot technique, one can immobilize either the probes or the sample mixture to solid support and then hope that there is only one biochemical entity binding to the probe (Maniatis et al., 1982). In traditional biochemistry, this approach was rather marginally applied – results always looked more convincing when supported by the correct size information. Interestingly, this method started a second, spectacularly successful life in the form of microarray technology.

10.5 High Throughput Measurements and Low Accuracy—A Necessary Compromise?

10.5.1 High Throughput Gene Expression Measurements

In principle, given enough manpower and financial support, every biochemical measurement can be scaled up in a massively parallel fashion even to genomic scale. In fact, the current interest in system level modeling was started by the introduction of massively parallel measurement techniques, such as gene expression microarrays (Schena et al., 1995). These are essentially highly efficient multiplexed dot blots, enabled by microfabrication and automatization. Varying and patenting essential experimental details led to the development of a large number of alternative microarray platforms (Hardiman, 2004). Nevertheless, microarray based RNA or DNA quantification methods are all based on the same basic principles. Nucleotide probes of varying lengths, from 25 base pairs up to hundreds of base pairs, are either immobilized or in situ synthesized on solid support. The RNA or DNA mixture to be quantified is labeled, most often by fluorescent dyes, and then hybridized to the microarray chips. An enormous amount of work, thousands of publications, went into working out both the experimental and computational analytical details of optimal microarray analysis. These efforts were often hindered by legal interference from manufacturers (Rouse and Hardiman, 2003) and the unavailability of essential information, such as microarray probe sequences, for the research community (Mecham et al., 2004). The inordinate number of relevant technical publications suggests a rather limited satisfaction with the accuracy of microarray technology. Here we can summarize only the most relevant concerns, and we have to refer to appropriate reviews for further details (Jordan, 2004; Draghici et al., 2006).

Due to low cost efficiency, the estimated accuracy of microarray measurements is only sparsely supported by independent verification data. A typical microarray platform contains thousands or tens of thousands of probes, but most studies will verify the quantification provided by microarrays only for a much smaller number of genes, typically less than one hundred (see for example (Gold et al., 2004; Holland, 2002)). A few recent studies quantified gene expression levels by quantitative RT-PCR from several hundred to over a thousand genes and a couple of interinstitutional efforts are also underway to perform similar validation (Czechowski et al., 2004; Holland, 2002). However, the overall level of accuracy by microarray measurements is far from being established(Draghici et al., 2006).

The lack of independent verifications was intended to be replaced by cross–platform comparison of RNA aliquots, which is, however, an imperfect tool with which to validate microarray platforms. Lack of consistency can be caused by the inferior performance of either one or both platforms, without clear indication of their relative merit. On the other hand, highly similar results across platforms could be simply caused by consistent cross-hybridization patterns without either platform measuring the true level of expression. Current experience in the field suggests that short oligo based microarray platforms show a rather good correlation, with a Pearson correlation coefficient of about 0.7 or better (Bammler et al., 2005; Woo et al., 2004; Yauk et al., 2004). cDNA microarray–based results, however, have a more limited correlation with short oligonucleotide based platforms, around 0.5 on average (Bammler et al., 2005; Mecham et al., 2004; Woo et al., 2004). It must be noted that these correlations are always based on gene expression ratios between two different RNA samples. Despite some optimistic reports (Hekstra et al., 2003), absolute levels of gene expression can hardly be estimated using only microarray data. This problem is best exemplified when looking at microarray data produced by a platform, for example, the Affymetrix gene chip, that uses multiple probes against the same transcript. Probes that are producing signals by hybridizing to the same transcript may show orders of magnitude variations in their signal intensity(Draghici et al., 2006). Surface chemistry, significantly different free energy binding values between probes, cross hybridization, or the efficiency of labeled nucleotide incorporation probably each have an effect on the poorly understood correlation between signal intensity and target concentration. As a consequence, while gene expression ratios can be estimated with reasonable certainty for a significant number of genes, measuring absolute concentrations in a comprehensive fashion is currently beyond the capabilities of microarray technology.

Microarray analysis is based on a strong, although rarely discussed, assumption: most microarray probes on a given platform produce sufficiently specific signals under a single, rather permissive hybridization condition(see Figure 10.3). Increasing evidence suggests that this might be true only for a subset of probes on any microarray platform (Zhang et al., 2005). Consequently, while gene expression ratios

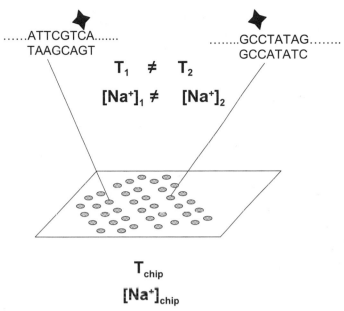

......ATTCGTCA.......
TAAGCAGT

........GCCTATAG........
GCCATATC

$T_1 \neq T_2$

$[Na^+]_1 \neq [Na^+]_2$

T_{chip}

$[Na^+]_{chip}$

Figure 10.3 Massively parallel measurement technologies, such as gene expression microarrays, are usually run under a single experimental condition, which is not ideal for most of the individual reactions. Each microarray probe can be associated with an ideal hybridization condition determined by the hybridization temperature, salt concentrations, etcetera. This is expected to produce the most specific signal with minimal cross-hybridization. The ideal hybridization temperatures (T_i) and salt concentrations ($[Na^+]_i$) are usually probe-sequence dependent, producing different values for the two microarray probes highlighted in the figure. Nevertheless, gene chips are routinely run under a single set of conditions (T_{chip}, $[Na^+]_{chip}$). A similar one-size-fits-all strategy is often implemented for economic reasons.

might be estimated for some genes with an error of less than 20–30%, the error for other genes may be far greater than that. There is only little, if any, guidance in the literature that would help with predicting the accuracy of a microarray measurement for a given gene in a particular experimental setting. The general expectation is that two-fold changes in gene expression can be reliably measured across the board. The detection limit of current microarray technology is around 10 copies of mRNA per cell (Holland, 2002; Kane et al., 2000). It should also be noted, that this level of sensitivity may be insufficient to detect relevant changes in low abundance genes, such as transcription factors (Holland, 2002).

A possible compromise between the relatively inaccurate microarray technology and low throughput, high precision methods is running a large number of quantitative real-time PCR reactions in a parallel fashion using, for example, 384-well optical plates. A recent study quantified 1,400 genes in Arabidopsis with high reproducibility and high sensitivity (0.001 mRNA copies per cell) over a six-orders-of-magnitude dynamic range (Czechowski et al., 2004). It remains to be seen to what extent the increased accuracy and sensitivity may be able to offset the impact of lower coverage and higher initial labor cost of PCR primer design.

10.5.2   High Throughput Protein Quantification

Soon after the runaway success of gene chip technology, antibodies were arrayed on solid phase support in order to develop protein microarrays (Haab et al., 2001). For proteins, unfortunately, neither sample labeling nor probe preparation is as straightforward as in the case of nucleic acids. This resulted in a much slower development, when researchers had to try several alternative detection methods (MacBeath, 2002). The detection limits of protein microarrays are perhaps not much worse or are similar to that of gene chips (1 part in 1,000,000), especially when preceded by fractionation, in which the relative concentration of the protein of interest is increased at the expense of other proteins. For example, high-speed centrifugation can remove the abundant cytoskeletal, structural proteins, which, this way, will be prevented from interfering with the quantification of soluble proteins. There has not been a large enough body of experience published yet that would provide a comprehensive estimate on the accuracy antibody microarray measurements. A consistent and reliable detection of 2-fold changes would perhaps satisfy most current users.

Detecting proteins with high specificity on protein microarrays depends on "luck" whether an appropriate antibody can be developed for a given protein. This and other, detection–related, problems associated with protein microarrays were probably not lost on experts who were developing alternative technologies for high throughput quantification of protein mixes. By far the most powerful and most widely used of these methods is mass spectrometry–based proteomics. Identification of proteins is based on measuring the mass-to-charge ratio of ionized protein fragments, and their quantification is based on counting the numbers of a given ionized fragment reaching the detector. Mass spectrometry requires by far the most complex sample processing and instrumentation of all the methods discussed so far. First, protein mixes are usually fractioned in order to reduce the complexity of the sample. This is rather important, because abundant proteins may outnumber rare proteins by four to five orders of magnitude, thus obscuring the signals obtained from less abundant proteins. Then, the fractionated protein mix is subjected to tryptic digestion in order to produce smaller fragments that are later ionized. The peptide fragments are then separated by liquid chromatography. These fragments are ionized and then analyzed by, usually, two tandem mass spectrometry analyses. The second mass spectrometry is run on fragments derived from a single mass-to-charge peak derived from the first mass spectrometry step. For each of these steps a multitude of techniques exist with their relative advantages and disadvantages reviewed by (Aebersold and Mann, 2003). Here we will review only those aspects of mass spectrometry that are relevant for system level modeling.

Standard mass spectrometry is not appropriate for accurate quantifications without the further modifications discussed below. Qualitative modeling methods, such as graph theoretic interpretations of intracellular regulatory networks, can take advantage even of these non-quantitative data. For example, a significant portion of the yeast "interactome" was mapped out recently by the mass spectrometric analysis of protein complexes that were isolated by several thousand various tagged "bait" proteins (Gavin et al., 2002; Ho et al., 2002). (In this case the tag is a short peptide sequence attached to the end of the native amino acid sequence that allows a high affinity separation of the bait protein and its interacting partners. While "protein-tagging" is a widely used and efficient technology it must be noted that the tag may influence the behavior of the tagged protein.) The mass spectrometry–based interactomes were combined with other types of interactome data sets yielding a probabilistic functional network of yeast, mapping out potential modules or clusters for further analysis (Lee et al., 2004). Although the principle of protein identification may sound deceptively simple, its associated difficulties become apparent upon closer inspection. Tryptic digestion fragments rarely have a unique mass-to-charge ratio (Alterovitz et al., 2006). Hence the need for a second mass spectrometry step on the fragment ions of a given peptide peak identified in the first step. However, the fragment ion spectra cannot be readily converted into peptide sequences solely based on theoretically expected distributions. Instead, the spectra generated during the second mass-spectrometry step are usually compared to comprehensive protein sequence databases. Therefore the success of protein sequencing will highly depend on the quality of reference databases. A large number of "machine learning"–inspired methods have been suggested to overcome this problem with varying success (Alterovitz et al., 2006). Nevertheless, considering the speed of the development of mass spectrometry based proteomics, there is little doubt that for organisms with comprehensive lists of sequence information, protein identification by this technology will be achieved within the foreseeable future.

Quantifying ratios of protein concentrations by mass spectrometry involves another dimension of technical challenges and usually relies on stable isotope labeling. In this, one protein mixture (for example, cancer tissue), is isotope labeled by one of several appropriate methods (Aebersold and Mann, 2003) while the other sample (for example, normal tissue), is left unchanged. The normal and cancer samples are then mixed. Since the chemical properties of a given peptide are still the same after isotope labeling, a mixture of the isotope labeled and native peptide can be co-analyzed by the mass analyzer. The difference in mass due to isotope incorporation will yield two different mass-to-charge peaks and the difference in the area under those peaks will provide an estimate for the relative expression level of that protein in the two samples. A wide variety of ingenious isotope tagging techniques have been developed, targeting various peptide side-chains such as sulfhydril groups (Gygi et al., 1999), amino groups (Munchbach et al., 2000), etcetera. The actual side-chain targeted and the protein purification techniques preceding the mass spectrometry will restrict the number of different proteins quantified in a given experiment (Gygi et al., 1999).

The accuracy of these technologies usually allows the reliable identification of at least 1.5- to 2-fold changes, although this estimate probably applies only to more abundantly expressed proteins (Gygi et al., 1999). The actual sensitivity of this technology is highly dependent on the purification steps preceding the actual mass spectrometry analysis. In absolute terms, the detection limit of proteins isolated from polyacrilamide gel electrophoresis bands and then subjected to mass spectrometry is in the femtomole range, or on a weight basis, it is in the low nanogram range (1–5 ng). This would roughly translate into a detection limit of 1 part per million. This ratio may reflect very different sensitivity levels in terms of copy number per cell depending on the pre-electrophoresis fractionation.

10.5.3   Further Uses of Mass-spectrometry

The universal principle underlying this technology offers itself to a wide variety of exploitations. In fact, given appropriate sample preparation methods and adequate mass spectrometry databases (to which the mass–to–charge peaks can be compared), any biologically relevant molecule can be, at least in principle, identified and quantified. Whether mass spectrometry is applied to a given task, it is highly dependent on cost efficiency, ease of use, and other practical considerations. For example, although nucleic acids can be just as well analyzed by mass spectrometry as proteins (Jurinke et al., 2004), for the average end user microarray analysis offers a cheaper and easier alternative. While the mass spectrometric analysis of a single biopolymer (that is a single band after gel purification) costs around one hundred dollars, a cheaper microarray platform may quantify several thousand genes for the same cost. When no such easily scaleable alternative technique exists, mass spectrometry provides an excellent general tool for high throughput biochemical measurements. It has been used for the quantitative analysis of the metabolome (Allen et al., 2003), identification and to some extent the quantification of tyrosine phosphorylation (Rush et al., 2005), protein ubiquitination (Peng et al., 2003), etcetera. These recently developed applications, however, currently belong to the realm of semi-quantitative methods.

10.6   Detecting Regulatory Interactions and Quantifying Kinetic Parameters

As we saw in section 10.4, concentration measurements can be rather accurately performed on a wide variety of intracellular molecules. Under certain experimental conditions this accuracy can be transferred to other types of measurements as well, since measuring kinetic parameters can be reduced into a time-series of concentration measurements, and detecting regulatory interactions can be reduced into a combination of appropriately modified concentration measurements. However, in many cases seemingly accurate measurements may provide misleading data about the actual intracellular conditions.

10.6.1 Detecting Regulatory Interactions

These measurements are usually based on simple modifications of several, above–discussed methods. One member of the interacting molecules is designated as the "bait," which is used to separate the entire macromolecular complex. The success of the method depends on whether the bait molecule can be targeted with sufficient specificity. If yes, for example, by using an appropriate antibody, then the whole complex is isolated and its members are identified or quantified by standard methods. Protein-protein or chromatin co-immunoprecipitation are such techniques. In the former, by using an appropriate antibody, a specific protein, such as a transcription factor, is isolated, and then other proteins interacting with it can be determined. In the latter, the same antibody can be used to pull down the transcription factor and with it the DNA regulatory regions to which the transcription factor is binding. The nucleotide sequence of this regulatory region can then be determined. In order to preserve the intracellular regulatory interactions, chemical cross linkers are often applied. These molecules have two highly reactive moieties. Within a certain distance, called "spacer arm length" (measured in Ångströms), these molecules tend to cross-link their target macromolecules by covalent binding. Therefore, when the cross-linkers are applied to the cell, interactions between proteins, for example, are fixed and carried over to subsequent analytic steps performed in free solution (Agou et al., 2004).

10.6.2 Quantifying Kinetic Parameters

Dynamic, for example ODE-based, models require accurate kinetic parameters that are usually determined by direct biochemical experimentation. In most published models these parameters are usually extracted from the literature. These models are usually based on individual gene derivatives and not, for example, on functional modules. Therefore, the kinetic parameters have to reflect the dynamic interaction between individual genes and proteins. For example, an ODE-based model of the epidermal growth factor (EGF) receptor pathway requires measuring the affinity between EGF and its receptor, the kinetic parameters of the tyrosine phosphorylation of the EGF receptor, etcetera (Schoeberl et al., 2002). Most often these kinetic parameters are determined in experiments containing a more or less purified population of the interacting proteins in free solution (that is not measured inside the cell.) Parameter optimization (Mendes and Kell, 1998) acknowledges the fact that the thus obtained kinetic parameters may often be misestimated relative to the true values reflecting intracellular conditions. In order to obtain more accurate estimates on these parameters, direct measurements on intracellular protein levels, without disrupting the cellular structure, were introduced. The actual tool, as so often the case in biology, was provided by nature itself in the form of fluorescent proteins (Tsien, 1998). The first of these, the green fluorescent protein was isolated from a jellyfish that, for reasons not entirely clear, produces this highly efficient protein fluorophore. The emphasis here is on the fact that this is an ordinary protein in terms of production and processing. Therefore, when cloned and fused to a protein in an experimental organism, the thus labeled protein starts to glow, emitting measurable signals allowing direct quantification. A remarkable diversity of fluorescent proteins have been developed recently, emitting signals at various wavelengths (Hawley et al., 2001). It should be noted that fusing the rather sizable fluorescent proteins to the target may have profound effects on the production rate or function of the labeled protein. The ease and efficiency with which one can label proteins in prokaryotes has led to several interesting applications, such as reverse engineering of bacterial regulatory networks (Ronen et al., 2002).

10.6.3   High Throughput Detection of Regulatory Interactions

By now the observant reader must have realized that novel methods in biological data acquisition more often involve the ingenious combination of existing technologies than the introduction of truly novel measurement principles. The ChIP-chip technology is an elegant combination of chromatin immunoprecipitation and gene expression microarrays (hence the name) (Ren et al., 2000). In the first step, a tagged transcription factor is used to isolate the upstream DNA regulatory regions it is binding to. Then a DNA microarray containing probes for a large number of upstream gene regulatory regions, in the case of yeast for the entire genome (Ren et al., 2000), is used to determine which of the regions have been enriched during the first step. This will then constitute a microarray–based approach that will determine which regions a given transcription factor is binding to under a given experimental condition. This will effectively map out a network of putative regulatory interactions between gene expression regulators and regulated genes (Lee et al., 2002). Obviously, the experimental noise of the two technologies will be compounded, requiring various computational methods to produce reliable measurements by using independent supporting evidence from other data sources, such as the coregulatory patterns of genes (Bar-Joseph et al., 2003).

10.7   Population Averaged versus Single Cell Measurements

Although single molecule measurements may just be around the corner (Chan et al., 2004), the detection limits of currently applied measurement techniques in systems biology, especially the massively parallel technologies, require the presence of millions of a given molecular species for reliable quantification. Such a high number of molecules can be derived from at least tens of thousands or more cells. Consequently, most of the above-described methods will yield population averaged values. The potential risk of using population averaged data is well exemplified by an interesting study by the group of Ferrell (Bagowski and Ferrell Jr., 2001). In this study they showed that in individual Xenopus oocytes the activation of the kinase JNK shows a steep dose response curve with a Hill coefficient of around 100. In population averaged measurements using hundreds of oocytes, the way these experiments are usually performed, the same dose response curve shows no cooperativity with an apparent Hill coefficient of 1. This means that the ultrasensitive bistable switch of JNK activation can be detected only in single cell measurements, and a very different kinetic model would be built based on the population averaged data. In such cases, single cell measurements produce a more accurate description.

The experiments described above involve the isolation of material for Western blot analysis from individual oocytes and thus require a significant amount of meticulous bench-work to obtain information about each individual cell. Fortunately, fluorescent proteins (see section 10.6) provide a convenient experimental tool for relatively high throughput single cell measurements. Sometimes the genomic sequence of a given fluorescent protein is simply inserted behind a transcriptional promoter of interest. In these cases the expression level of the reporter protein serves as a surrogate marker for promoter activity and can be quantified in hundreds or thousands of individual cells by direct measurement of fluorescent intensity (Elowitz et al., 2002; Raser and O'Shea, 2004). This relatively simple experimental arrangement has already produced interesting results by directly demonstrating the stochasticity of gene expression in individual cells (Elowitz et al., 2002) and also by quantifying the noise of transcriptional and translational activity (Blake et al., 2003; Raser and O'Shea, 2004). A further level of experimental complexity can be achieved by fusing fluorescent proteins to other proteins of biological interest. For example, in a rather intriguing study the two members of the p53-Mdm2 feedback loop were labeled individually by fluorescent proteins of different colors (Lahav et al., 2004). Surprisingly, the study showed that in human cancer cells p53 was expressed in discrete pulses after DNA damage, with the number of pulses differing between individual cells. This is a potentially relevant and rather unexpected observation regarding the function of one of the most studied tumor suppressor genes, p53, which could not have been detected by population averaged measurements.

The number of various fluorescent proteins and the procedure required to fuse them to other proteins of interest limits the number of proteins that can be studied simultaneously in a single cell. A recent approach combining the application of fluorescently labeled antibodies with multiparameter flow cytometry (Irish et al., 2004) may increase the number of quantified parameters of single cell measurements by one to two orders of magnitudes. A causal protein-signaling network was reconstructed from such measurements using Bayesian network inference (see chapter 11) on eleven key signaling proteins in T lymphocytes (Sachs et al., 2005). Again, the ability to measure changes in the signaling proteins in single cells as opposed to population averaged measurements was essential to the success of this approach.

10.8   Conclusions: A Final Look at Experimental Design

Researchers applying ordinary differential equations based models to intracellular networks are well aware of the importance of accurate parameter estimations.

Therefore, these studies usually apply high accuracy, low throughput measurements for parameter optimization (Hoffmann et al., 2002; Schoeberl et al., 2002). However, not every chemical species can be measured accurately, therefore, these studies rely on measurements of only a subset of the variables represented in the model. Consequently, a set of interesting theoretical questions arises with practical implications. How many variables should be quantified for reliable parameter optimization of a network with a given size? By what strategy should this subset of parameters be selected? How does the inaccuracy of measurements propagate back to parameter estimation in a robust dynamic network (see chapter 11)? Problems like these demonstrate well the intricate relationship between theory, modeling and experimental biology.

11   Methods to Identify Cellular Architecture and Dynamics from Experimental Data

Rudiyanto Gunawan, Kapil G. Gadkar, and Francis J. Doyle III

A system-level understanding of the functioning behavior of a cell requires an accurate representation of the underlying complex networks of gene and protein interactions. Advances in molecular biology have provided a glimpse of such complexity through diverse measurements of cellular activities. In systems biology, the goal of network inference or reverse engineering problems is to reconstruct the complex network of regulatory interactions from available measurements using a mathematical framework. Here, the reverse engineering effort faces two daunting problems: network size and complexity, and incomplete and inaccurate measurements. In addition, complete knowledge of a cellular network entails the identification of not only the network architecture (topology) but also its dynamics. Indeed, implicit in the term *regulation* is the importance of dynamics of these interactions. Network inference from experiments has been extensively investigated in the field of engineering, which is known as system identification. In addition, many concepts in engineering, such as robustness (see chapter 2), modularity (see chapter 3), and optimality, have been observed in many biological systems. For these reasons, engineering approaches have been instrumental in the reverse engineering effort. This chapter highlights the methodologies and challenges in the reverse engineering of cellular networks, in particular the identification of network dynamics using engineering approaches.

11.1   Introduction

At the turn of the century, scientists successfully sequenced the human (International Human Genome Sequencing Consortium, 2001) and other genomes, enabled by advances in high throughput measurements in molecular cell biology. The complete human genome provides the blueprint of human cells and creates opportunities

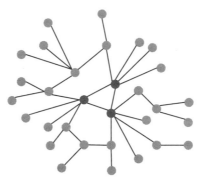

Figure 11.1 A hypothetical biological network topology. Each node in the graph can represent a biological entity such as genes, transcripts, or proteins. The edges show the interactions, such as activation/inhibition. In contrast, the dynamics describes the nature of each interconnection, for example, a Hill-type kinetics. The topology and dynamics completely characterize a network behavior.

to advance our understanding of cellular functions. The success of genomics ushers in a new era that is characterized by a shift from a reductionist approach of molecular cell biology research in the past, to a systemic or integrated approach: systems biology. The emphasis in a systemic approach is to ascertain the complex interactions in the network of genes and proteins that produce the observed cellular phenotypes under different conditions and/or stimuli. Here, the function of a gene or protein is described in the context of its dynamical interactions with other elements in the network. This chapter introduces the methods and challenges in one aspect of systems biology, namely the identification of the cellular networks from experiments. This area of research is also known as *reverse engineering* or *network inference.*

Biological networks can be categorized according to the cellular functions that they describe, such as protein-protein network, transcriptional network, metabolic network, and signal transduction pathway. There exist two primary facets in a typical biological network, its topology and dynamics (kinetics). The former describes the interconnections among the parts of the network (genes, transcripts, proteins), while the latter gives the nature of these interactions. The dynamics can be as simple as a linear function, or a nonlinear function such as Michaelis-Menten kinetics. A schematic of a hypothetical biological network is shown in figure 11.1. The goals of reverse engineering such complex networks are also multi-faceted, including: (i) hypothesis generation, (ii) design of experiment, (iii) understanding of cellular function, and (iv) unraveling design principles.

The sources of experimental data for the reverse engineering problems include large scale deletion projects, high throughput DNA microarray experiments, and chromatin immunoprecipitation assays (ChIP-on-chip) (see chapter 10 for more complete discussions on the data acquisition techniques for systems biology). The utility of these data, however, is limited by many factors, such as high level of noise, low sampling frequency, type of experimental protocol, and other issues as highlighted in chapter 10. The noise inversely correlates with the amount of information in the data, while the low sampling frequency restricts the identification of dynamics in the network. Because of the wide variety of modeling objectives and the heterogeneous sources of data, there exist a wide spectrum of modeling approaches in the reverse engineering of cellular networks, as described in these reviews (D'haeseleer et al., 2000; Ideker and Lauffenburger, 2003; Stelling, 2004; Barabási and Oltvai, 2004). The highest level of the spectrum, and hence the most abstract, involves models that mostly describe the network topology with little or no dynamics, such as signed directed graphs or Bayesian networks. The identification of network topology benefits greatly from these models as they can efficiently handle highly complex networks. The lower level models incorporate the physicochemical details into the network topology, which greatly increases the difficulty of reverse engineering problems (Ronen et al., 2002). Such models typically consist of differential equations such as those described in chapter 6, though they can be as simple as Boolean networks.

One of the simplest representations of a cellular network is a directed graph (similar to the network shown in figure 11.1, but the edges have directions/arrows). The directed edges convey the flow of influence in the network. For example, a node A with a directed edge to a node B implies that A directly influences the activity of B. Such model structure mostly captures the network topology, which can be effectively reconstructed from gene perturbation data (Wagner, 2001). A step down in the modeling spectrum is a signed directed graph (SDG), which is also a graph node with directed edges. However, the edges here can assume positive or negative values based on the influences, that is, activation or inhibition, respectively. The network inference problem of this model structure uses comparative methods on gene expression of wild-type and mutants created from deletion experiments (Kyoda et al., 2004).

Another model structure with a directed graph architecture is a Boolean network, which is also a graph node with directed edges. Here, the nodes assume binary numbers representing high or low levels (1 or 0, respectively). In a gene network, high level represents activated/expressed, while low indicates inactive/suppressed. Each directed edge corresponds to a Boolean logic function describing the influence of one gene to another. The inference problem of this model structure utilizes steady state gene expression levels from perturbation experiments of gene deletion or overexpression (Ideker et al., 2000).

Bayesian networks use a probabilistic approach to modeling cellular networks with a directed acyclic graph, in the same spirit as chapter 4. Here, each node represents a random variable characterized by a conditional probability with respect to its immediate parent nodes (that is, the start nodes of incoming edges). Thus, the state of the network is described by the joint probability

$$P(X_1, ..., X_N) = \prod_{i=1}^{N} P(X_i | X_{\mathbf{J}}) \tag{11.1}$$

where N is the total number of nodes in the network, $P(X_i|X_{\mathbf{J}})$ is the conditional probability of the i-th node to assume the value X_i given the values of its parents nodes X_j, $j \in \mathbf{J}$ such that $\mathbf{J}$ is the set of indices of the parent nodes. This steady state model structure does not provide the dynamics of the network but can explicitly account for the noise from experimental measurements, protocols, and the inherent stochastic nature of gene expression. As in the previous model structures, the reverse engineering using Bayesian networks utilizes data from perturbed gene expression profiles (Pe'er et al., 2001).

The more detailed model structures involve detailed dynamics of each interaction, such as the S-systems (Savageau, 1988). This model structure is based on mass action kinetics, in which the dynamics of interaction is described using nonlinear polynomial functions:

$$\frac{dx_i}{dt} = \alpha_i \prod_{j=1}^{N} x_j^{g_{i,j}} - \beta_i \prod_{j=1}^{N} x_j^{h_{i,j}} \tag{11.2}$$

where x_i is the state variable describing the concentration of cellular molecules (genes, proteins), α_i and β_i are the rate constants, and $g_{i,j}$ and $h_{i,j}$ are the kinetic orders. This framework is flexible enough to capture common dynamics in cellular functions such as Michaelis-Menten kinetics. As expected, the inference problem of this model structure is computationally intensive because of the need to simulate highly nonlinear differential equations (Kimura et al., 2005). In addition, the reverse engineering of network dynamics (that is, estimating the model parameters α_i, β_i, $g_{i,j}$, and $h_{i,j}$) requires time-series data, whenever available.

One model structure, based on Petri nets, attempts to combine the graph representation of the network and the detailed dynamics of differential equations. This hybrid functional Petri net (HFPN) architecture supports different cellular entities using various primitive data types (Boolean, string, real), types of interactions (discrete/stochastic, continuous, generic), and prior knowledge of the system (Matsuno et al., 2000; Nagasaki et al., 2004). Here, the nodes are connected to each other by connectors (arcs), and the dynamics are described by mappings associated with each connector. Because of its flexibility, the reverse engineering of this model structure can potentially accommodate any type of data, including gene expression and biological facts.

The complete reverse engineering of a cellular network needs to identify both the topology and dynamics of interactions. The challenges in this problem are multiple, starting from the selection of model structures to the identification of model parameters from noisy measurements. In particular, the inference of network dynamics is difficult due to the data quantity and quality and the parameter identifiability issues, which will be discussed in greater detail in section 11.3. The underlying reason for the difficulty is the mismatch between the available and the required data to uniquely identify a model structure. Indeed, the selection of model structure determines the types and amount of data necessary for a complete reconstruction of the network (Selinger et al., 2003). For example, to identify p

number of parameters in a set of nonlinear differential equations, one theoretically needs $2p + 1$ number of randomly chosen experiments (assuming zero measurement noise) (Sontag, 2002). When only the network topology is desired, the number needed reduces to $r + 1$ experiments, where r is the total number of possible connections.

The purpose of this chapter is to provide a conceptual overview of the issues involved in the reverse engineering of cellular networks with emphasis on the dynamical characterization. This chapter complements and extends a previous review that emphasized the inference of network topology (D'haeseleer et al., 2000). The next section gives a motivating example, which highlights some of the difficulties in a cellular network inference problem. Section 11.3 discusses the different issues and methodologies in reverse engineering with respect to both the topology and dynamics. Tutorials are presented in the form of case studies involving a metabolic network in *E. coli* and a signal transduction pathway in a caspase-activated apoptosis. Finally, open research problems in this area are identified based on the analysis of the case studies.

11.2   A Motivating Example

High throughput gene expression profiles can provide system-wide level measurements for reverse engineering of genetic regulatory networks in the cells. The efficacy of these measurements for inferring the network information, such as the kinetic parameters, depends on the complexity of the underlying gene network as well as the quality and quantity of measurement data. These issues were addressed using a formal identifiability analysis by Zak et al. (2003), whose results will be summarized here. In particular, the analysis considers two types of identifiability: *a priori identifiability* and *practical identifiability*, as a function of the input perturbations and the fluctuations in the gene expression due to the inherent stochastic nature of the process. A priori identifiability is concerned with the ability to uniquely identify model parameters from noise-free experimental data, given a particular model and a particular input-output experiment. On the other hand, practical identifiability is concerned with the accuracy of parameter values that can be estimated from noisy measurements.

In the aforementioned study, an *in silico* genetic regulatory network was constructed from an arrangement of common regulatory motifs: cascade, mutual repression, auto-activation and sequestration, and agonist-induced receptor downregulation (Zak et al., 2003). The model consists of 44 species with a total of 97 parameters involved in 118 reactions, including promoter binding/unbinding, transcription, transcript degradation, translation, protein monomer degradation, protein dimerization/undimerization, and dimer degradation processes. The network exhibits multiple steady state behavior depending on the presence of a ligand input (see figure 11.2). The identifiability analyses were performed on the *in silico* network as a function of the ligand perturbations: a step, a 1-hour pulse, and

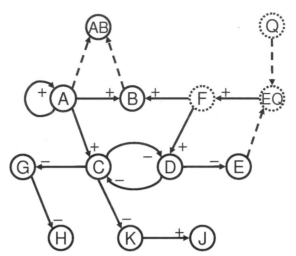

Figure 11.2 *In silico* genetic regulatory network. Dashed arrows represent chemical reactions (not regulation). The dotted nodes (F, EQ, Q) exist only when the ligand Q is present in the system. In the presence of the ligand Q, the genes B, D, F, G, J, and K are fully expressed (HIGH state), while genes A, C, E, and H are suppressed (LOW state). On the other hand, the absence of the ligand drives the genes A, C, E, and H to HIGH and the genes B, D, G, J, and K to LOW states.

two 1-hour pulses 1 day apart. In addition, as single-cell gene expression profiling is realizable (Hemby et al., 2002), the number of cells collected in each sampling time is also treated as an experimental variable. Two approaches, deterministic and stochastic, were used in the simulations of the *in silico* network with the parameter estimates from published values for genes and proteins with similar roles to those in the network.

11.2.1   Methodologies

A priori identifiability analysis utilizes the correlation matrix of the parameters, $\mathbf{M}_c$ (Beck and Arnold, 1977)

$$\mathbf{M}_c(i,j) = \mathbf{V}_p(i,j)\left(\mathbf{V}_p(i,i)\mathbf{V}_p(j,j)\right)^{-1/2} \tag{11.3}$$

where $\mathbf{V}_p(i,j)$ is the (i,j)-th element of the parameter covariance matrix. The covariance matrix quantifies the degree of (co)variability in random variables (such as noise in the measurements, parameter inaccuracies), which is given by the expected value:

$$\mathbf{V}_w = E[(\mathbf{w} - \bar{\mathbf{w}})(\mathbf{w} - \bar{\mathbf{w}})^T] \tag{11.4}$$

where $\mathbf{w}$ denotes the vector of random variables, $\bar{\mathbf{w}}$ denotes the mean values of $\mathbf{w}$, and $E[\cdot]$ represents the expected value operator. The (i,j)-th element of the symmetric $\mathbf{M}_c$ conveys the degree of correlation between the i-th and j-th parameters where a value of 1 (-1) implies a perfect (opposite) correlation (the diagonal elements of $\mathbf{M}_c$ are exactly 1 since a parameter is perfectly correlated with itself). For example, consider the following simple system:

$$y = (p_1 + p_2)\, u \tag{11.5}$$

where y is the output and u is the input. From the measurements of y (given u), one can only identify the sum of parameters $(p_1 + p_2)$, but not p_1 and p_2 independently. Here, the two parameters are said to have a perfect correlation (in this case, a correlation coefficient of 1). Thus the parameters that have correlations between -1 and 1 (that is, $-1 < M_c(i, j) < 1$) can be independently identified from experimental data (assuming perfect measurements). Thus a parameter that has a perfect correlation only to itself is said to be a priori identifiable. Further, the parameters that do not satisfy this condition can be reduced to a smaller set of identifiable parameters through an iterative parameter reduction process (Zak et al., 2003).

Practical identifiability analysis uses the Fisher information matrix (FIM) as a measure of the informativeness of noisy measurement data for estimating the model parameters. The inverse of FIM provides the lower bound for the variances of the parameter estimates (or the upper bound for accuracy) based on the Cramer-Rao inequality (Ljung, 1999a). If the noise in the data follows the Gaussian distribution, the FIM reduces to

$$\mathbf{FIM} = \sum_{i=1}^{N} \mathbf{S}_I^T(t_i)\mathbf{V}_\mu^{-1}\mathbf{S}_I(t_i) \tag{11.6}$$

where $\mathbf{S}_I$ is the sensitivity matrix with respect to the a priori identifiable parameters (Varma and Palsson, 1994) and $\mathbf{V}_\mu$ is the covariance of the measurements (a measure of noise in the data). The lower bounds of the parameter variances are given by

$$\sigma_{p_i}^2 \geq [\mathbf{FIM}^{-1}]_{ii}, \tag{11.7}$$

from which the 95% confidence interval for each parameter p_i can be defined as

$$[p_i - 1.96\sigma_{p_i}, p_i + 1.96\sigma_{p_i}] \tag{11.8}$$

In this example, a parameter is called practical identifiable when its estimated value is non-zero within a 95% confidence. The level of confidence interval for practical identifiability can be varied to include or exclude more parameters.

The Gaussian assumption may not apply for gene expression as this process involves very low copy numbers of chemical species, which makes it behave as a discrete stochastic system (see chapter 8 for the mathematical description of such system). With a lower bound of zero for the number of copies, for instance, the distribution will not be symmetric and the noise in the system can become non-

Gaussian (such as log-normal or bimodal distributions). Nevertheless, the FIM can still be evaluated using a direct analysis of the chemical master equation (Gunawan et al., 2005). Here, the Fisher information matrix is expressed as the variance of the score function (Cover and Thomas, 1991). For discrete stochastic systems, the FIM can be evaluated by simulating the master equation for the joint probability density function of the states. This simulation uses a Monte Carlo approach such as the stochastic simulation algorithm (Gillespie, 1976) or its approximate accelerated algorithm as discussed in chapter 16. The score function is equivalent to the normalized sensitivity of the joint distribution function with respect to the model parameters.

11.2.2   Insights from Identifiability

A priori identifiability analysis applied to the *in silico* network revealed that one-third to over half of the parameters in the network are not a priori identifiable, where the step ligand input performed the worst among the three perturbations. A major fraction of these parameters belonged to promoter binding/unbinding and transcription factor dimerization/undimerization, of which many exhibited perfect correlations. This result suggested that some of these parameters can be combined by equilibrium assumption for these processes by setting the forward and reverse rates equal. After removal of the parameters that can be combined or measured directly from experiments such as mRNA degradation rates (Wang et al., 2002), the step ligand input only allowed 3/4 of the parameters to be a priori identifiable, while the pulse experiments allowed 8/9 of all parameters.

Further analysis of the model using the FIM showed that only about half of the parameters were practically identifiable, of which the double pulse experiment performed the best among the three. In addition, the number of cells sampled in each experiment, which was captured using discrete stochastic simulations of the network, affected the fraction of identifiable parameters in a nonlinear manner. Increasing cell count in each sample reduced the noise in the measurements and improved the practical identifiability in a diminishing return trend. The term *noise* here relates to the inherent (discrete) stochastic nature of gene expression, which differs from the more common data noise arising from the measurement devices. The impact of cell sampling was especially pronounced in the identifiability of transcriptional interactions (differences between bound and unbound transcriptional parameters). Here, the step perturbation gave a higher fraction of identifiable parameters at a lower number of cell sampling while the double pulse input became more efficient at higher cell sampling.

This example highlights a number of difficulties in reverse engineering a cellular network. The first analysis showed that even with prior structural knowledge of the network and noise-free experimental data, a priori identifiability of the full set of parameters remained elusive. This finding signified the importance of obtaining good prior estimates of the model parameters and avoiding over-parametrization of the network through model reduction such as the equilibrium assumption.

In addition, the practical identifiability analysis underscored the importance of designing the experiment protocol, such as the type of ligand perturbation used in this study, to produce the most informative experimental data (see section 11.4.2). In general, a perturbation that is rich in dynamics can more effectively excite the system for accurate estimation of the kinetic parameters. Also, the intrinsic stochastic nature of gene expression can play an important role in the practical identifiability of the parameters when only few cells are collected at each time point. This effect, however, was diminished with increasing number of collected cells.

11.3   Issues for Network Inference

The problem of network inference or reverse engineering has long been an active research area in control theory, known as system identification (Ljung, 1999a). In addition to control theoretic approaches, research in other fields such as computer science and statistics (known as machine learning and statistical learning, respectively) have also made significant contributions to this problem (Bock and Gough, 2003; Perrin et al., 2003). However, the reverse engineering of cellular networks pushes the envelope on many approaches in these fields because of the characteristics of these networks: large size and high nonlinearity. As such, the modeling efforts have focused on capturing both aspects: (i) network complexity, and (ii) level of detail (Stelling, 2004). Unfortunately, identification of models that embody both high complexity and details of a cellular network is an untenable problem, and thus, one major issue in reverse engineering as well as in data acquisition is to select the appropriate model structure that balances the network complexity and the detail of interactions. This selection depends on the type of network and organism, the available experimental data, and the intended use of the resulting model.

Many conceptual approaches from system identification have found appropriate uses in the identification of cellular networks. For example, a singular value decomposition was used to identify all possible networks that are consistent with given gene expression profiles (Yeung et al., 2002). When choosing the solution among the candidate networks, this approach also assumed that the biological networks are sparse. Using a similar assumption, Gardner and colleagues (Gardner et al., 2003) proposed a network inference algorithm based on linear regression of gene expression profiles. Here, each gene is assumed to have only k connectivities, where k is (much) fewer than the total number of genes in the network. The solution is then chosen to minimize the mismatch between the model prediction and experimental data. Further, an iterative approach was proposed by Tegner et al. (2003) to identify the network connections from gene perturbation data. At each iteration, the algorithm ranks the genes based on the variance of predicted connectivities from all consistent solutions. The gene that has the highest variance, that is, the most uncertainty, will be selected for the next perturbation experiment. Again, the network was assumed sparsely connected as in the other approaches.

A significant challenge of constructing the cellular network from experiments, especially for a gene network model, is the large number of nodes, on the order of $10,000$, that renders the inference problem practically intractable (for example, determination of 10^8 parameters of interactions). Fortunately, cellular networks are tremendously sparse and highly structured (Wagner, 2002), such that the actual interactions to be identified are orders of magnitude fewer. In addition, these networks are not randomly connected, but highly modular and structured with regular hierarchies, motivating the use of a structured approach to the identification of such networks (Zak et al., 2005). One hierarchical decomposition of the network is to call the top level a *network* which is comprised of regulatory *motifs* of 2-4 genes (Lee et al., 2002; Shen-Orr et al., 2002; Zak et al., 2003). By searching through biological networks for common motifs, one can find the frequencies with which each simple motif occurs in the network. The much higher occurrences of these motifs in cells than in randomized networks (Shen-Orr et al., 2002) give support to a postulation that these are the basic building blocks of cellular networks. Many of these motifs have direct analogs in system engineering architectures, such as the three dominant motifs in *E. coli*: (i) feedforward loop, (ii) single input module, and (iii) densely overlapping regulon (Shen-Orr et al., 2002). At the lowest level of the hierarchy is the *module* that represents transcriptional regulation, of which a nice example is given by Barkai and Leibler (2000). The existence of structures in the complex cellular network creates an opportunity for reverse engineering methods to incorporate this knowledge by constraining the search methods or exploiting prior knowledge in Bayesian frameworks.

The interconnections between the nodes in a cellular network are not static. In fact, dynamic behavior is an essential property of complex biophysical networks (Zaslaver et al., 2004) that must be captured in the modeling efforts. There exist preliminary ideas in capturing network behavior using dynamic models in both discrete time (Hartemink et al., 2002) and continuous (Zak et al., 2004). The problems associated with the curse of dimensionality as noted above are more pronounced when one augments the dynamics with the network interconnections, especially for a full continuum model. Here, one major issue is the challenge of uniquely identifying the kinetic parameters from experimental data, typically gene expression profiling. This issue, known as parameter identifiability in control theory (Ljung, 1999a), deals with the informativeness of the data: the quantity and quality of the measurements with respect to the model parameters. The example in the previous section revealed that full knowledge of gene interconnections and perfect measurements still could not guarantee full identifiability of gene interactions.

Coupled to this, the noise in measurements and the inherent stochastic nature of gene expression make practical identification of genetic regulatory networks difficult. In practice, the reverse-engineering of a gene network should involve a careful design of experiment using prior knowledge of the system to obtain the most informative measurements. As described by the cycle of knowledge in chapter 1, this process should be iterative, in which the result from each trial is used to better design the next experiment. Here, a measure of informativeness of data, such as the Fisher

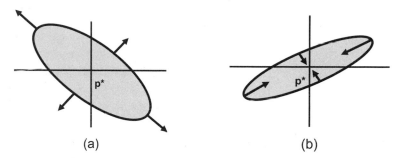

Figure 11.3 Fisher information matrix-based optimality criteria. The axes represent the model parameters where the origin describes the best parameter estimates. For simplicity, only two parameters are shown. In a system with three or more parameters, these ellipses are projections of the higher dimensional ellipses (hyperellipsoids) onto two-parameter axes. (a) The ellipse of information. The ellipsoidal axes are defined by the FIM eigenvalues and eigenvectors. The area quantifies the amount of information, while the shape indicates the distribution of information for each parameter. D-optimality design aims to maximize the area/volume of information (as indicated by the arrows), which is proportional to the determinant of FIM. (b) The ellipse of parameter uncertainty The lengths of the ellipsoidal axes are equal to the inverse of the eigenvalues of FIM. A-optimal design aims to reduce the region of parameter uncertainty (shown by the arrows), which is measured by the sum of the parameter variances.

information matrix, can help in formulating the optimal experiment design into a (nonlinear) optimization problem. The Fisher information matrix (FIM) takes into account the noise in the measurements and also gracefully handles the stochastic effect of gene expression. In addition, the FIM allows flexibility in choosing the appropriate criterion for optimality depending on the goal of model identification. Figure 11.3 illustrates the two most effective FIM-based optimality criteria, D-optimal and A-optimal, in designing experiments (Emery and Nenarokomov, 1998). D-optimal design aims to maximize the degree of informativeness in data by maximizing the determinant of FIM, which corresponds to the area/volume of the information hyperellipsoid (figure 11.3a). On the other hand, A-optimal design is equivalent to reducing the hyperellipsoid of uncertainty in parameter estimates (figure 11.3b).

The iterative nature of this framework for model development and refinement of experimental protocol necessitates a termination criterion, which typically consists of a model validation test. The selection of tests to use still remains an open research problem because of the difficulty in comparing the performance of different algorithms. In the application domain of systems engineering, it is understood that for certain experimental data, it is not possible to absolutely confirm whether a model is valid. Typically, the converse test is implemented, that is, whether the given data contradict the model prediction (Poola et al., 1994). Such model (in)validation tests for reverse engineering problems can be formulated based on the difference between predicted and observed output with some statistics about

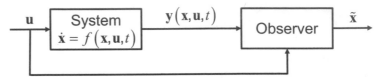

Figure 11.4 A schematic diagram of an observer. In a physical system, the complete state information is usually not available. Further, the measurements typically represent functions of only part of the states. An observer uses (inexact) knowledge of the system dynamics $f(\mathbf{x}, \mathbf{u}, t)$ and the input $\mathbf{u}$ to "guess" the states from the available measurements $\mathbf{y}$.

these differences. These statistics limit the degree of model errors using, for example, maximum absolute value, mean value, and variance.

Aside from the aspect of the quality of data, another practical limitation in most (if not all) attempts to reverse engineer cellular networks is the limited quantity of data, both in terms of sampling frequency and number of independent measurements. For example, although gene expression profiling can provide high throughput data to estimate interactions among thousand of genes, this method still does not depict the protein-mediated regulatory effects. As noted in chapter 10, current system level modeling efforts face the challenge of compromising data quantity and quality (low throughput, accurate measurements versus high throughput, relatively inaccurate measurements). In many cases, parameter estimation from limited measurements suffers from stringent computational requirement and degeneracy, where many parameter combinations give similar agreement to the observed behavior. Here, measurement selection procedures can help identify the combination of measurements that give the best identifiability. Also, an observer can provide estimates of all system states (gene, transcript, protein levels) from limited measurements.

The concept of an observer is described in figure 11.4. The purpose of an observer is to infer the states of a system (for example, internal energy, entropy) from the measurements (such as temperature, pressure). For this reason, an observer is also known as a state estimator in control systems theory. There exist multiple approaches for designing an observer for biological systems, including extended Kalman filters (Stephanopoulos and San, 1984; Gee and Ramirez, 1996), artificial neural network (Glassey et al., 1997; Simutis and Lübbert, 1997), and state regulator problem (SRP) (see Section 11.4,2) (Gadkar et al., 2005b). The state regulator problem approach builds on dynamic flux balance analysis (dFBA) described in this chapter to estimate the unknown variables in a biological system from dynamical measurements. The dFBA extends traditional flux balance analysis (Varma and Palsson, 1994) to allow the estimation of dynamic fluxes in a given metabolic network. The SRP observer is formulated as a constrained optimization problem where the gene network is assumed to operate optimally by minimizing unnecessary accumulation of intermediates (states) and fluxes (reactions) in the framework of dFBA. Given the full estimates of the network states and fluxes, the parameter estimation becomes decoupled and thus computationally efficient with lower probability for degeneracy.

11.4 Case Studies

The case studies derive from the applications of engineering approaches to the reverse engineering of cellular networks, in particular to identify the dynamics of biological networks. In the first example, an optimization-based approach is demonstrated for estimating the dynamic behavior of a metabolic network in *E. coli* from experiments (Mahadevan et al., 2002). The second example introduces a framework for iterative network inference to identify a signal transduction pathway in a caspase-activated apoptosis (Fussencgger et al., 2000).

11.4.1 Dynamic Flux Balance Analysis

As recent developments in genomics provide information of the cellular architecture, the logical next step is to study the dynamic behavior of the cellular network. A primary bottleneck for this is the lack of kinetic information of the intracellular reactions. The flux balance analysis approach in chapter 5 utilizes the known stoichiometry to predict the flux distributions in the network without requiring the kinetic information (Varma and Palsson, 1994). However, the approach can be used to study only the steady state operations of the network, preventing its applicability in situations where dynamic reprogramming of the metabolic network is important. In this case study, a dynamic Flux Balance Analysis (dFBA) approach is discussed that is capable of predicting the dynamics of the metabolic network with modest requirements of experimental data.

To motivate the concept, we consider the diauxic growth of *Escherichia coli*. Using the metabolic network of *E. coli*, the extreme pathways are identified with glucose, acetate, and oxygen as input and acetate and biomass as output. From the extreme pathways, four primary pathways are determined, based on the biomass yield, to represent both aerobic and anaerobic growth on glucose and aerobic utilization of acetate. These pathways are expressed as a simplified network shown in figure 11.5. A dynamic model for the prediction of the time profiles for the batch bioreactor based on the simplified network is represented in the equations,

$$
\begin{aligned}
\frac{dGlcxt}{dt} &= \mathbf{St}^{Glcxt}\mathbf{v}X \\
\frac{dAc}{dt} &= \mathbf{St}^{Ac}\mathbf{v}X \\
\frac{dO_2}{dt} &= \mathbf{St}^{O_2}\mathbf{v}X + k_la(O_2^* - O_2) \\
\frac{dX}{dt} &= (v_1 + v_2 + v_3 + v_4)X
\end{aligned}
\tag{11.9}
$$

where X represents the biomass concentration, $\mathbf{St}^{Glcxt}$, $\mathbf{St}^{Ac}$, $\mathbf{St}^{O_2}$ are the rows of the stoichiometric matrix associated with glucose, acetate, and oxygen, respectively, $\mathbf{v}$ is the vector of reaction fluxes, and k_la is the mass transfer coefficient for oxygen ($7.5 \ hr^{-1}$) and O_2^* is the oxygen concentration in the gas phase (0.21 mM).

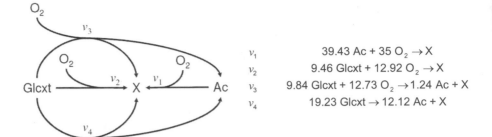

Figure 11.5 The simplified metabolic network of the diauxic growth in *E. coli* with glucose, acetate, and oxygen as inputs and biomass and acetate as outputs.

To determine the dynamic profiles of the metabolite levels, the dynamic fluxes need to be determined. With the absence of kinetic information that relates the fluxes to concentrations, dynamic optimization is proposed for determining the fluxes and metabolite concentrations. It is based on an assumption that the cellular processes are performed optimally in order to achieve a cellular objective. Similar assumptions are made in the FBA approach (Edwards et al., 2001a) and in the cybernetic modeling approach (Varner and Ramkrishna, 1999). For the dFBA approach considered here, maximizing the instantaneous growth rate is proposed as the built-in cellular objective. Other candidate objective functions which have shown good fit of experimental data include maximization of biomass (Burgard and Maranas, 2003) or minimization of total fluxes (known as the principle of flux minimization) (Holzhütter, 2004).

The dFBA approach involves an optimization over the entire time period of interest to obtain time profiles. The optimization problem is shown below:

$$\max_{\mathbf{v}(t)} \quad \sum_{j=0}^{M} \int_{t_0}^{t_f} \frac{X(t)}{X_0 e^{\mu^{avg} t}} \delta(t - t_j) \mathrm{d}t \tag{11.10}$$

such that

$$\frac{\mathrm{d}\mathbf{z}}{\mathrm{d}t} = F(\mathbf{v}, \mathbf{z}) \tag{11.11}$$

$$|\dot{\mathbf{v}}| \leq \dot{\mathbf{v}}_{max}; \quad \mathbf{z} \leq \mathbf{z}_0; \quad c(\mathbf{v}, \mathbf{z}) \leq 0 \qquad \forall \ t \in [t_0, t_f] \tag{11.12}$$

$$\mathbf{z}(t_0) = \mathbf{z}_0 \tag{11.13}$$

$$t_j = t_0 + j \frac{t_f - t_0}{M} \quad j = 0, \cdots, M \tag{11.14}$$

The time period of interest is divided into finite number of intervals (equation 11.14). The optimization maximizes the growth rate at each of these intervals. The objective function is scaled such that all points are equally weighed. Equation 11.11 is the matrix representation of equation 11.9 and represents the mass balance

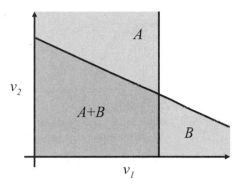

Figure 11.6   An example of a system with two fluxes and two constraints A and B (aside from the positivity of fluxes). Each inequality constraint is shown in each shaded region. The combination of the two constraints $A + B$ limits the feasible space of flux pairs for the dFBA optimization.

and continuity constraints. The rate of change of flux constraint, the non-negative metabolite level constraint, and the additional nonlinear constraints are imposed by equation 11.12. As discussed in chapter 5, these constraints reduce the feasible search space for the fluxes that maximize the objective function, as illustrated in figure 11.6. Equation 11.13 represents the initial conditions of all species.

In most cases, limited fermentation data are available. Substrate and oxygen uptake rates and product formation rates are usually calculated. These limited experimental data are used as additional constraints to the dynamic optimization problem. For this case study, the glucose uptake rate, and oxygen uptake rates are bound by the additional constraints shown below:

$$\mathbf{St}^{Glcxt}\mathbf{v} \leq \frac{v_{max}^{Glcxt}Glcxt}{K_m + Glcxt} \tag{11.15}$$

$$\mathbf{St}^{O_2}\mathbf{v} \leq v_{max}^{O_2} \tag{11.16}$$

The glucose uptake is bounded by the Michaelis-Menten kinetic involving the glucose concentration, and the oxygen uptake is bounded by a maximum possible flux. The unknown constants in the above equations are determined from the available experimental data (v_{max}^{Glcxt}=10 mmol/gdw-h (Varma and Palsson, 1994); K_m=0.015 mM (Wong et al., 1997); $v_{max}^{O_2}$=15 mmol/gdw-h (Varma and Palsson, 1994)).

The dynamic optimization is solved by parameterizing the dynamic equations through the use of orthogonal collocation on finite elements (Cuthrell and Biegler, 1987). Details of solving the dynamic optimization problem are discussed by Mahadevan et al. (2002). Figure 11.7 shows the profiles of the the metabolite levels suggesting that the dFBA approach accurately predicted the dynamics of the diauxic growth on glucose and acetate. The dFBA also correctly predicted the re-

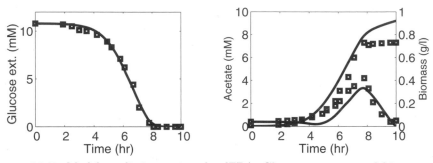

Figure 11.7 Model predictions using the dFBA. Glucose, acetate, and biomass concentrations from the model predictions (solid lines) are compared to experimental data (squares) (Varma and Palsson, 1994).

utilization of the acetate, which was not possible with the classical FBA approach (Varma and Palsson, 1994).

The dFBA approach does not require kinetic information of the intracellular reactions; it could, however, incorporate available kinetic information into the constraints of the dynamic optimization. Further, it allows the dynamic formulation of the objective function describing characteristics, such as, reduction of transition time between two steady states or end-point optimization into the rigorous mathematical framework. The primary drawback of the approach is that it typically requires solving a nonlinear optimization problem. As the size of the network increases the computation burden could become infeasible. However, the use of a simplified form of the important pathways as done here assists in capturing the dynamics of the crucial components of the network. In summary, the dFBA approach provides a useful tool for the quantitative study of the dynamic reprogramming of metabolic networks to obtain a better understanding of the behavior of the network.

11.4.2 Iterative Model Identification

As mentioned in the previous section, the reverse engineering of a cellular network should involve an iterative process. One possible framework for this process is depicted in figure 11.8. The model identification step is decoupled into two parts. The first part uses the limited measurements to give estimates of time profiles for all concentrations and reaction rates. These full estimates of system variables allow for an efficient parameter estimation in the second part. When a model (in)validation step necessitates further model refinement, an optimal experiment design and/or an optimal measurement set is determined to guide the next experiment.

The application of the framework is demonstrated for the model identification of caspase function in cell apoptosis. The schematic of this system is shown in figure 11.9, which was developed by Varner and co-workers (Fussenegger et al., 2000). This model with the published parameters is assumed as the "real" system.

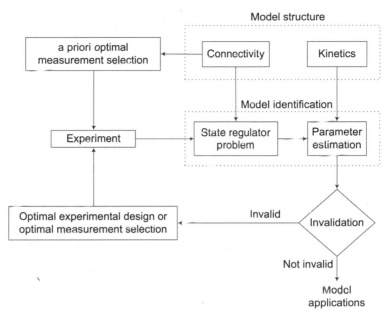

Figure 11.8 Iterative scheme for model identification.

The network topology and the mechanism of the interactions are assumed to be known. The model can be represented in a very general form:

$$\dot{\mathbf{x}} = \mathbf{A}\mathbf{x} + \mathbf{B}\mathbf{r} + \mathbf{C} \qquad (11.17)$$

$$\mathbf{r} = f(\mathbf{x}, \mathbf{p}) \qquad (11.18)$$

where the vectors $\mathbf{x}$ and $\mathbf{r}$ represent the states and reaction rates, respectively. The matrices $\mathbf{A}$ and $\mathbf{C}$ describe the degradation and auto-generation respectively, whereas the matrix $\mathbf{B}$ represents the stoichiometry of the network. The nonlinear function $f(\mathbf{x}, \mathbf{p})$ represents the reaction rate equations. Further details of this model representation are included in Gadkar et al. (2005b). A discrete version for the continuous time invariant affine system is derived using a standard technique known as the zero-order hold (Brogan, 1991). The discrete model equation is represented as:

$$\mathbf{x}(k+1) = \bar{\mathbf{A}}\mathbf{x}(k) + \bar{\mathbf{B}}\mathbf{r}(k) + \bar{\mathbf{C}} \qquad (11.19)$$

where

$$\bar{\mathbf{A}} = e^{\mathbf{A}\Delta T},$$
$$\bar{\mathbf{B}} = (e^{\mathbf{A}\Delta T} - 1)\mathbf{A}^{-1}\mathbf{B},$$
$$\bar{\mathbf{C}} = (e^{\mathbf{A}\Delta T} - 1)\mathbf{A}^{-1}\mathbf{C}.$$

The goal of this case study is to identify the kinetic parameters $\mathbf{p}$ in the nonlinear reaction rates.

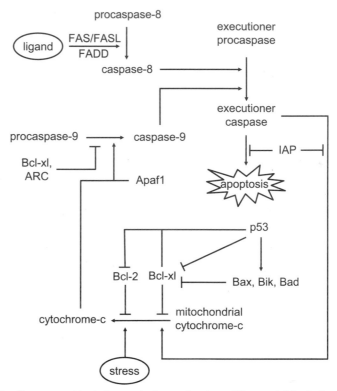

Figure 11.9 Caspase-activated apoptosis mechanism. The model includes two triggers for the activation of cell suicide mechanism, extracellular death ligand and stress-related factor (Fussenegger et al., 2000). The cell death occurs when executioner caspase is activated by caspase-8 (ligand effector) or caspase-9 (stress-related effector).

A possible first step in the model identification framework is the measurement selection. Parameter identifiability is crucial prior to the parameter estimation from experimental data. Practical identifiability of parameters discussed in section 11.2 is used for the selection of measurements that minimize the confidence interval (equation 11.8) for the model parameters. In this case study, the efficacy of model refinement by changing the experiment design or by improving the measurement set is compared. Thus, the first iteration in this case study is carried out with a suboptimal measurement set. The details of parameter confidence intervals for both optimal and suboptimal sets are included in (Gadkar et al., 2005a).

The model identification is decoupled into two parts: a state regulator problem (SRP) based estimator and a parameter estimation step. The SRP estimator uses the limited measurements to determine the time profiles of all unknown concentrations and reaction rates. It is based on a premise, similar to the dFBA approach, that cellular processes have evolved regulatory structures to optimally use the cellular resources. This translates into two postulates: (1) network flows are managed to minimize intracellular accumulation and (2) networks are managed to minimize the number of edges carrying flux. The estimator is formulated as a quadratic optimization problem as shown below:

$$\min_{\mathbf{r}(k),\cdots,\mathbf{r}(k+h-1)} \sum_{j=0}^{h-1}[\mathbf{x}(k+j+1)^T\mathbf{W}_x\mathbf{x}(k+j+1) + \mathbf{r}(k+j)^T\mathbf{W}_r\mathbf{r}(k+j)] \quad (11.20)$$

subject to:

$$\mathbf{x}(k+j+1) = \bar{\mathbf{A}}\mathbf{x}(k+j) + \bar{\mathbf{B}}\mathbf{r}(k+j) + \bar{\mathbf{C}} \quad \forall \quad j = 0,\cdots,h-1 \quad (11.21)$$

$$\mathbf{x}(k+j+1) \geq 0 \quad \forall \quad j = 0,\cdots,h-1 \quad (11.22)$$

$$|\mathbf{x}(k+j+1) - \mathbf{x}^*(k+j+1)| \leq \Delta_{tol} \quad \forall \quad j = 0,\cdots,h-1 \quad (11.23)$$

The objective function consists of two terms: the first represents the accumulation of the intracellular species, and the second describes the flux utilization. The terms $\mathbf{W}_x$ and $\mathbf{W}_r$ are the matrices of weights associated with these two terms. The SRP optimization is subject to constraints of mass balance (equation 11.21), non-negativity of concentrations (equation 11.22), and constraints imposed by the available measurements (equation 11.23). The term $\mathbf{x}^*$ represents the measurements and Δ_{tol} denotes the tolerance around the measurement describing the measurement error. Finally, the variable h denotes the prediction horizon of the SRP estimator. The optimization problem is solved for each sampling time to determine the profiles of all fluxes and species concentrations.

The SRP estimates of all system variables allow for efficient determination of the parameter values by decoupling the full parameter estimation into multiple sets, each with fewer parameters. The kinetic parameters associated with a reaction rate are determined independently from the others using a Bayesian approach, known as maximum *a posteriori* estimation (Gunawan et al., 2003). In this formulation, the difference between the SRP rate estimate and that predicted by the rate equation (equation 11.18) is minimized. Further, the deviations of parameter values from those obtained in the previous iteration are penalized. The formulation is represented as:

$$\min_{\mathbf{p}} \left[\left(\hat{\mathbf{r}}^i - \mathbf{r}^i(\hat{\mathbf{x}},\mathbf{p})\right)^T \mathbf{V}_\epsilon^{-1} \left(\hat{\mathbf{r}}^i - \mathbf{r}^i(\hat{\mathbf{x}},\mathbf{p})\right) + (\mathbf{p}-\mathbf{p}^0)^T\mathbf{V}_p^{-1}(\mathbf{p}-\mathbf{p}^0)\right] \quad (11.24)$$

$$\forall \quad i = 1,\cdots,N_R$$

where $\hat{\mathbf{r}}^i$ and $\hat{\mathbf{x}}$ are the SRP estimates of the i-th reaction rate and the concentrations, respectively, N_R represents the total number of reactions in the network, $\mathbf{p}^0$ is the vector of parameter values obtained in the previous iteration, and $\mathbf{V}_\epsilon$ and $\mathbf{V}_p$ are the variances of the reaction rates and the parameters, respectively. The parameter variances are determined using equation 11.7, and the reaction rate variances are determined from the noise in the measurements from which the rates were estimated. The second term in the objective function of equation 11.24 is zero in the first iteration.

An important step in the iterative approach is the model refinement method. In this work, the model refinement is achieved by an optimization of the experiment protocol or an optimal selection of the measurement set. The optimal experiment design maximizes the number of identifiable parameters, which is determined using the orthogonal procedure of McAuley and colleagues (Yao et al., 2003). When there exist multiple experiment designs with the same number of identifiable parameters, the selection is done to maximize the data informativeness by maximizing the D-optimality criterion. Mathematically, the optimal experiment design determination is given by:

$$\max \quad \det(\mathbf{FIM}) \tag{11.25}$$

$$\text{s.t.} \left[\max_{E \in \mathbf{E}} \ N_{\tilde{p}} \right]$$

where $\mathbf{E}$ denotes the parameterized space of experiment protocol and $N_{\tilde{p}}$ denotes the number of identifiable parameters.

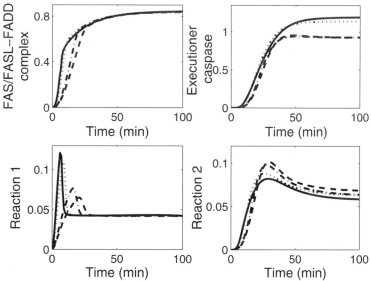

Figure 11.10 Model predictions of a few of the concentrations and reaction rates of the apoptosis model. Reaction 1 involves the binding of the FADD to the FAS-FASL complex; reaction 2 involves the activation of executioner procaspase by caspase-8. Solid line: real system; dashed line: prediction with estimated parameters after first iteration (suboptimal experiment with suboptimal measurements); dash-dotted line: prediction with estimated parameters after second iteration (suboptimal experiment with optimal measurements); dotted line: prediction with estimated parameters after second iteration (optimal experiment with suboptimal measurements).

Figure 11.10 presents the time profiles of a few species concentrations and reaction rates predicted by the models identified by the iterative framework, which show the improvements in model predictions with each iteration. As model identification is closely related to the parameter identifiability, the improvements are better when using the optimal experiment design than the optimal measurement set (Gadkar et al., 2005a).

This case study demonstrates an iterative framework for model identification to study quantitatively the dynamics of cellular pathways. It shows tremendous potential in improving the predictive capabilities for biological systems, especially in cases where experimental data is available but the kinetic parameters involved in the pathway reactions are unknown.

11.5   Summary and Future Directions

The reverse engineering of cellular networks represents a crucial aspect of a systemic approach for biological discovery in the post-genomic era. The major hurdle in this task is the high complexity of cellular networks, implying models with large numbers of nodes and interactions. Fortunately, the cellular networks appear to have structures (that is, they are not random) that are shared with engineered systems (see chapter 3). This may be one reason for why the application of engineering methods, in particular systems identification, has shown to be fruitful in approaching these problems.

In brief, the challenges in the cellular network inference condense to the integration between experimental and modeling efforts. Advances in molecular biology have allowed high throughput measurements of the interactions, but the data may not carry sufficient information to (uniquely) identify the network interactions, as shown in one of the examples above. In addition, the large size of network models renders the inference problem practically intractable. The case studies in the chapter demonstrate attempts to solve these problems by building an estimator for the network and formulating an iterative model identification framework. Here, the experiments and models are coupled together through a model-based experiment design and a Bayesian approach to incorporate prior knowledge of the network. Such engineering method and other approaches from engineering, computer science, and statistics have found great successes in their domains and will likely find greater application in systems biology as experimental methods are refined and closer collaborations are developed between modelers and experimentalists.

There still exist many open research problems in the reverse engineering of cellular networks. On the experimental front, the challenges are: (i) to improve the signal-to-noise ratio in the measurements, (ii) to develop new tools for measuring the cellular concentrations, fluxes, and interactions in both space and time, and (iii) to incorporate model-based design of experiment protocol. All of these will allow efficient and accurate dynamical modeling of the networks. The efforts here can benefit from existing models to identify the most useful type of measurements, for example, using information from sensitivity analysis. As experimental data will come from different measurements, data preprocessing may become necessary to extract relevant information before the inference step.

On the modeling front, the main challenge still remains in the formulation of model structures that can exploit the characteristics of cellular networks: sparsity, hierarchy, robustness, and optimality. Decomposition methodologies can exploit these characteristics and reduce the network inference problem to a reasonable scope. Categorically, these methods fall into either horizontal or vertical decomposition. The horizontal approaches focus on the topology of the network by dividing the network into building blocks (such as the aforementioned motifs and modules). The vertical approaches decompose the network based on the time scale of interconnections (that is, dynamics). There exists a need to integrate the two approaches in systems biology to obtain integrated system models. As noted above, the goal is to strike a balance between the size and level of detail, that is, a model structure that can sufficiently capture the dynamical behavior of a cellular network and is also amendable for numerical simulation and analysis in model identification. There may be no universal model for all cellular systems and purposes, but rather a tailored model structure for each system and use. Again, the modeling research should be integrated with the experimental efforts such that advances in each area will improve the other.

Acknowledgments

The authors would like to acknowledge the role of Daniel Zak in developing the motivating example at the beginning of the chapter, and Radhakrishnan Mahadevan and Prof. Jeremy Edwards for the dFBA methodology as described in the corresponding case study in this chapter. Funding from the Institute for Collaborative Biotechnologies through grant DAAD19-03-D-0004 from the U.S. Army Research Office and separate funding from the DARPA BioCOMP program are gratefully acknowledged.

12   Using Control Theory to Study Biology*

Brian P. Ingalls, Tau-Mu Yi, and Pablo A. Iglesias

* Reprinted in part with permission from *J. Phys. Chem.* B **2004**, 108, 1143–1152. Copyright 2004 American Chemical Society.

Much work has been devoted to determining the responses of biochemical networks to changes in their environment or their internal components. These studies have been motivated both by direct application to metabolic engineering and pharmaceutics as well as by the desire to improve our understanding of the behavior of these systems. This sensitivity analysis has focused primarily on the steady state (asymptotic) response of a system to constant (step) changes in parameters; see also chapter 1. However, there are cases in which a dynamic analysis of system response is crucial. This is clearly the case for mechanisms whose nominal behavior is time-varying, for example, the cell cycle. Moreover, investigations of the transient behavior invoked in signal transduction networks or the role of Ca^{2+} oscillations as a second messenger demand a dynamic analysis. This chapter presents a framework which is ideally suited to analysis of dynamic systems. Tools from control theory can be applied to elucidate the functioning of self-regulating (homeostatic) systems and to predict the effect of perturbations.

12.1   Linear Systems and the Frequency Response

We begin with an introduction to the framework of linear systems and one of the primary tools for describing their behavior: the frequency response. These ideas can be seen as a natural extension of a standard approach to analysis of biochemical systems: parametric sensitivity analysis.

Analytic tools for the study of the sensitivity of biochemical systems have been developed within the fields of Metabolic Control Analysis (MCA) (Kacser and Burns, 1973; Fell, 1992; Hofmeyr, 2000) and Biochemical Systems Theory (BST) (Savageau, 1976; Voit, 2000). This analysis is carried out in a linear (or log-linear) regime in which only small perturbations are addressed. This restriction is necessary since it is only after linearization that the analysis becomes tractable.

The same approach is taken here – the linearized response of a biochemical system is considered. The sensitivity analysis is extended by considering the response not just to constant parameter changes, but also to time-varying perturbations. This is achieved through a *frequency domain* analysis that describes the response of the system to a canonical set of inputs (sinusoids). The response to arbitrary perturbations can be reconstructed by use of the Fourier transform. This analysis can be interpreted as an extension of MCA as presented by Ingalls (2004).

12.1.1   Linear Input/Output Models of Biochemical Networks

A network consisting of n chemical species involved in m reactions is modeled. The n-vector $\mathbf{s}$ is composed of the concentrations of each species. The r-vector $\mathbf{p}$ is composed of the (external) parameters of interest in the model. The m-vector valued function $\mathbf{v} = \mathbf{v}(\mathbf{s}, \mathbf{p})$ describes the rate of each reaction as a function of species concentrations and parameter values. Finally the n by m stoichiometry matrix $\mathbf{N}$ describes the network: component $\mathbf{N}_{i,j}$ is equal to the net number of individuals of species i produced or consumed in reaction j; see chapter 5. The network can then be modeled by the ordinary differential equation

$$\frac{d}{dt}\mathbf{s}(t) = \mathbf{N}\mathbf{v}(\mathbf{s}(t), \mathbf{p}(t)) \qquad \text{for all } t \geq 0 \tag{12.1}$$

The vector $\mathbf{p}$ contains any external parameters which have a direct effect on the rates of the reactions (including, for example, kinetic constants of enzymes and external effectors).

For the purposes of this presentation, we will assume that the species concentrations are not constrained by any structural conservations (as when there are conserved moieties), and so the matrix $\mathbf{N}$ has full row rank. For a treatment of the general case, see (Ingalls, 2004).

Local analysis of system 12.1 will be carried out in the neighborhood of a steady state $(\mathbf{s}^0, \mathbf{p}^0)$ of interest. This point is brought to the origin by a change of variables in the states: $\mathbf{x}(t) = \mathbf{s}(t) - \mathbf{s}^0$, and in the parameters: $\mathbf{u}(t) = \mathbf{p}(t) - \mathbf{p}^0$. The n-vector $\mathbf{x}$ and the m-vector $\mathbf{u}$ indicate the deviation from the nominal state and parameter values of system 12.1, respectively. The linearized system then takes the form

$$\dot{\mathbf{x}}(t) = \left[\mathbf{N}\frac{\partial \mathbf{v}}{\partial \mathbf{s}}\right]\mathbf{x}(t) + \left[\mathbf{N}\frac{\partial \mathbf{v}}{\partial \mathbf{p}}\right]\mathbf{u}(t) \tag{12.2}$$

where the derivatives are taken at $(\mathbf{s}^0, \mathbf{p}^0)$. By construction, this linearized system has steady state $(\mathbf{x}, \mathbf{u}) = (\mathbf{0}, \mathbf{0})$.

The behavior of the original system 12.1 is approximated by that of the linearized system 12.2 near the nominal operating point. In particular, the linearized model faithfully represents the response of the original system to small changes in the parameters (for which the function $\mathbf{u}(\cdot)$ remains near zero). Standard sensitivity analysis involves gauging the response of system 12.2 to *constant* (step) changes in the parameter levels. In extending this analysis to nonconstant perturbations, it is useful to introduce the notations used in systems and control theory for analyzing such systems.

The standard model of a linear time-invariant input-output system has the form

$$\dot{\mathbf{x}}(t) = \mathbf{A}\mathbf{x}(t) + \mathbf{B}\mathbf{u}(t) \tag{12.3}$$

$$\mathbf{y}(t) = \mathbf{C}\mathbf{x}(t) + \mathbf{D}\mathbf{u}(t) \tag{12.4}$$

where $\mathbf{x}$ is an n-vector, $\mathbf{u}$ is an r-vector, $\mathbf{y}$ is a q-vector, and $\mathbf{A}$, $\mathbf{B}$, $\mathbf{C}$, and $\mathbf{D}$ are matrices of appropriate dimensions. The dynamics of the linearized model 12.2 take this form with

$$\mathbf{A} = \mathbf{N} \left. \frac{\partial \mathbf{v}}{\partial \mathbf{s}} \right|_{\mathbf{s}=\mathbf{s}^0, \mathbf{p}=\mathbf{p}^0} \qquad \text{and} \qquad \mathbf{B} = \mathbf{N} \left. \frac{\partial \mathbf{v}}{\partial \mathbf{p}} \right|_{\mathbf{s}=\mathbf{s}^0, \mathbf{p}=\mathbf{p}^0} \tag{12.5}$$

The components of the *input* vector $\mathbf{u}$ can play a number of roles in the system. In control engineering, three of the most common are: reference input (providing an external signal to which the system should respond), control input (by which one subnetwork might regulate the activity of another), and disturbance (to incorporate the effect of perturbations).

The vector $\mathbf{y}$ is referred to as the system *output* and represents a function of the state and input which is of specific interest. In addressing biochemical systems, there are several outputs which may be of interest, including species concentrations, reaction rates, pathway fluxes, transient times, and rates of entropy production (cf. section 5.8.1 of (Heinrich and Schuster, 1996)). In what follows, two output vectors of primary interest will be addressed.

The first is the vector of independent species concentrations, or more precisely, the deviations of these concentrations from the nominal level. In the linearized model 12.2, these deviations are described by the state $\mathbf{x}$. This choice of output is thus characterized by

$$\mathbf{y}(t) = \mathbf{x}(t) \tag{12.6}$$

which correspond to the choice $\mathbf{C} = \mathbf{I}$ (the $n \times n$ identity matrix) and $\mathbf{D} = \mathbf{0}$.

The second output of interest is the vector of reaction rates. Again, it is the deviation from the nominal rates which is the natural choice for $\mathbf{y}$. This is approximated by the linearization of the reaction rate function $\mathbf{v}(\cdot, \cdot)$ at the nominal point as follows:

$$\mathbf{y}(t) = \frac{\partial \mathbf{v}}{\partial \mathbf{s}} \mathbf{x}(t) + \frac{\partial \mathbf{v}}{\partial \mathbf{p}} \mathbf{u}(t) \tag{12.7}$$

where the derivatives are evaluated at $(\mathbf{s}^0, \mathbf{p}^0)$. This output takes the form of equation 12.4 with $\mathbf{C} = \frac{\partial \mathbf{v}}{\partial \mathbf{s}}$ and $\mathbf{D} = \frac{\partial \mathbf{v}}{\partial \mathbf{p}}$.

12.1.2   Frequency Response

Sensitivity analysis is concerned with determining the steady state response of a system to constant disturbances, for example, to an instantaneous change in the activity of an enzyme from one constant level to another. Extending that analysis to determination of the asymptotic response to arbitrary time-varying perturbations may seem a daunting task. Indeed, this is an intractable problem in general. However, when restricting to linear systems, a satisfactory result can be achieved.

There are two features of linear systems that can be exploited in this analysis. The first is simply the linear nature of their input-output behavior which implies an *additive property*: provided the system starts with initial condition $\mathbf{x}(0) = \mathbf{0}$ (which corresponds to the nominal steady state of the biochemical network), the output produced by the sum of two inputs is the sum of the outputs produced independently by the two inputs. That is, if input $\mathbf{u}_1(\cdot)$ elicits output $\mathbf{y}_1(\cdot)$ and input $\mathbf{u}_2(\cdot)$ yields output $\mathbf{y}_2(\cdot)$, then input $\mathbf{u}_1(\cdot) + \mathbf{u}_2(\cdot)$ leads to output $\mathbf{y}_1(\cdot) + \mathbf{y}_2(\cdot)$.

The additive property allows a reductionist approach to the analysis of system response: if a complicated input can be written as a sum of simpler signals, the response to each of these simpler inputs can be addressed separately and the original response can be found through a straightforward summation. This leads to a satisfactory procedure provided one is able to find a family of "simple" functions with the following two properties: 1) the family has to be "complete" in the sense that an arbitrary signal can be decomposed into a sum of functions chosen from this family; and 2) it must enjoy the property that the asymptotic response of a linear system to inputs chosen from the family is easily characterized. The family of sinusoids (sines and cosines) satisfies both of these conditions.

The decomposition of a signal $f(t)$ into a combination of sinusoids is the foundation of *Fourier analysis* (Strang, 1986; Lynn, 1982), which allows the description of $f(t)$ in terms of its *Fourier transform* $F(\omega)$ defined as a function of frequency ω by

$$F(\omega) = \frac{1}{2\pi} \int_{-\infty}^{\infty} f(t)e^{-i\omega t}\, dt. \tag{12.8}$$

The transform provides a record of the frequency content of $f(t)$ and is an alternative characterization of the original function. While complete recovery of a signal from its transform is difficult to achieve, important aspects of the nature of the signal can be gleaned directly from the graph of the transform. In particular, one can determine what sort of variations dominate the signal (low frequency or high frequency) by comparing the content at various frequencies. Quickly-varying signals have transforms with most of their content at high frequencies, while slowly-varying functions show primarily low-frequency content.

The second crucial property of linear systems that will be used is that, as mentioned above, their response to sinusoidal inputs can be easily described. Indeed, it is this property of sines and cosines which makes Fourier analysis a useful tool for analyzing linear time-invariant systems.

Consider the case of a system for which the input and output are scalars, referred to as Single-Input-Single-Output (SISO) systems. For such systems, a sinusoidal input of frequency ω, for example, $u(t) = \sin(\omega t)$, generates an output which is, after an initial transient, a sinusoid of the same frequency: $y(t) = A\sin(\omega t + \phi)$. This response can be described by two numbers: A, the amplitude of the oscillatory output, known as the *system gain*; and ϕ, the phase of the oscillatory output, known as the *phase shift*. For systems which are not SISO, there is one such pair of numbers which characterizes the response of each output channel (or component) to each input channel. The particular gain and phase shift which correspond to each frequency ω can be conveniently described by the assignment of a single complex number $Ae^{i\phi}$ (with modulus A and argument ϕ) to each frequency. The resulting complex-valued function is called the *frequency response*. The frequency response for system 12.3 is

$$\mathbf{H}(i\omega) = \mathbf{C}(i\omega\mathbf{I} - \mathbf{A})^{-1}\mathbf{B} + \mathbf{D}, \qquad \text{for all real } \omega \qquad (12.9)$$

This function will in general be matrix-valued but is scalar-valued in the SISO case. The frequency response can be derived through an algebraic calculation involving the Laplace transform of the system (Morris, 2001). The Laplace transform is a standard tool in the analysis of linear systems. It allows a linear differential equation, stated in the *time domain*, to be restated as an algebraic equation, in the *Laplace domain* – the complex plane. The behavior of the system in the Laplace domain is characterized by its *transfer function*, which is recovered from equation 12.9 by replacing the purely imaginary argument $i\omega$ with a general complex variable s.

In addressing biochemical networks, system response can be described as in equation 12.9. Recall, the matrices $\mathbf{A}$ and $\mathbf{B}$ describing the dynamics were derived in equation 12.5. If the independent species concentrations are chosen as output we have (from equation 12.6) $\mathbf{C} = \mathbf{I}$ and $\mathbf{D} = \mathbf{0}$, and so the frequency response takes the form

$$\mathbf{H}_s(i\omega) = (i\omega\mathbf{I} - \mathbf{N}\frac{\partial\mathbf{v}}{\partial\mathbf{s}})^{-1}\mathbf{N}\frac{\partial\mathbf{v}}{\partial\mathbf{p}} \qquad (12.10)$$

Alternatively, for the reaction rate output, equation 12.7 gives $\mathbf{C} = \frac{\partial\mathbf{v}}{\partial\mathbf{s}}$ and $\mathbf{D} = \frac{\partial\mathbf{v}}{\partial\mathbf{p}}$, so that

$$\mathbf{H}_v(i\omega) = \frac{\partial\mathbf{v}}{\partial\mathbf{s}}(i\omega\mathbf{I} - \mathbf{N}\frac{\partial\mathbf{v}}{\partial\mathbf{s}})^{-1}\mathbf{N}\frac{\partial\mathbf{v}}{\partial\mathbf{p}} + \frac{\partial\mathbf{v}}{\partial\mathbf{p}} \qquad (12.11)$$

Each element of these matrix-valued frequency responses is a scalar-valued function which describes the response of one output channel to one input channel. For each such input/output channel pair, the complex-valued function which describes the system behavior can be plotted in a number of ways. Perhaps the most useful of these visualizations is the *Bode plot*, in which the magnitude and argument of the frequency response are plotted separately. The magnitude of the function value (the system gain) is plotted on a log-log scale, where the gain is measured in decibels (dB) (defined by $r = 20 \log_{10} r$ dB; note that $0\,\text{dB}$ corresponds to a gain of one). The argument of the function value (the phase shift) appears on a semi-log plot, with log frequency plotted against phase in degrees. Bode plots will be used to illustrate frequency responses in the remainder of this chapter.

The response of a system to a constant input (which can be thought of as a sinusoid with frequency zero) is characterized by the frequency response at $\omega = 0$. Making this substitution into equations 12.10 and 12.11 the response of the system is found as

$$\mathbf{H}_s(0) = -(\mathbf{N}\frac{\partial\mathbf{v}}{\partial\mathbf{s}})^{-1}\mathbf{N}\frac{\partial\mathbf{v}}{\partial\mathbf{p}} \quad \text{and} \quad \mathbf{H}_v(0) = -\frac{\partial\mathbf{v}}{\partial\mathbf{s}}(\mathbf{N}\frac{\partial\mathbf{v}}{\partial\mathbf{s}})^{-1}\mathbf{N}\frac{\partial\mathbf{v}}{\partial\mathbf{p}} + \frac{\partial\mathbf{v}}{\partial\mathbf{p}} \quad (12.12)$$

These expressions can be derived from a standard sensitivity analysis of the system, such as that provided by MCA.

12.1.3   Illustration of the Frequency Response

The effect of negative feedback will be illustrated by an analysis of the bacterial *trp* operon, which is responsible for tryptophan production. A number of models of bacterial tryptophan biosynthesis have appeared in the literature, originating with the work of Goodwin (1965). The model of Xiu et al. (1997) will be considered here. A more complete model, including explicit time delays, has also appeared (Santillán and Mackey, 2001).

The Xiu model involves three state variables: the concentration of tryptophan P, the concentration of mRNA transcribed from the *trp* operon M, and the amount of expressed enzyme E. (It is an abstraction of the model that tryptophan synthesis is catalyzed by a single enzyme.) The dynamics of the model describe production of mRNA, enzyme, and tryptophan, as well as the degradation and dilution (due to cell growth) of each of these species. Cellular consumption of tryptophan is also included. In addition, two negative feedbacks are incorporated. The first is the inhibition of enzyme E by tryptophan. The second is the repression of transcription of mRNA, also tryptophan dependent. This genetic regulation is achieved through the activity of a repressor molecule R which, when bound to two units of tryptophan, interacts with an operator region of the operon, thus blocking transcription.

The dynamics, indicated in figure 12.1 can be described by the equations

$$\frac{dx}{dt} = \frac{z+1}{1+(1+r)z} - (\alpha_1 + u)x \tag{12.13}$$

$$\frac{dy}{dt} = x - (\alpha_2 + u)y \tag{12.14}$$

$$\frac{dz}{dt} = y\frac{k_i^2}{k_i^2 + z^2} - (\alpha_3 + u)z - \alpha_4\frac{z}{z+1} - \alpha_5(1 + \alpha_6 u)u\frac{z}{z+k} \tag{12.15}$$

$$\alpha_1 = 0.9, \ \alpha_2 = 0.02, \ \alpha_3 = 0, \ \alpha_4 = 0.024, \ \alpha_5 = 430,$$
$$\alpha_6 = -7.5, \ u = 0.00936, \ k_i = 2283, \ k = 0.05.$$

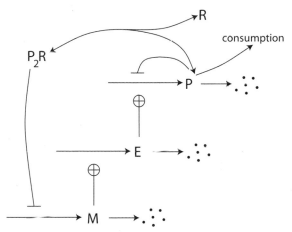

Figure 12.1   Tryptophan biosynthesis: reaction scheme.

where x, y, and z are dimensionless concentrations of mRNA, enzyme, and tryptophan respectively. The dimensionless parameters are described by Xiu et al. (1997).
The behavior of the system under changes in the value of α_5 will be addressed, with
a nominal value of $\alpha_5 = 430$. The effect of the enzyme inhibition on this response
will be illustrated by considering two values of the parameter r: strong feedback is
exhibited with $r = 10$, while weaker feedback will be addressed by taking $r = 5$.
The concentration of tryptophan (z) is taken as the output of the system. The
magnitude frequency responses to changes in α_5 are shown in figure 12.2.

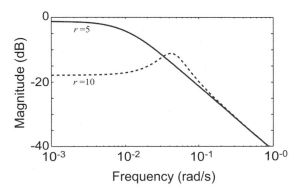

Figure 12.2   Frequency response for *trp* operon model.

In this model, α_5 describes the effect of cellular demand for tryptophan (Xiu et al.,
1997). The behavior shown in the figure is typical of a negative feedback system.
With weak feedback ($r = 5$), the effect of the input on asymptotic tryptophan
levels decreases monotonically as the frequency grows larger. Strengthening the

feedback (to $r = 10$) has two effects. The first is that the low frequency response is improved: as a standard sensitivity analysis would show, increasing the feedback reduces the effect of perturbations on the output. The other feature of stronger negative feedback is an increase in sensitivity at higher frequencies – to the point that the feedback actually makes the system *more* sensitive to disturbances over a certain frequency range.

The knowledge that negative feedback can introduce such resonance effects is crucial to the design of feedback systems. The trade-off between improved response at low frequencies and increased sensitivity at higher frequencies can be made explicit (for certain linear systems) by a constraint known as *Bode's integral formula* (Bode, 1945). System designers work around this "performance constraint" by implementing feedback that introduces increased sensitivity only at frequency ranges over which the system is unlikely to be excited. One could postulate that the same is true of feedback mechanisms within the cell: they have been crafted by natural selection in such a way that a trade-off is made between improved response to common low-frequency inputs and amplification of rare disturbances at higher frequencies.

Having illustrated the effect of negative feedback on the frequency response, we now turn to a more complete description of feedback strategies, highlighting the critical role of integral feedback.

12.2   Integral Feedback Control: From Homeostasis to Chemotaxis

Homeostasis is the dynamic self-regulation of a system to maintain essential variables within limits necessary for acceptable performance in the presence of unexpected disturbances. It is one of the defining features of living organisms. Homeostasis is achieved through countless control systems that regulate the multiplicity of biological processes. This intricate control network ensures robustness in the constantly changing real world; see chapter 2.

A related phenomenon is that of sensory adaptation in which the sensory system adjusts itself to changing environmental conditions for peak performance. For example, one's vision can adapt to the ambient background light intensity (bright or dim) so that there is sufficient contrast to detect objects. In signal transduction pathways, negative feedback regulation causes the output to return toward its prestimulus value after the application of a step increase in the input. During movement toward a chemical signal (chemotaxis), this type of adaptation facilitates the sensing of chemical gradients over a wide range of concentrations.

In this section we will discuss how one particular type of control system, integral feedback control, plays a crucial role in both homeostasis and sensory adaptation.

12.2.1 Negative Feedback Control

The most fundamental control system is negative feedback control. In such a system, the controller measures the difference between the current output and the desired output and based on this error takes some control action that reduces the error (see figure 12.3A). Negative feedback promotes regulation around a set point, stability, and robustness when performing some task. An important goal of the controller is to minimize the effect of the disturbances d_1 and d_2 on the output y. As shown in section 12.3, positive feedback increases the difference between the current and previous output, and thereby acts as an amplifier, but with potentially destabilizing consequences.

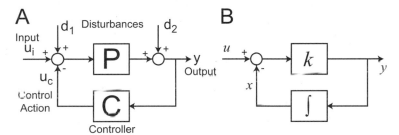

Figure 12.3 A) Block diagram of typical control system. The system to be controlled, which is usually referred to in control engineering as the "plant," P, takes an input u_i and converts it into an output y, which is typically normalized to represent the deviation between the current output and the desired output (that is, the error). The disturbances d_1 and d_2 perturb the input and output, respectively. The controller measures the error and takes an appropriate control action to reduce y. B) Block diagram of integral feedback control system. The input is u, the gain of the plant is k, and the controller is an integrator. From this diagram, it is clear that the feedback term $x = \int y\,dt$. Thus, $dx/dt = y$ and at steady state, $y \to 0$ as long as the system is stable.

One can classify controllers according to the mathematical operations used to convert the error signal into a control action. In today's world of ultrafast computers, one can design fancy digital controllers that implement arbitrarily complex strategies. In the past, however, control engineers resorted to three basic types of feedback control: (1) proportional control: the error term is multiplied by a constant before being fed back; (2) integral control: error is integrated; or (3) derivative control: error is differentiated. Each type of feedback has beneficial features. Proportional control corrects for "current" errors. One can adjust the amount of feedback by increasing or decreasing the constant factor. Higher feedback gain is better at rejecting disturbances, but it also causes the system to become less stable. Integral control eliminates steady state errors. Finally, derivative control provides "anticipation" of upcoming changes, which increases damping, improves

Controller Type	Controller (time domain)	(Laplace domain)	Closed-loop transfer function
Proportional	$u_c(t) = k_P\, y(t)$	$U_c(s) = k_P\, Y(s)$	$\frac{P(s)}{1+k_P P(s)}$
Integral	$u_c(t) = k_I \int y(t)dt$	$U_c(s) = \frac{k_I}{s} Y(s)$	$\frac{sP(s)}{s+k_I P(s)}$
Derivative	$u_c(t) = k_D\left(\frac{dy}{dt}\right)$	$U_c(s) = sk_D Y(s)$	$\frac{P(s)}{1+sk_D P(s)}$

Table 12.1 Controller types with time- and frequency-domain descriptions. The last column shows the transfer function from $D(s)$ to $U(s)$.

stability, and decreases transient errors. Mathematically, we can represent these controllers as in table 12.1.

By substituting these controllers into the feedback system shown in figure 12.3A, one can calculate the relationship (transfer function) between the disturbance inputs and the output using the block diagram and some simple algebra. The transfer function from d_1 to y, found in table 12.1, represents the sensitivity of the system to the disturbance.

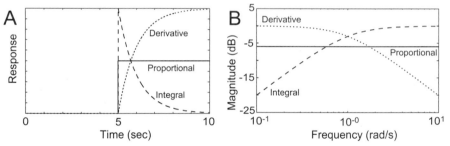

Figure 12.4 A) Time course of response of three control systems to a unit step disturbance ($d_1 = 1/s$, $k_P = k_I = k_D = 1$). Proportional (solid), integral (dashed), and derivative (dotted) control systems are depicted. (B) Bode plot of the sensitivity function $Y(s)/D_1(s)$. Recall that 0 dB corresponds to an output that has the same magnitude as the input; that is, a system with unity gain.

From these transfer functions one can run simulations of the output responses to an input disturbance signal and compare the three controllers. Applying a unit step disturbance at $d_1(t)$ ($D_1(s) = 1/s$) produces the time histories shown in figure 12.4A. Proportional control attenuates the disturbance at both short and long time scales; integral control completely neutralizes step disturbances at steady state but has little effect early on; derivative control is the opposite, blocking the immediate change but showing no attenuation at steady state. An alternative representation of these dynamics in the frequency domain is possible by taking the Fourier transform of the time domain signals. This frequency response of the transfer function or Bode plot offers perhaps a simpler depiction of the above comparison: one readily observes the disturbance attenuation at low frequencies by integral control, attenuation at high frequencies by derivative control, and attenuation at all frequencies by proportional control.

To capture the properties of all three controllers, one can combine them into a proportional-integral-derivative (PID) control system. The transfer function for a PID controller is written as $PID(s) = k_P + k_I/s + k_D s$. One can obtain the desired performance by tuning the parameters k_P, k_I, and k_D to obtain the best balance of steady state error, transient behavior, and stability. From a Bode plot perspective, manipulating the three coefficients allows one to shape the Bode plot to obtain the optimal disturbance attenuation at the critical frequencies. Thinking in terms of transfer functions, the PID controller offers sufficient flexibility to place the dominant pole(s)[1] of the system at appropriate location(s) for the desired system behavior.

Although the above analysis, based on transfer functions, is for linear feedback systems, the general lessons also apply to nonlinear systems. Indeed, it is helpful to think about complex digital controllers in these simpler terms in order to gain intuition. In addition, PID controllers are still widely used for systems possessing slower dynamics, such as in process control.

12.2.2   Primer on Integral Control

Now we will focus our attention on integral feedback control because of the remarkable property of perfect regulation at steady state against step disturbances. More importantly, this regulation is robust to internal and external perturbations. For example, the presence of an additional disturbance d_2 does not affect the perfect regulation of the step disturbance d_1. Likewise, the steady state output is robust to variations in the parameters of the plant. Thus, integral control ensures the robust tracking of a specific steady state value so that the error approaches zero despite uncertainty in internal and external conditions. Exceptions arise in the case of higher-order unstable disturbance inputs (for example, ramp inputs) or when the controller itself is perturbed.

Integral controllers are ubiquitous in man-made systems. For example, the cruise control in a car uses integral control to maintain robustly the speed of the vehicle at the set point despite disturbances such as the wind or a hill. In an airplane, integral control loops are found at every level from CPUs to instruments to the entire vehicle. A single oil refinery possesses more than 10,000 integral feedback loops.

A block diagram of a simple linear system with integral feedback illustrates its chief features; see figure 12.3B. The plant or network, represented by the block with gain k $(P(s) = k)$, takes the input u and produces the output y_1. The difference between the output y_1 and the desired steady state output y_0 is the error term y. Then, y is integrated and fed back into the system. The key to integral control is that the feedback term $x = \int y$ so that

$$\frac{dx}{dt} = y \qquad\qquad (12.16)$$

At steady state the time derivatives of the variables go to 0, so that $y \to 0$ as $t \to \infty$ independent of the values of the input u and the gain k. Hence, the error asymptotically approaches zero as long as the system is stable. It is important to note that this analysis does not depend on the fact that the system is linear; the perfect regulation property of integral feedback also applies to nonlinear systems.

Transfer function and state-space interpretations. As we described above, for a typical linear feedback system with plant $P(s)$ and controller $C(s)$, the sensitivity transfer function from the disturbance input $D_1(s)$ to the output is $S(s) = P(s)/(1 + P(s)C(s))$. We can then prove that if the input signal is a step of size A, then the output will approach zero asymptotically in time if and only if the sensitivity function has a zero at the origin (Doyle et al., 1992).

In transfer function form, a step input of size A is described by $U_i(s) = A/s$. This leads to the output $Y(s) = U_i(s)S(s) = AS(s)/s$. If the feedback system is stable, then by the final value theorem (Doyle et al., 1992), $y(t) \to AS(0)$ as $t \to \infty$. Clearly, the right-hand side is zero if and only if $S(0) = 0$. An integral feedback system possesses such a zero at the origin: $S(s) = sP(s)/(s + P(s))$.

Alternatively, one can represent the dynamics of a system in state-space form as a set of first order differential equations:

$$\frac{dx}{dt} = f(x, u), \quad \text{and} \quad y = g(x, u) \tag{12.17}$$

The vector x is the state of the system (typically describing the concentrations of the species) and u is the input. For a linear system, we can simplify this description to the matrix form of equation 12.3. One can introduce a new integral feedback state z with the dynamics

$$\frac{dz}{dt} = Cx + Du = y \tag{12.18}$$

In this manner, integral feedback is implemented, and these dynamics guarantee that the steady state error approaches zero no matter what the values of u, A, B, C, and D as long as the system is stable.

12.2.3 Necessity of Integral Control and the Internal Model Principle

The previous section demonstrated the sufficiency of integral control for robust perfect regulation. What about necessity? Is it true that any system exhibiting robust perfect regulation must contain integral feedback? A simple necessity proof for linear systems is provided that relies on the state-space description given above.

At steady state, $dx/dt = 0$, so that $x = -A^{-1}Bu$ and $y = (D - CA^{-1}B)u$. Thus, $y = 0$ at steady state for all constant u, if and only if either

$$\begin{bmatrix} C & D \end{bmatrix} = 0, \quad \text{or} \quad \det \begin{bmatrix} A & B \\ C & D \end{bmatrix} = 0 \tag{12.19}$$

The former is the trivial case when $y(t) = 0$ for all t, and the latter holds if and only if there exists a $k \neq 0$ such that

$$k \begin{bmatrix} A & B \end{bmatrix} = \begin{bmatrix} C & D \end{bmatrix} \qquad (12.20)$$

Thus, defining $z = kx$, we have $dz/dt = k\dot{x} = k(Ax + Bu) = Cx + Du = y$, which is the standard integral control equation.

This necessity statement suggests that integral control is prevalent at all levels of biology from cellular regulation to organismal physiology to ecosystem balance. Just as integral feedback is used ubiquitously in man-made systems, it must also be a common control strategy in biological systems given the requirement that internal variables maintain constant steady state values despite step disturbance changes.

The necessity of integral control applies to step changes. The internal model principle (IMP) generalizes this notion. The principle states that the robust tracking of an arbitrary signal requires a model of that signal to be in the controller. The intuition is that the internal model counteracts the external signal so that $y(t) \to 0$ as $t \to \infty$ even in the presence of parameter perturbations. For example, a controller containing an integrator ($C(s) = 1/s$) is necessary to track robustly a step signal ($U(s) = 1/s$). Francis and Wonham (1976) proved IMP for linear systems. Isidori and Byrnes (1990) have established a general framework for IMP in nonlinear systems using techniques from differential geometry. Sontag (2003) has provided a succinct statement of IMP relevant to biological systems. However, these topics are beyond the scope of this chapter. It is important to appreciate that living systems are subject not only to constant, or step, changes, but also to perturbations that involve steadily rising or falling signals, and to even more complex disturbance behaviors (for example, neural signals). In order to maintain homeostasis, the feedback control system implemented by the biological network must contain an internal model of the disturbance according to IMP. An area for future research is cataloging these control structures and addressing the question of how biology builds these internal models.

12.2.4   Examples of Integral Control in Biology

Here we illustrate two biological examples of integral control. One is in the area of blood calcium homeostasis and the other is in the area of sensory adaptation.

Blood calcium regulation. The level of calcium in the blood is carefully regulated against disturbances in calcium utilization and uptake. The two compounds parathyroid hormone (PTH) and vitamin D (VitD) play a central role in this regulation. They control how much calcium is introduced into the blood from the intestine (vitamin D) and from the bone (PTH). El-Samad and Khammash have formulated a model, illustrated schematically in figure 12.5A, of these dynamics in mammals (El-Samad et al., 2002).

A disturbance d_1 affects the rate at which calcium is taken up or removed from the blood; this disturbance is compensated for by the action of PTH and vitamin

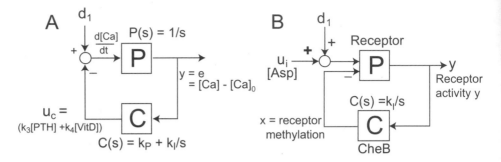

Figure 12.5 A) Model of blood calcium regulation (El-Samad et al., 2002). A disturbance, d_1, in calcium dynamics is attenuated by the control action u_c, mediated by PTH and vitamin D, which is produced by the block C representing a PI controller. The error y is the current level of calcium, [Ca], minus the set point level of calcium, $[\text{Ca}]_0$. B) Model of regulation of bacterial chemoreceptors. Receptor activity depends on the ligand aspartate as well as the degree of receptor methylation (feedback term). An integral controller is implemented because CheB only demethylates active receptors. This control system ensures that the steady state level of receptor activity is constant despite changes in ligand concentration or in receptor numbers.

D, $u_c : d[\text{Ca}]/dt = u_c + d_1$. The error is the deviation from the steady state blood calcium level ($y = [\text{Ca}] - [\text{Ca}]_0$). It is known from physiological measurements that the level of PTH is proportional to this error ($y \propto [PTH]$). In addition, the rate of production of vitamin D is proportional to the concentration of PTH, and assuming that vitamin D has a slow degradation rate on the time scales of interest, we have $d[\text{VitD}]/dt = k[\text{PTH}]$. Thus, we can calculate the error in terms of [PTH] or [VitD]: $y = k_1[\text{PTH}] = k_2 d[\text{VitD}]/dt$. Finally, if we approximate the rate of calcium absorption from the intestine or bone as linear functions of [VitD] and [PTH], respectively, we have the following equation for the control action: $u_c = k_3[\text{PTH}] + k_4[\text{VitD}] = k_P y + k_I \int y$. Thus, this system exhibits proportional-integral (PI) control.

Bacterial chemotaxis signaling pathway. Bacteria are able to sense gradients of attractants and repellents. The signal transduction pathway responsible for this behavior possesses several special features to ensure both exquisite sensitivity and wide dynamic range. One such feature is perfect adaptation: the output of the pathway (flagellar rotation) exactly returns to its prestimulus value even in the presence of continuous stimulation so that the steady state level of output activity asymptotically approaches a constant value independent of the attractant concentration. The bacterial chemotaxis system is a two-component signaling system (Stock et al., 1991). The receptor complex, which consists of the receptor, the histidine kinase CheA, and the adaptor protein CheW, phosphorylates the response regulator CheY. Phosphorylated CheY interacts with the flagellar motor to induce clockwise (CW) rotation and tumbling behavior. The attractant inhibits the receptor complex resulting in counterclockwise (CCW) flagellar rotation and straight runs. Receptor complex activity is regulated by methylation, which mediates adaptation. Methylation by CheR increases receptor activity; demethylation by CheB decreases activity. Although there is no direct evidence, we assume that CheB senses the activity state of the receptor by only demethylating active receptor complexes. This assumption results in an important negative feedback loop; see figure 12.5B.

Robust versus non-robust perfect adaptation. An important question is whether perfect adaptation in bacterial chemotaxis is robust to changes in internal and external conditions. Alon, Leibler, and colleagues experimentally tested the robustness of perfect adaptation to dramatic changes in the concentration of key components of this pathway (Alon et al., 1999). They demonstrated that as the methylase CheR was varied over a 50-fold range, the adaptation precision remained close to perfect. They went on to show that perfect adaptation was robust not only to changes in levels of CheR, but also to changes in the concentration of CheB, receptor, and CheY.

Is it possible to model perfect adaptation in bacterial chemotaxis? Most models in the literature indeed were able to reproduce perfect adaptation, but only through fine-tuning of the model parameters. Perfect adaptation is nonrobust in these models because altering a parameter disrupts perfect adaptation. This can be evaluated by systematically varying the model parameters and testing for perfect adaptation using continuation methods (Yi et al., 2000). For example, varying the total receptor concentration over a 100-fold range in a particular model of bacterial chemotaxis (Spiro et al., 1997), one observes perfect adaptation for only one particular receptor concentration, $8\,\mu\text{M}$. This is an example of non-robust perfect adaptation.

Alternatively, one can imagine that perfect adaptation is a structural property of the system, insensitive to parameter variation, perhaps resulting from a particular feedback control mechanism. For example, perfect adaptation was robust to a 100-fold change in receptor concentration in another model by Barkai-Alon-Leibler (BAL) (Barkai and Leibler, 1997). Varying the levels of several other components in the model did not disrupt perfect adaptation. The necessity of integral control argues that an integral control mechanism must be present in the BAL model to explain this robust regulation.

Implementation of integral control in the bacterial chemotaxis system. How is integral control implemented in the BAL model of the chemotaxis system? A simplified version of the derivation is shown here. The variable x represents the methylation state of the receptor. The change in x: dx/dt, equals the methylation rate r minus the demethylation rate. Using the assumption that CheB only demethylates active receptor complexes so that the demethylation rate is proportional to A, we obtain the following: $dx/dt = r - bA$. At steady state, $x = 0$, $r = bA$, and hence the steady state activity level $A_0 = r/b$. We can rewrite this as the familiar $dx/dt = -b(A - A^0) = -by$. The key point is that if r and b are independent of u, then this system will exhibit perfect adaptation that is robust to changes in the system parameters.

12.3   Feedback in Cellular Communication

In the previous section we saw how integral control can be used to achieve robust perfect adaptation in the signaling pathway regulating chemotaxis in *E. coli*. We now consider other uses of feedback mechanisms in cell signaling pathways.

12.3.1   Signal Detection: Fast Excitation, Slow Inhibition

One of the roles of the integral feedback mechanism employed in *E. coli* chemotaxis is that it allows the cell to determine the rate at which the chemoattractant concentration is varying temporally around the cell. Thus, the objective of the integral feedback control is not necessarily to reject the step disturbances, but to generate signals that mirror the rate of change in the external signal.

An alternative role of the mechanism can be envisioned (Koshland Jr. et al., 1982; Sontag, 2003). A cell may need to monitor the external environment for sudden changes, to which it can then adapt. A mechanism for achieving signal detection is then required so that the cell can alter its behavior in response to this change.

A digital logic mechanism demonstrating how this monitoring can take place is shown in figure 12.6. The current state of the environment is sensed continuously and compared to the previous state through an EXCLUSIVE-OR gate (X-OR). This circuit has a "low" output if the two inputs coincide, but a "high" if they differ. Thus, if the present and past states of the environment differ, a transient signal is generated that can trigger a response.

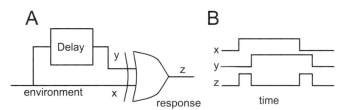

Figure 12.6   Signal detection scheme. A) Changes in the external environment can be detected by the scheme outlined here. The environmental signal, x, is compared with stored copies of this signal, y, using an X-OR signal. This generates a pulse whenever the present state does not match the previous one. B) Sample signal levels. The two changes in the environment lead to two response pulses.

For biological signaling, a similar transient response can be effected by a mechanism in which the current state of the environment generates a fast excitory signal that stimulates a response regulator (Koshland Jr. et al., 1982; Levchenko and Iglesias, 2002); see figure 12.7. The environment also generates a slower inhibitory signal on this same response regulator. Whenever the state of the environment is constant, the positive and negative influences balance and the response returns to basal levels. Recently, this scheme was used to create a synthetic gene network to allow cell-to-cell communication in *E. coli* (Basu et al., 2004).

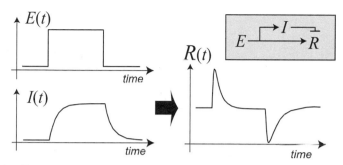

Figure 12.7   Fast excitation, slow inhibition. Biochemical scheme for implementing the signal detection scheme of figure 12.6 (Insert). As seen in the time courses, a rise in the excitation signal, E, stimulates an inhibitory signal, I, whose level rises slowly. The excitation leads to a response, R, which is then attenuated by the inhibitor. Together this leads to a short pulse in the response.

A mathematical model that effects this general mechanism is given by

$$\frac{dI}{dt} = -k_{-1}I + k_1 E$$
$$\frac{dR}{dt} = -k_{-2}IR + k_2 E$$

A change in the environmental signal, E, leads to both a fast increase in the response, R, as well as a slower buildup in inhibition, I; see figure 12.7. At steady state,

$$I = (k_1/k_{-1})E \quad \text{and} \quad R = (k_2/k_{-2})\frac{E}{I} \qquad (12.21)$$

Together, these two equations imply that, at steady state,

$$R \to R^\star = \frac{k_2 k_{-1}}{k_{-2} k_1} \qquad (12.22)$$

ensuring that the level of response is independent of that of E.

Though the system looks like a purely feedforward control mechanism, integral feedback is still being employed. Rewriting the differential equation as

$$\frac{dR}{dt} = -k_{-2}I\left(R - \frac{k_2}{k_{-2}}\frac{E}{I}\right)$$
$$= -k_{-2}I\left(R - R^\star + \frac{k_2}{k_{-2}}\left(\frac{k_{-1}I - k_1 E}{k_1 I}\right)\right)$$
$$= -k_{-2}I\left(R - R^\star + D\right)$$

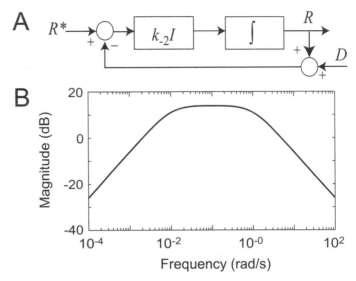

Figure 12.8 Integral control mechanism. A) The fast excitation, slow inhibition mechanism described in the text can be expressed in block-diagram form. A signal consisting of the response, R, plus a transient signal, D, is compared against a constant reference signal, R^*. After scaling by the time-varying gain $k_2 I(t)$, the error is integrated. B) The magnitude Bode plot of the linearized model of this mechanism exhibits the low frequency rise proportional to frequency that is characteristic of a closed-loop system that has integral control. Parameters used: $k_2 = 5$, $k_{-1} = 0.01$, and $k_{-2} I_0 = 1$.

where

$$D = \frac{k_2}{k_1 k_{-2}} \left(\frac{dI/dt}{I} \right) \tag{12.23}$$

approaches zero as $t \to \infty$ provided that $I > 0$. Hence, the system can be redrawn as in figure 12.8A where the integral control feedback is evident.

This can also be observed by computing the transfer function of the linearized model of this system. In particular, the closed-loop transfer function between the environment and the output is

$$\frac{R(s)}{E(s)} = \frac{k_2 s}{(s + k_{-1})(s + k_{-2} I_0)} \tag{12.24}$$

where I_0 is the inhibitor concentration about the operating point. The frequency response of this transfer function is shown in figure 12.8B. This transfer function demonstrates that the system has two poles, corresponding to the off-rates for the inhibitor (k_{-1}) and response-regulator ($k_{-2} I_0$) equations. It also has a zero at $s = 0$, which is a consequence of a closed-loop system that has integral feedback.

12.3.2 Amplifying the Signal through Positive Feedback

The adaptive property of the fast excitation, slow inhibition mechanism is not affected by model parameter values; it is clear from the discussion above that the kinetic coefficients can be altered, but that the response, after a change in the stimulus, will return to its prestimulus concentration. These concentrations, however, will be affected by the parameter values, echoing the results obtained in bacterial chemotaxis models (Alon et al., 1999; Yi et al., 2000). However, it is also known that the external gradient is not amplified with this single mechanism and hence cannot explain all the observed response of chemotaxing eukaryotic cells, such as neutrophils and the amoeba *D. discoideum*. To amplify the effect of the stimulus, several mechanisms have been proposed (Iglesias and Levchenko, 2002; Levchenko and Iglesias, 2002).

One means of amplifying the response is to add a positive feedback loop downstream, as shown in figure 12.9A. Suppose that the response of the sensing mechanism activates a downstream autocatalytic effector according to:

$$\frac{dX}{dt} = k_{-3}X + k_3 R + \frac{k_f X^n}{1 + k_s X^n} \tag{12.25}$$

The parameter k_f, as well as the Hill coefficient, n, denote the strength of the positive feedback. For now, assume that $n = 1$. In the absence of this feedback ($k_f = 0$) the concentration of X is proportional to that of R, with proportionality constant equal to $1/k_{-3}$. Now, assume that $k_f \geq 0$ and that $k_s X \ll 1$. Then X is, once again, proportional to R, but the proportionality constant is now $1/(k_{-3} - k_f) > 1/k_{-3}$. This can be arbitrarily large if $k_f \approx k_{-3}$. In this situation, of course, saturation conditions exist; see figure 12.9B.

12.3.3 Positive Feedback and Cooperativity: Hysteretic Behavior

More interesting behavior can arise if the Hill coefficient of the feedback term is greater than one. In this case, the response does not vary significantly for small changes in the stimulus; see figure 12.9C. However, once a threshold value is reached, the response changes abruptly to a higher level. At this point, the response is once again relatively insensitive of the stimulus. To return to prestimulus levels, a significant decrease in the stimulus is needed. This hysteretic response arising from a bistable system is common in engineering circuits. For example, the Schmitt trigger implements a bistable circuit by closing a positive feedback loop around an operational amplifier (Sedra and Smith, 2004). There is experimental evidence that cells also rely on bistability for regulation (Xiong and Ferrell Jr., 2003). Synthetic biological switches have also been designed and built based on these principles (Hasty et al., 2000; Ozbudak et al., 2004).

Several models based on bistable signaling systems have been proposed to account for the large chemoattractant-induced responses in cells (Meinhardt, 1999; Narang et al., 2001; Postma and Van Haastert, 2001). However, because of the hysteretic

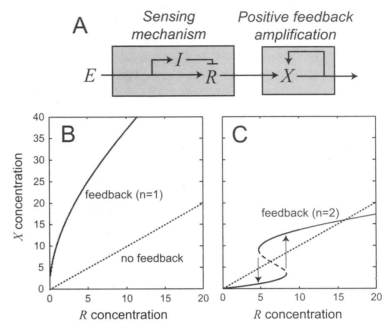

Figure 12.9 Amplification through positive feedback. A) To amplify the response to an external environmental signal, the sensing mechanism of figure 12.7 can be followed by an amplification mechanism that relies of positive feedback. B) Steady state concentration of X for the system described by equation 12.25 as a function of the input concentration R. For comparison, the response of the system assuming linear feedback, $n = 1$ (solid) is plotted alongside that of the open-loop system, $k_f = 0$, (dotted line). Coefficients used are $k_{-3} = k_3 = k_f = 1$ and $k_s = 0.01$. C) Hysteretic behavior observed when $n = 2$ (solid). When the input concentration is low, $R \lesssim 5$, or high, $R \gtrsim 8$, the system exhibits only one steady state response. For intermediate values, two stable (solid lines), as well as one unstable (dashed line), steady states are present. This bistable system can lead to discontinuous behavior. When the input is increased beyond the transition level, a sudden rise in the output can be observed, as the system moves from the low level of response to the higher level, shown by the arrow. To return to the lower level, the input signal must be reduced significantly. The response in the absence of feedback is shown for comparison (dotted).

nature of their response, these models cannot account for the behavior seen in unpolarized *D. discoideum* cells. These unpolarized amoebae are equally sensitive around the whole membrane, and so when they are subjected to sudden changes in the concentration gradient, these cells can rapidly respond (Iglesias and Levchenko, 2002; Devreotes and Janetopoulos, 2003). A hysteretic switch, however, could account for the response of polarized cells, which "remember" their polarization. These cells, when subjected to a change in the chemoattractant gradient, tend to turn.

More recently, a means for amplifying the response to an external signal has been suggested in which parallel sensing mechanisms, acting independently, cooperate to produce an amplified response to the chemoattractant stimulus (Ma et al., 2004b). The advantage of such a mechanism for amplifying the signal is that it employs a redundant mechanism. If, for some reason, one of the two pathways is impeded, the cell can still detect stimuli. Such mechanisms have been demonstrated experimentally in *D. discoideum* cells in which one of the two pathways is disrupted, either through knockout mutations or pharmacological inhibitors. These cells are still able to sense external stimuli, though chemotaxis is partially impaired (Iijima and Devreotes, 2002; Funamoto et al., 2002).

12.3.4   Oscillations: Positive and Negative Feedback Work Together

Besides providing for strong amplification, one of the uses of positive feedback is as a means of obtaining oscillatory behavior. This was first used in engineering around 1915 (Bennett, 1979).

In biology, oscillatory systems are ubiquitous, from the circadian rhythms to genetic oscillators to the wave pattern observed in *D. discoideum* cell-to-cell communication (Goldbeter, 2002; Kruse and Jülicher, 2005). Many of these systems rely on the interplay of positive and negative feedback loops.

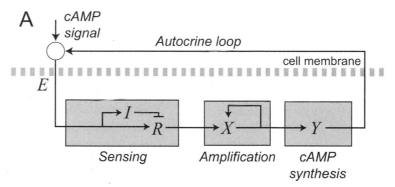

Figure 12.10   Positive feedback through autocrine loop. In *D. discoideum* cells, the pathway that senses extracellular cAMP also stimulates the production of intracellular cAMP, which is then secreted from the cell. In doing so, a positive feedback loop is closed.

Autocrine loops arise when a cell secretes a chemical that stimulates the secretory cell itself. For example, receptor binding of extracellular cAMP in the amoeba *D. discoideum* induces the activation of adenylyl cyclase of aggregation (ACA). This leads to the synthesis of intracellular cAMP from ATP. This cAMP is then secreted into the extracellular medium where it can diffuse away and thereby signal nearby cells. However, the secreted cAMP may also find its way back to the cell. In doing so it closes a positive feedback loop involving the chemoattractant sensory system; see figure 12.10.

A thorough analysis of autocrine loops requires that the stochastic nature of the diffusion of the signaling molecule be considered (Batsilas et al., 2003). However, if we ignore the spatial considerations of the diffusion, the analysis of the autocrine loop is relatively straightforward. The positive feedback path is coupled to the negative feedback mechanism, described above, that provides adaptation. Together, these intertwined positive and negative feedback loops lead to the formation of cAMP waves that can propagate as circular or spiral wave forms (Kessin, 2001).

Several models have been proposed to describe the oscillatory behavior found in cAMP signaling in *D. discoideum* (Halloy et al., 1998; Laub and Loomis, 1998; Nagano, 2000; Iglesias, 2003). While these models differ in the biochemical identities of activators and inactivators, they all rely on an interplay between positive and negative feedback to achieve this periodic oscillation.

Here we take a systems-level approach to the analysis of the autocrine loop and thereby demonstrate the use of several control-theoretic analysis techniques. Using the scheme described in figure 12.10, we assume that the production of intracellular cAMP is governed by

$$\frac{dY}{dt} = -k_{-4}Y + k_4 X \qquad (12.26)$$

and that this changes the extracellular concentration as:

$$\frac{dE}{dt} = -k_{-5}E + k_5 Y \qquad (12.27)$$

The parameter k_5 can be used to describe the strength of the feedback loop.

The system can now be treated as in figure 12.11A. Here, a linear system is found in feedback with a discontinuous nonlinear element, which can serve as an approximate model for the hysteretic bistable system described in section 12.3.3. This type of feedback system has been studied extensively in the control literature, where it is sometimes known as a relay or relaxation oscillator (Tsypkin, 1984; Varigonda and Georgiou, 2001). In these cases, the *describing function method* can be used (Khalil, 2002) to determine whether oscillatory behavior is possible.

The analysis is predicated in computing an equivalent gain through the nonlinear system. Suppose that the input to the nonlinearity is the sinusoid $x(t) = A \sin(\omega_0 t)$ and assume that the output $y(t)$ is also periodic. It can then be described by the Fourier series. For example, if the nonlinearity involved is the hysteresis function described in figure 12.11A, then the sinusoidal input leads to a square wave output with Fourier series

$$y(t) = \frac{\epsilon}{\pi} \sum_{k=0}^{\infty} \frac{\sin\left([2k+1][w_0(t-t_0)]\right)}{2k+1}, \quad \sin(\omega_0 t_0) = \frac{\delta}{A} \qquad (12.28)$$

If we focus on the fundamental frequency, $k = 0$, the nonlinearity has a gain equivalent to:

$$\eta(A) = \frac{\epsilon}{A\pi} e^{-i\omega_0 t_0} \qquad (12.29)$$

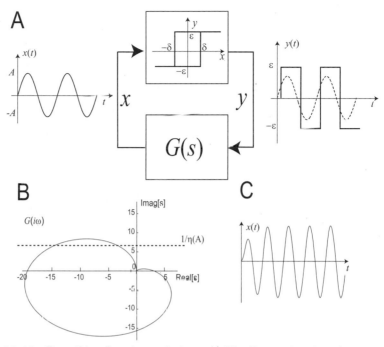

Figure 12.11 Describing function technique. A) The linear, time-invariant system with transfer function $G(i\omega)$ is placed in feedback with a nonlinear system. If the system exhibits oscillatory behavior, then a sinusoidal input to the nonlinear system, x, gives rise to the square wave output $y(t)$. The frequency of this output matches that of the input, though the hysteretic nature of the nonlinearity gives rise to a phase shift. In the describing function analysis technique, the square wave is expressed as a Fourier series, and an effective gain, $\eta(A)$ is computed. B) To determine whether oscillatory behavior is predicted, a polar plot of the linear system's transfer function is obtained. Here, the real and imaginary parts of $G(i\omega)$ are plotted as a function of ω (solid). On the same axes, the function $1/\eta(A)$ is plotted as a function of A (dotted). Points of intersection correspond to frequencies ω_0 and amplitudes A where $1 - G(\omega_0)\eta(A) = 0$. These are predicted magnitudes and frequencies of oscillation. C) Limit cycle oscillation arising from the system described in panel B.

If there is oscillatory behavior, then the loop gain at the frequency ω_0 must be one; that is $1 = G(i\omega_0)\eta(A)$. Thus, by plotting $G(i\omega)$ as a function of ω and $1/\eta(A)$ as a function of A, we can seek for points where this equation is satisfied; see figure 12.11B. These points predict the existence of an oscillation with the given frequency and magnitude.

The describing function technique is one means of studying oscillatory behavior analytically. Other techniques and methods include the use of bifurcation analysis (Ma and Iglesias, 2002).

12.4   Discussion

Control theory originated to meet the needs from a variety of engineering disciplines. In its essence, it facilitates the analysis and design of dynamical systems that are used to regulate the performance of a larger system. Almost always, these systems involve considerable feedback loops which can endow them with excellent robustness properties, but can also make them vulnerable; see chapter 2.

In this chapter we have attempted to introduce readers to several of these tools. Because many of the techniques used in the analysis and design of control systems are based on linear analysis, we have emphasized this. While it is true that "real systems" are not linear, it is also true that considerable insight can be obtained regarding the dynamical behavior of nonlinear systems near equilibria by considering their linearizations (Khalil, 2002). This approach is especially fruitful when applied to systems whose architecture ensures they will spend most of their time near a steady state, including systems governed by homeostasis. We have shown how, by considering linear systems, powerful frequency-domain and transfer-function tools are available for analysis.

We have also tried to illustrate how understanding of engineering control systems can lead to some intuition as to the system behavior of biological systems. For example, knowledge of the internal model principle helps evaluate models of perfect adaptation in biology. Similarly, understanding how hysteretic switches and amplifiers arise out of positive feedback may lead to a better understanding of how these behaviors arise in biology.

Finally, we note that, historically, control theory first arose out of a need to understand the behavior of systems (Bennett, 1979). This theory was then used to design and engineer better systems. It is not difficult to foresee that, in biology, control theory may follow the same path. At first, we expect that both existing and new tools will be used to analyze existing biological systems. However, we expect that these tools will later allow us to design and implement synthetic biological systems. In fact, we now see the first steps in this process (Hasty et al., 2002).

Acknowledgements

This work was supported by the Natural Sciences and Engineering Research Council of Canada (BPI), the Whitaker Foundation (PAI), and the National Science Foundation's Biocomplexity program, through grant number DMS-0083500 (PAI) and NIGMS grant 71920 (PAI).

Notes

1.   For a rational transfer function $G(s) = N(s)/D(s)$ where $N(s)$ and $D(s)$ are polynomials with no common roots, the poles (respectively zeros) of the transfer function are the roots of the denominator (respectively numerator) polynomials.

For example, the transfer function $G(s) = (s + 1)/(s^2 + 1)$ has one zero at -1 and two poles at $\pm i$. Much of classical control theory deals with the use of feedback to manipulate the location of the poles of the closed-loop transfer function. See (Doyle et al., 1992) for examples.

13   Synthetic Gene Regulatory Systems

Mads Kærn and Ron Weiss

In parallel with the development of high-throughput technologies fueling systems biology, advances in modeling of biological systems and in synthesis of long DNA fragments with arbitrary nucleotide sequences have fostered the emergence of a nascent field termed synthetic biology. At its core, this field uses recombinant DNA manipulation techniques to design and embed complex "programmed" functions into living organisms. An important notion that pervades most of the work in synthetic biology is the use of mathematical models for forward design. As such, systems and synthetic biology can be viewed as being two sides of the same coin. While systems biology attempts to unravel how the set of instructions encoded by an organism's DNA orchestrates its phenotypical complexity, synthetic biology aims to create cells with desirable behaviors through the integration of additional instructions. This can be achieved by first investigating which network architectures support the desired outcome and then augmenting the genotype accordingly. The construction of synthetic gene regulatory systems can thus help understand natural systems by complementing approaches in which quantitative analysis is used to elucidate "design principles" underlying the functioning of natural intracellular networks. Moreover, synthetic systems provide excellent examples of the direct link between theoretical modeling and biological reality.

13.1   Introduction

During the last few decades, the ability to isolate, sequence, and manipulate DNA has led to tremendous advances in genetic engineering with numerous benefits to science, agriculture, and medicine. Typically, genetic engineering is used to endow a genetically modified organism with a novel trait, such as resistance to certain pesticides or the ability to efficiently synthesize pharmacological molecules, for example by transferring a gene from another organism. Gene therapy is another example. There, a trait lost due to a nonfunctional endogenous gene is typically recovered by inserting a normal copy of the gene at a non-specific location within the genome. Synthetic biology can be viewed as a natural extension of such single-gene approaches in the sense that entire systems are inserted into the genome of the host cell.

So far, efforts in synthetic biology have included the construction of novel gene regulatory networks, signal transduction pathways, metabolic pathways, synthetic multicellular systems, engineered sensory proteins, and the regulation of proteins that control intrinsic cell functions. An excellent introductory review of the many different aspects of synthetic biology is given by Benner and Sismour (2005). Here, we focus on the mathematical design and experimental implementation of selected synthetic gene regulatory networks that embody important architectural properties. Prerequisites for designing and implementing synthetic gene regulatory networks include understanding how transcriptional regulation works, how transcription factor proteins regulate the expression of each other within networks, and knowledge of recombinant DNA technologies. An excellent introduction to the latter is given by Nicholl (1994).

General aspects of transcriptional regulation and how transcription is modeled are discussed in sections 13.2 and 13.3, respectively. The remaining sections highlight how synthetic gene regulatory systems have been designed and implemented in the bacterium *Escherichia coli* based on network models constructed from phenomenological mathematical descriptions of transcriptional regulation. In sections 13.4 and 13.5, we discuss linear transcriptional networks and feedforward networks, respectively. In section 13.6, we provide examples of networks that support bistability and oscillations by incorporating feedback control. These systems demonstrate how some of the principles investigated in chapter 6 have been used to create living cells with complex dynamical properties.

13.2 Transcriptional Regulatory Modules

In order to engineer gene regulatory systems, it is necessary to appreciate some of the basic elements of gene regulation. Natural genetic circuits are typically described as circuits of interconnected modules consisting of interacting proteins, DNA, RNA, and small molecules that regulate the transcription of genes into mRNA, the translation of mRNA into polypeptides, and the biological activity of the expressed proteins. While the abundance of an expressed protein can be controlled by many different mechanisms, the regulation of gene transcription is one of the most common. In prokaryotes, this type of control is often mediated through transcription factor proteins that alter the ability of the RNA polymerase to bind to and initiate transcription from promoter regions located upstream of the regulated genes.

Prokaryotic transcriptional regulatory modules often consists of four elements: a promoter region, the gene (or genes) expressed from that promoter, the transcription factor proteins that regulate the expression level, and additional regulatory

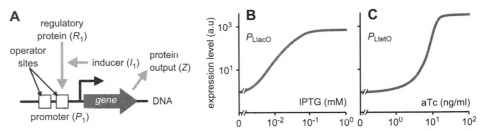

Figure 13.1 **A**. Architecture of a prototypical transcriptional regulatory module. **B−C**. Population-averaged signal-response curves measured in *E. coli* cells expressing a reporter gene from LacI/P_{LlacO} and TetR/P_{LtetO} modules with fixed concentrations of LacI or TetR and varying concentrations of the inducers isopropyl-β-D-thiogalactoside (IPTG) and anhydrotetracycline (aTc), respectively. Based on experimental data from (Lutz and Bujard, 1997).

molecules that modulate the activity of the transcription factors. A schematic layout of such modules is shown in figure 13.1A. The expression from the regulated promoter can be measured in single cells by expressing a reporter gene, such as the *gfp*, *yfp*, or *cfp* genes encoding green (GFP), yellow (YFP) and cyan (CFP) fluorescent protein, respectively (see chapter 10).

Transcriptional regulatory proteins increase (activators) or decrease (repressors) the probability that a gene is transcribed into mRNA by binding to stretches of DNA within or near promoter regions referred to as operators or *cis*-regulatory elements (figure 13.1A). While activators may facilitate the binding of RNA polymerase to the promoter, repressors often exert their function by competing with the RNA polymerase for promoter access. Transcription from a promoter containing appropriate *cis*-regulatory elements can thus be controlled by up- or down-regulating the cellular abundance of the corresponding transcription factor proteins. In some cases, external control over such *in vivo* signals is provided by small molecules called inducers. These molecules typically function by modulating the activity of a transcription factor protein. Specifically, when the inducer binds to the protein, it causes an alteration in its three-dimensional structure that increases or decreases the affinity between the protein and its cognate *cis*-regulatory elements. Varying the inducer concentration thus provides a means of regulating transcription without altering cellular protein abundances directly.

Figures 13.1B and 13.1C illustrate how expression of a reporter protein from two engineered transcriptional regulatory modules, LacI/P_{LlacO} and TetR/P_{LtetO}, is modulated by the inducers isopropyl-β-D-thiogalactoside (IPTG) and anhydrotetracycline (aTc), respectively. The P_{LlacO} and P_{LtetO} promoters are obtained by inserting lacO and tetO operator sequences, corresponding to the binding sites of LacI and TetR, respectively, into the P_{L} promoter normally repressed by the protein CI. In both cases, the signal-response curve, in other words, the relationship between the regulatory input signal (the inducer concentration) and the output signal (the abundance of the reporter protein), is highly nonlinear and sigmoidal. The endogenous *E. coli* promoters P_{lac} and P_{tet}, which are repressed by LacI and

Table 13.1 Transcription regulatory modules used frequently to construct synthetic gene regulatory networks in *E. coli*. TetR and LacI are repressors that are inactivated by their inducers. LuxR is an AHL-dependent activator. CI is generally a repressor, but activates transcription from the P_{RM} promoter.

Regulatory protein	Regulated promoters	Inducer
TetR	P_{tet}, P_{LtetO}	tetracycline
LacI	P_{lac}, P_{LlacO}, P_{trc}	lactose, IPTG
CI	P_L, P_R, P_{RM}, P_{luxOR}	
LuxR	P_{lux}, P_{luxOR}	acyl-homoserine lactones (AHL)

TetR, respectively, respond to induction in a similar fashion. Additional transcriptional regulatory modules used frequently in synthetic biology are summarized in table 13.1.

13.3   Modeling Transcriptional Modules

In the remaining sections of this chapter, we discuss how simulation and analysis of mathematical models have been employed to forward engineer *E. coli* cells with novel characteristics and sophisticated computational capabilities by interconnecting the modules in table 13.1 into larger networks. We use a convenient abstraction to model these biochemical networks with ordinary differential equations (ODEs) that include basal expression of a protein, protein decay, and Hill function descriptions of gene regulation (see chapter 6).

In general form, the ODE that models the output Z of a genetic module given the regulatory input S is given by:

$$\frac{d[Z]}{dt} = k' + \frac{k \cdot (S^n/K^n)^\mu}{1 + (S^n/K^n)} - d \cdot [Z] \tag{13.1}$$

where the parameter μ is used to distinguish between the cases of repression ($\mu = 0$) and activation ($\mu = 1$) of transcription by S (see for example, Kuznetsov et al. (2005) for details). The constants K and n are the Hill constant and Hill coefficient, respectively. The Hill constant gives the value of the input signal that yields 50% response, and the Hill coefficient gives the slope of the signal-response curve at this input signal. The parameter d is the rate constant associated with the decay of the output reporter protein. Additionally, the parameters k and k' are the rate constants associated with signal-independent (basal) and signal-dependent gene expression. The values of k and k' are typically correlated, and this interdependence is often modeled by setting $k' = a \cdot k$ with $0 \le a < 1$. With this relationship, the steady state solution of equation 13.1 is given by:

$$[Z]_{ss} = \frac{k}{d} \left(a + \frac{(S^n/K^n)^\mu}{1 + (S^n/K^n)} \right) \tag{13.2}$$

Hence, in steady state, the cellular abundance of the reporter protein reflects the relationship between the regulatory signal and the transcription rate modeled by the Hill function.

It is noted that equation 13.1 models transcription and translation as a single step. Because the separation of transcription and translation introduces response delays, it can be important in models of temporal dynamics to include mRNA as an independent variable. In this case, a single-input transcriptional regulatory module is described by the ODEs:

$$\frac{d[M]}{dt} = a \cdot k_{tr} + \frac{k_{tr} \cdot (S^n/K^n)^\mu}{1 + (S^n/K^n)} - d_M \cdot [Z]$$

$$\frac{d[Z]}{dt} = k_{tl}[M] - d \cdot [Z] \tag{13.3}$$

where $[M]$ is the mRNA concentration, k_{tr} is the rate constant associated with transcription, k_{tl} is the rate constant associated with translation and d_M is the mRNA decay constant. Equation 13.1 is obtained from equation 13.3 by invoking a steady state assumption for $[M]$ and defining the constant k by $k = k_{tr}k_{tl}/d_M$. Hence, modeling transcription and translation as a single step does not change the steady state solution in equation 13.2.

Equation 13.1, or equation 13.3 when mRNA is included, are used to model both the effect of changing the intracellular concentration of a regulatory protein and the extracellular concentration of an inducer. For example, the effect of varying the concentration of a repressor R is modeled by setting the input signal equal to the repressor concentration, $S = [R]$, with $\mu = 0$. When the concentration of the repressor is constant, the effect of varying the concentration of its inducer I is modeled by setting $S = [I]$ and $\mu = 1$. For example, the steady state signal-response curves in Figs. 13.1B and 13.1C for induction of the LacI/P_{LlacO} and the TetR/P_{LtetO} modules can be modeled using equation 13.2 with $\mu = 1$ and the concentrations of IPTG and aTc defining the signal S, respectively. Other input-output functions are also possible depending on the regulatory role of the protein and how the inducer affects the activity of this protein. In cases where both repressor and inducer concentrations vary, the signal S is the concentration of active repressor molecules. This signal is modeled by setting $S = [R_T]/(K_I^{n_I} + [I]^{n_I})$ with $[R_T]$ being the total repressor concentration, and K_I and n_I the Hill constant and coefficient associated with the repressor-inducer interaction, respectively.

For the purpose of network modeling, we will use the following notations: Each transcriptional regulator protein is given an index $i = 1, 2, \ldots N$. The concentration of the transcription factor protein is given by $[R_i]$, and its inducer, if present, by $[I_i]$. The gene and mRNA that encode the regulatory protein R_i are denoted $r(i)$ and M_i, respectively. The rate constant associated with the decay of protein R_i is given by d_i. Promoters are identified as follows: P denotes a constitutively active promoter, P_i a promoter regulated by the protein R_i, and P_{ij} a promoter regulated by the proteins R_i and R_j. The parameters characterizing the transcription from each promoter are identified by the same index as the promoter for the parameters

k and a (or k'), and by the protein index for the Hill constant K and the Hill coefficient n.

13.4   Linear Networks

Linear transcriptional regulatory networks consist of modules placed in series with the output of one module acting as the input to the next module. In the simplest case, a linear network is composed of two modules and one regulatory step. The LacI/P_{LlacO} and TetR/P_{LtetO} modules discussed in section 13.2 are examples of such one-step transcriptional cascades because the transcriptional regulator (R_1 in figure 13.1) is expressed at high constant levels from a constitutively active promoter. For clarity, the constitutive promoter (that is, the first transcription module) is omitted from the diagram in figure 13.1. The construction and analysis of longer transcriptional cascades, which will be discussed next, is useful for determining how information flows through transcriptional networks and can help better understand the rules of module composition. For example, cascades comprised of two and three regulatory steps have been engineered with the purpose of investigating time delays, ultrasensitivity in signal-response relationships and stochacticity in transcriptional regulation (see, for example, Blake et al. (2003); Hooshangi et al. (2005); Rosenfeld et al. (2005); Pedraza and van Oudenaarden (2005)).

13.4.1   Two-Step Cascades

Figure 13.2 depicts the schematics of a two-step linear repressor cascade obtained by adding a third transcriptional module to the one-step cascade. The first module comprises the promoter P with no regulatory inputs; the second module the repressor R_1, its inducer I_1, and the promoter P_1. The third module comprises the R_2 repressor and the P_2 promoter. This configuration provides a mechanism to measure the behavior of the R_2/P_2 module. The constitutive promoter P drives expression of the R_1 repressor, which in turn, inhibits the expression of R_2. The inducer I_1 can thus be used to determine the input signal to the R_2/P_2 regulatory module by modulating the cellular abundance of repressor R_2.

Using equation 13.1 with $\mu = 1$ and $S = [I_1]$ to model the R_1/I_1-dependent expression of R_2 and with $\mu = 0$ and $S = [R_2]$ to model the inhibition by R_2 of expression from P_2, the ODEs describing the two-step linear network are given by:

$$\frac{d[R_2]}{dt} = a_1 \cdot k_1 + \frac{k_1 \cdot ([I_1]/K_1)^{n_1}}{1 + ([I_1]/K_1)^{n_1}} - d_2 \cdot [R_2]$$

$$\frac{d[Z]}{dt} = a_2 \cdot k_2 + \frac{k_2}{1 + ([R_2]/K_2)^{n_2}} - d \cdot [Z] \qquad (13.4)$$

where the meaning of the parameters were defined in section 13.3. Notice that it is not necessary to include an equation for R_1 because its steady state level is constant. As discussed in section 13.3, the combined regulatory activity of R_1 and I_1 can be

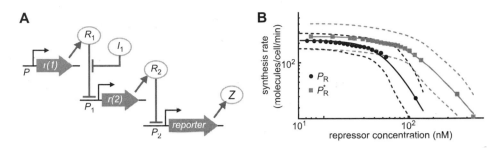

Figure 13.2 **A**. Architecture of a two-step repressor cascade. **B**. Population-averaged rates of reporter protein synthesis from the P_R promoter (black points) and the P_R^* promoter (grey points) measured at the single-cell level using a two-step repressor cascade. The broken curves give the standard deviation associated with the measured synthesis rate, and full curves the fit to a Hill function with $\mu = 0$. The fitted parameter values are: $n = 2.4 \pm 0.3$, $K = 55 \pm 10$ nM, $k = 220 \pm 15$ min^{-1} for the P_R promoter, and $n = 1.7 \pm 0.3$, $K = 120 \pm 25$ nM, $k = 255 \pm 40$ min^{-1} for the P_R^* promoter (Rosenfeld et al., 2005).

captured phenomenologically in one Hill function to model the relationship between the inducer concentration and the expression from the regulated promoter.

The steady state solution of equation 13.4 is given by:

$$
\begin{aligned}
[R_2]_{ss} &= \frac{k_1}{d_2}\left(a_1 + \frac{([I_1]/K_1)^{n_1}}{1 + ([I_1]/K_1)^{n_1}}\right) \\
[Z]_{ss} &= \frac{k_2}{d}\left(a_2 + \frac{1}{1 + ([R_2]_{ss}/K_2)^{n_2}}\right)
\end{aligned}
\tag{13.5}
$$

In terms of the overall response of this network, the steady state solution predicts that the presence of inducer (high input) results in repression of P_2 (high R_2, low output) and the absence of inducer (low input) allows transcription from the P_2 promoter (low R_2, high output).

13.4.2   Characterizing Module Input-Output Functions

The network illustrated in figure 13.2 and described by the model in equation 13.4 can be implemented using different repressor/promoter pairs. Elowitz and colleagues (Rosenfeld et al., 2005) implemented a version using the aTc-inducible TetR/P_{tet} module to characterize a CI/P_R repressor module driving CFP ($R_1 =$ TetR, $I_1 =$ aTc, and $R_2 =$ CI, $Z =$ CFP). In this implementation, the *cI* gene is fused with the *yfp* gene to synthesize a yellow-fluorescent variant of the CI protein. This dual-color labeling allows for simultaneous measurements of the input and output signals in single cells. Additionally, using time-lapse microscopy to determine the rate of change in fluorescence, the dependency of the rate of protein synthesis on the repressor concentration can be determined at the level of single cells. The system thus enables a direct investigation of the suitability of the Hill function in equation 13.1 as a model of the most fundamental signal-response relationship in gene regulatory systems.

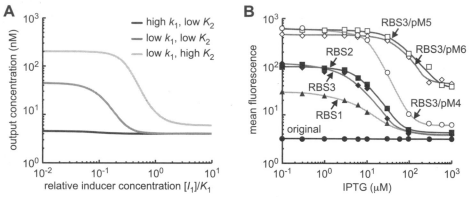

Figure 13.3 **A**. Predicted effects on the signal-response curve of the two-step repressor cascade of decreasing the rate of repressor R_2 synthesis (from $k_1 = 1000$ nM/min to $k_1 = 100$ nM/min) and increasing the Hill constant (from $K_2 = 10$ nM to $K_2 = 100$ nM). Other parameter values are $a_1 = a_2 = 0.02$, $d_1 = d_2 = 0.1$ nM/min, $k_2 = 20$ nM/min, and $n = 2$. **B**. Experimentally measured signal-response in two-step repressor cascades containing altered ribosome binding sites (RBS) of repressor-encoding mRNA to change k_1, or mutations in the regulated promoter (pM) to change K_2. Based on experimental data from (Weiss and Basu, 2002).

Figure 13.3B illustrates the experimentally observed relationship between the concentration of the CI-YFP protein and the population-averaged rate of CFP synthesis for the P_R promoter and a variant of this promoter, designated P_R^*, where one of the CI binding-sites is mutated. Also included are the standard deviations associated with the average protein synthesis rates and the signal-response curves obtained by fitting the data to Hill functions.

13.4.3   Matching Kinetic Characteristics

While the two-step cascade composed of the TetR/P_{tet} and CI/P_R modules exhibits a useful inverse sigmoidal signal-response relationship, it is often the case that coupling transcriptional regulatory modules does not yield the desired behavior. Another version of the same network uses the LacI/P_{lac} pair as the inducible module to control the input to the CI/P_R module. However, when initially assembled, no fluorescence was observed from cells harboring the network regardless of whether the inducer, in this case IPTG, is absent or present. Apparently, even with maximum repression of the P_{lac} promoter, CI is synthesized at a sufficiently high level to fully repress transcription from the P_R promoter. Unfortunately, our models are presently not sufficiently accurate to predict such mismatch problems partially because accurate *in vivo* parameter values are difficult to obtain. Hence, it is often necessary to first construct a network, and then use modeling tools to guide the correction and fine-tuning of its behavior.

In order to overcome impedance mismatch problems, one can mutate genetic elements until the desirable network response is obtained. Starting with a nonfunctional or non-optimal network, such mutations can be introduced to affect biological parameters identified by model analysis as most likely to yield the desired behavior. For example, Feng et al. (2004) showed how to use global sensitivity analysis to determine the best genetic targets for mutations that could make the two-step $LacI/P_{lac}$, CI/P_R cascade functional. The steady state model in equation 13.5 predicts, as shown in figure 13.3A, that decreasing the value of the maximal repressor synthesis rate k_1 or increasing the Hill constant K_2 should confer a non-responsive network with the desired network properties. The Hill constant can be modified by mutating one of the CI-binding sites within the P_R promoter to lower the CI-binding affinity, and the maximal CI synthesis rate k_1 can be changed by mutating the ribosome-binding site (RBS) on the CI-encoding mRNA.

That the model correctly predicts the genetic mutations required to obtain a functional network is shown in figure 13.3B. The experimental results are obtained with three different cI-RBS sequences yielding lower translation efficiencies than the original RBS (Weiss and Basu, 2002). The plots show that the systems with the weakened RBS are able to respond to induction with IPTG, in agreement with the model predictions. Also shown are the effects of introducing mutations into the CI-binding site within the P_R promoter. These mutations are combined with the weakest RBS in order to optimize the response.

13.4.4   Interfacing Transcriptional Modules

Once the kinetic characteristics of the individual transcriptional regulatory modules are appropriately matched, they can be coupled together into larger networks. This can be accomplished by combining modules at random (Guet et al., 2002) or rationally to achieve a specific network property. Perhaps the simplest extension of the two-step cascade is to add an additional repressor module to form a linear three-step network (figure 13.4A). The experimental investigation of this cascade highlights interesting properties that are important for the understanding of the more complex systems discussed in the sections to follow.

An implementation of the three-step linear repressor cascade uses the $TetR/P_{LtetO}$ module as the inducible input component and the $LacI/P_{lac}$ and CI/P_R modules as the first and second repressor module, respectively (Hooshangi et al., 2005). Figure 13.4B shows the experimentally measured population-averaged steady state network outputs at varying concentration of the aTc inducer when a fluorescent reporter is expressed from the P_{LtetO} (P_1), the P_{lac} (P_2), and the P_R (P_3) promoter, respectively. These population-averaged protein abundance curves have the correlations expected for the network. While the expression from the P_1 and P_3 promoters show a positive correlation with the input aTc concentration (that is, high-pass detection), expression from P_2 shows a negative correlation (that is, low-pass detection). When fitted to a Hill function, the Hill coefficients for the steady state response in the cascades of length one, two, and three are 2.3, 7.0, and 7.5, respectively. In other words, increased length of transcriptional regulatory cascades improves the sensitivity to the input signal by enabling more pro-

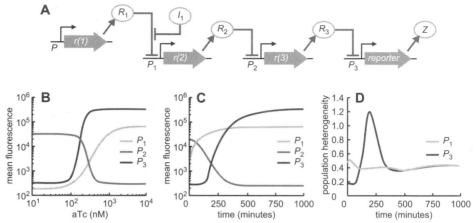

Figure 13.4 **A.** Architecture of the three-step repressor cascade. **B.** Population-averaged steady state expression levels obtained by expressing a fluorescence reporter gene at different steps in the cascade ($P_1 = P_{\text{LtetO}}$, $P_2 = P_{\text{lac}}$, $P_3 = P_{\text{R}}$) when the concentration of the inducer of the first transcriptional module (aTc) is varied. **C.** Time course of population-averaged expression levels at the different steps in the cascade following induction. **D.** Relative population heterogeneity (standard deviation over the mean) at steps one and three in the cascade following induction. Based on experimental data from (Hooshangi et al., 2005).

nounced all-or-nothing steady state responses. This phenomenon can also be found in naturally occuring regulatory motifs such as signal transduction phosphorylation cascades (Ferrell Jr., 1996).

It is also interesting to compare the time course of expression induction at the different steps in the cascade following aTc induction. This is done in figure 13.4C. While protein synthesis from the first promoter begins immediately after addition of aTc, there is a significant time lag in the repression and activation of the second and third promoter, respectively. The abundance of the protein expressed from P_{LtetO} (P_1) reaches the 50% of maximal abundance after $\sim$15 minutes, and it takes about 200 and 300 minutes for the proteins expressed from the P_{lac} (P_2) and the P_{R} (P_3) promoters to pass the 50% mark. While a model based on equation 13.1 predicts such delays, the experiments give an idea of the relative time scale involved in transcriptional regulation and the response-delay introduced as the regulatory signal propagates through the network. Specifically, the cell division time for *E. coli* is typically $\sim$45–120 minutes depending on the strain and the growth conditions, meaning it may take several generations for a full transcriptional response to be realized.

Another important observation that can be deduced from the time series experiment is that the regulatory signal propagates through the cascade at very different rates in individual cells. Figure 13.4D compares the relative variability in fluorescence among cells, measured as the standard deviation over the mean, and changes following induction in the cascades of length one and three. While the cell-to-cell variability changes little as time progresses for the one-step cascade, indicative of a fairly homogenous response, it changes significantly for the three-step cascade and reaches a peak value after about 200 minutes. At this time, which roughly corresponds to the point where expression from the P_R promoter is initiated, the cell population is highly heterogeneous. Hence, the increased steady state sensitivity in the longer cascade comes at the cost of a response that initially is highly asynchronous. An implication of this in terms of regulatory robustness is discussed further in section 13.6 in the context of feedback networks.

13.5   Feedforward Networks

Genetic feedforward networks are circuits in which transcriptional regulatory modules are configured with a common input that propagates through parallel cascades, and ultimately converge to regulate a shared downstream promoter. Several endogenous feedforward motifs have been documented (Lee et al., 2002) and three-gene networks with this architecture appear more frequently in cellular regulation than expected based on randomized networks (Shen-Orr et al., 2002). Modeling predicts that the three-gene feedforward networks support a variety of properties ranging from transcriptional response delay and filtering to the generation of transient pulses of gene expression (Maugan and Alon, 2003). Here, we limit our discussion to feedforward networks engineered in *E. coli* by interconnecting transcriptional regulatory modules in table 13.1. The first feedforward network (section 13.5.1) is composed of three genes and is designed to generate a transient pulse in response to a persistent inducing signal. The second network is composed of five genes and enables cells to respond to an inducing signal when the inducer concentration is within a specific range (section 13.5.2).

13.5.1   Pulse-Generating Network

When the downstream promoter in a feedforward network receives both an activating and a repressing signal, a transcriptional pulse can be generated if the repressing signal is delayed compared to the activating signal. Such a delay is realized if the repressing signal has to propagate through a higher number of transcriptional modules than the activating signal (see figure 13.4B). Hence, the feedforward network depicted in figure 13.5A should be able to generate a gene expression pulse. In this network, an inducing input signal (S_1) activates the transcription of a reporter gene from a multi-input promoter (P_{12}) and as well as the expression from the P_1 promoter of a repressor (R_2) of the P_{12} promoter.

Ignoring basal expression and modeling the expression from the P_{12} promoter as a product of an activating and a repressing Hill function, the feedforward network can be described by the following ODEs:

$$\frac{d[R_2]}{dt} = \frac{k_1 \cdot s^{n_1}}{1 + s^{n_1}} - d_2 \cdot [R_2]$$

$$\frac{d[Z]}{dt} = \frac{k_{12}}{1 + ([R_2]/K_2)^{n_2}} \cdot \frac{s^{n_1}}{1 + s^{n_1}} - d \cdot [Z] \tag{13.6}$$

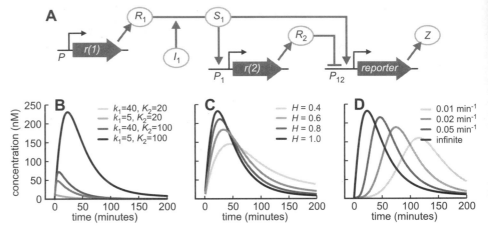

Figure 13.5   **A**. Architecture of the pulse-generating network. **B–D**. Simulations of the network model demonstrating the effect of changing parameters (panel **B**) at full induction ($H = 1$), the level of induction (panel **C**) and the rate of inducer accumulation (panel **D**). Unless otherwise indicated, parameter values are (in nM/min): $k_1 = 5$, $k_{12} = 20$, $d_2 = 0.01$, $d = 0.04$, (in nM): $K_1 = 1$, $K_2 = 100$, and $n_1 = n_2 = 3$. Inducer accumulation is modeled by setting $s(t) = k_s \cdot t$ with k_s being the rate of accumulation.

where $s = S_1/K_1$ is the inducing signal of P_1 (that is, dependent on the inducer concentration $[I_1]$). As before, it is not necessary to include the concentration of the R_1 protein because its concentration can be assumed constant.

Without resorting to computer simulations, let us see if we can generate intuition about the network dynamics directly from the ODEs. To do this, we define the induction level H as $H = s^{n_1}/(1 + s^{n_1})$ and find the steady states of the system. They are given by:

$$[R_2]_{ss} = \frac{k_1}{d_2} \cdot H$$

$$[Z]_{ss} = \frac{k_{12}}{d} \cdot \frac{K_2^{n_2} \cdot H}{K_2^{n_2} + [R_2]_{ss}^{n_2}} \tag{13.7}$$

Let us consider the case where the induction level is constant and the lifetime of the repressor is so long that its decay can be assumed negligible. In this case, the accumulation of repressor following induction at $t = 0$ is given by $[R_2](t) = k_1 \cdot H \cdot t$. This reduces equation (13.7) to a time-dependent ODE for the output concentration that is given by:

$$\frac{d[Z]}{dt} = \frac{k_{12}}{1 + (k_1 \cdot H \cdot t/K_2)^{n_2}} - d \cdot [Z] \tag{13.8}$$

This equation captures the initial high transcription rate from P_{12}, which leads to an overshoot of the steady state in equation (13.7), and the decrease in this rate as repressor accumulates with time. It also indicates that the duration of the pulse (and hence its magnitude) is linked to the maximal rate of repressor synthesis k_1, the Hill constant K_2 and the concentration of the induction level H. Specifically, a 50%

decrease in transcription occurs after a time period given by $t_{0.5} = K_2/(k_1 \cdot H)$. Hence, increasing K_2 or decreasing k_1 is predicted to cause a longer pulse and higher amplitude. This prediction is validated by the model simulations presented in figure 13.5B.

Suboptimal induction, that is, values of H less than one, is also predicted to increase the duration of the pulse. However, because the rate of output protein synthesis depends on the induction level, suboptimal induction is expected to decrease the amplitude. Additionally, because the length and the amplitude of the pulse depend on how fast the repressor accumulates, they should also depend on the rate at which the inducing signal accumulates. These predictions are validated by the model simulations presented in figure 13.5C and figure 13.5D, respectively.

The pulse-generating network is implemented experimentally by expressing the CI repressor from the AHL-activated P_{lux} promoter and GFP from the multi-input promoter designated P_{luxOR}. This promoter is obtained by inserting a CI binding site into the P_{lux} promoter to achieve repression of AHL-activated transcription by CI (Basu et al., 2004). The experimental observations reflect well the results of the above analysis. Figure 13.6A shows population-averaged temporal responses of four *E. coli* strains harboring networks constructed with different rates of CI synthesis (*cI*-RBS mutations) and different binding affinities of CI to the P_{luxOR} promoter (operator mutations) following induction with saturating AHL concentrations. It is seen that the effects of the mutations are in agreement with the model predictions. Due to a high repressor synthesis rate (high k_1) and strong repressor binding to the P_{luxOR} promoter (K_2), a pulse is not generated in the original network. Mutations that decrease the repressor synthesis rate or the operator binding strength yield a pulse with intermediate duration and amplitude. The best network performance is obtained when these mutations are combined.

In a second set of experiments, the temporal response was measured after induction with different AHL concentrations using the network with the best performance. As shown in figure 13.6B, at AHL concentrations below 47 nM, the pulse amplitude is decreased and its duration shortened. At an AHL concentration of 4.7 nM, the pulse can hardly be observed. In other words, the system responds differently at nonsaturating AHL concentrations as predicted by the model analysis. A third set of experiments measured the network response to different rates of AHL accumulation. The results are shown in figure 13.6C. As the rate of AHL accumulation is decreased, the onset of the pulse is delayed, and its amplitude decreased. This is also in agreement with the model prediction.

The experimental results in figure 13.6A−C are population-averaged responses obtained in a well-mixed environment. This leaves open the question of how cells harboring the feedforward network respond in an environment where signal diffusion plays an important role. Figure 13.6D illustrates the results of an experiment designed to determine the spatio-temporal response at the level of single cells. In these experiments, cells harboring the pulse-generating feedforward network are placed adjacent to *E. coli* cells that synthesize and emit AHL. The AHL-emitting "sender" cells harbor an aTc-inducible promoter controlling the expression of the

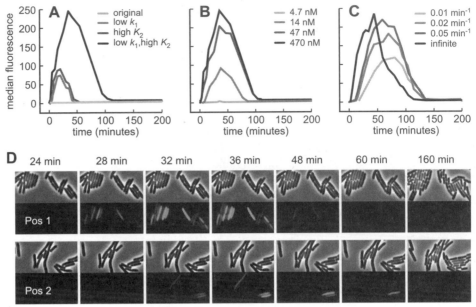

Figure 13.6 **A**–**C**. Experimental validations of the model predictions in a pulse-generating network activated by AHL. The effect of (**A**) introducing mutations to change the k_1 and K_2 parameters, (**B**) different inducer concentrations, and (**C**) changing the rate of inducer accumulation. Based on data from (Basu et al., 2004). **D**. Responses of pulse-generating cells placed at different distances from nearby sender cells to AHL synthesized by the senders. Notice that the response of cells in position 2, which is farther away from the senders, is delayed, and the maximum pulse amplitude is diminished.

enzyme (LuxI) that synthesizes AHL from common metabolites. As a result, the sender cells produce AHL when treated with aTc. The inducer subsequently diffuses into the environment and establishes an AHL concentration gradient. Figure 13.6D shows the phase-contrast and fluorescence microscopy images of "receiver" cells harboring the feedforward network taken at different time points and different distances from a colony of AHL-emitting senders. While there is distinct variability in the response from one cell to another, it is seen that single cells respond to increased AHL by generating a pulse of fluorescence. Moreover, AHL-induction elicits a response in receiver cells that depends on the distance from the senders. Because of AHL diffusion, the rate of AHL accumulation is slower farther from the AHL-emitting source. This allows receiver cells to differentiate between signals originating from nearby and distant senders.

13.5.2   Concentration Band Detection

The pulse-generating system discussed in the previous section is an example of the complex responses that can be generated from transcriptional networks combining one-step and two-step linear cascades in a feedforward architecture. In this section, we investigate a network in which a two-step and a three-step cascade are

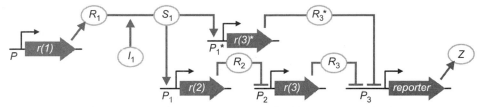

Figure 13.7 Architecture of the band detection feedforward network obtained by combining a high-pass two-step linear repressor cascade and a low-pass three-step linear repressor cascade. This combination enables the synthesis of the reporter protein only when the inducer I_1 is within a specific range.

activated by the same input and regulate the same output. The resultant five-gene feedforward system activates the expression of the output gene within a finite concentration range of the inducing signal, that is, concentration band detection, and supports the formation of spatial patterns in response to a gradient in the inducing signal. These experiments demonstrate a mechanism referred to as the "French flag model" in developmental biology (Wolpert, 2002) where cells read and respond to spatial information encoded in a "morphogen" gradient by having sharp induction thresholds.

Figure 13.7 illustrates the schematics of the five-gene feedforward network. Along the three-step branch, the inducing signal S_1, which is generated by a combination of the regulator R_1 and its inducer I_1, activates the expression of the R_2, which, in turn, inhibits the expression of the R_3 repressor. In the final step, the repressor R_3 inhibits the transcription of the reporter protein. Along the two-step branch, the inducing signal activates the expression of the repressor designated R_3^*, which is functionally equivalent to the R_3 repressor (that is, it also inhibits the expression of the reporter protein).

How will the system respond to different levels of the inducing signal? To answer this question, we look at the steady state concentration of the output reporter protein. Since R_3 and R_3^* are assumed to be functionally identical, it is given by:

$$[Z]_{ss} = \frac{k_3}{d} \cdot \frac{K_3^{n_3}}{K_3^{n_3} + ([R_3]_{ss} + [R_3^*]_{ss})^{n_3}} \tag{13.9}$$

The steady state concentrations of the three repressors are given by:

$$
\begin{aligned}
[R_2]_{ss} &= \frac{k_1}{d_2} \frac{s^{n_1}}{1 + s^{n_1}} \\
[R_3]_{ss} &= \frac{k_2}{d_3} \frac{K_2^{n_2}}{K_2^{n_2} + [R_2]_{ss}^{n_2}} \\
[R_3^*]_{ss} &= \frac{k_1^*}{d_3} \frac{s^{n_1}}{1 + s^{n_1}}
\end{aligned}
\tag{13.10}
$$

where k_1^* is the rate of R_3^* expression from the P_1^* promoter, and the inducing signal S_1 is expressed relative to the value that yields 50% response ($s = S_1/K_1$).

Along the two-step branch, the steady state concentration of R_3^* rises as the inducing signal increases and the expression of the output protein is inhibited when the inducing signal is high. This branch acts as a low-pass detector (see figure 13.4). The steady state concentration of R_3 in the three-step branch shows the opposite correlation and acts as a high-pass detector. For appropriately matched parameter values, this may leave a gap in the concentration of the repressors of transcription from P_3 at intermediate values of the inducing signal. It is in this gap that the output protein is synthesized.

The boundaries in the concentration of the inducing signal between which the output is expressed can be obtained from the steady states in equation 13.10 as the values S_{low} and S_{high} where R_3^* and R_3 have 50% of their maximal concentrations, respectively. They are given by:

$$S_{\text{low}} = K_1 \quad S_{\text{high}} = K_1 \sqrt[n_2]{\frac{d_2 K_2}{k_1 - d_2 K_2}} \tag{13.11}$$

Therefore, a gap in the total concentration of repressors may occur if $k_1 > 2 K_2 d_2$. If a sufficient gap exists, the range of band detection can be shifted by modifying the value of K_1.

The five-gene feedforward network is implemented experimentally (Basu et al., 2004) using the AHL-activated LuxR/P_{lux} module to regulate the expression of the CI protein (R_2), which, in turn, regulates the expression of the LacI protein (R_3) from the P_{R} (P_2) promoter. The system output protein is expressed from the P_{lac} (P_3) promoter. These transcriptional regulatory modules comprise the three-step branch of the network. Along the two-step branch, AHL activates transcription of a variant of the $lacI$ gene, designated $lacI^{\text{M1}}$, that differs in its DNA sequence from that of the $lacI$ gene, but encodes a protein with the same amino-acid sequence. The protein product encoded by $lacI^{\text{M1}}$ (R_3^*) is thus functionally identical to LacI.

The band detection network is implemented in different versions: one using the wild-type LuxR protein, designated BD2, the other, designated BD1, with a mutant variant of LuxR that is hypersensitive to AHL. In the latter, less AHL is required to achieve the same expression level from the P_{lux} promoter, corresponding to a decreased value of the Hill constant K_1. Accordingly, the range of AHL concentrations detected by the two versions should be different, with the mutated LuxR network expressing the output protein of the system at a lower inducer concentration. Figure 13.8A and figure 13.8B show that this differential response is also observed experimentally. In figure 13.8A, the measured steady state input-response of the two-step branch is shown at varying AHL concentrations. It is seen that the AHL concentration yielding 50% response is decreased more than 10-fold when the hypersensitive LuxR variant is employed. In figure 13.8B, the observed steady state input-response of the five-gene band network is shown at varying AHL concentrations. As predicted by the model, BD1 cells activate the expression of the system output at a lower range of AHL concentration than BD2 cells.

The different ranges of AHL detected by strains harboring the different band detection networks enable multicellular pattern formation. Figure 13.8C shows the

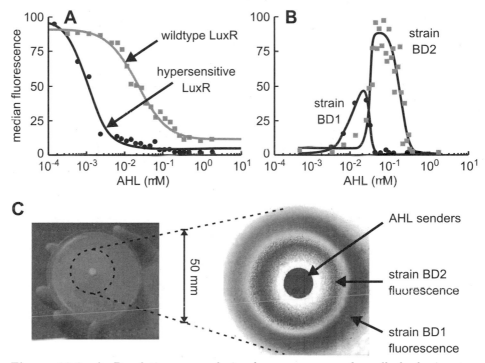

Figure 13.8 **A**. Population-averaged signal-response curves for cells harboring two versions of the two-step branches with a wildtype and a hypersensitive LuxR mutant, respectively. **B**. Signal-response curves for the five-gene feedforward network with the wildtype LuxR (strain BD2) and the mutant LuxR (strain BD1). **C**. Formation of a target pattern within a bacterial lawn containing a mixture of the BD1 and BD2 strains in the presence of AHL-emitting cells in the center of the lawn. Data from (Basu et al., 2004).

formation of a target pattern in an experiment where an AHL-emitting cell strain is grown at the center of a lawn containing a mixture of BD1 and BD2 cells. The BD2 cells turn on the expression of a fluorescent reporter gene at a short distance from the AHL-emitting cells, but remain quiescent farther away. On the other hand, the BD1 cells are quiescent near the center of the lawn and express a differently colored fluorescent reporter only at a distance from the AHL-emitting cells.

13.6   Feedback Networks

The experiments involving feedforward networks demonstrate how complex dynamics can be generated by combining linear signaling cascades. In natural regulatory systems, such behaviors are frequently generated in networks incorporating feedback loops as an additional control feature. As discussed in chapter 6, feedback control enables complex dynamics, such as bistability, hysteresis, and oscillations.

There are numerous examples of such behaviors in natural genetic circuits. For example, LacI is a key component of a natural genetic feedback network that exhibits bistability (Ozbudak et al., 2004). There are also many examples of gene regulatory feedback networks supporting dampened or sustained oscillations. They include, for example, the circadian clocks discussed in chapter 2 and chapter 12, and the Mdm2-p53 network discussed in chapter 6. A motivation for the implementation of synthetic gene regulatory feedback systems is to complement the analysis of mathematical models of natural circuits with investigations of how feedback networks behave *in vivo*. In this section, we discuss several genetic feedback networks implemented in *E. coli* to create cells capable of complex temporal dynamics and the mathematical models used to design or to understand the network properties.

13.6.1   Bistable Networks

Bistability and hysteresis are trademark features of networks that contain positive feedback or autocatalysis. Here, we investigate two single-gene positive feedback networks giving rise to hysteresis (Atkinson et al., 2003) and bimodal population distributions (Isaacs et al., 2003), respectively, and a two-gene system designed to operate as a bistable genetic toggle switch (Gardner et al., 2000; Kobayashi et al., 2004). In the single-gene positive feedback system depicted in figure 13.9A, a transcription activator R_1 binds to its own promoter and increases the rate of its own synthesis. This network can be described by the ODE:

$$\frac{d[R_1]}{dt} = a_1 \cdot k_1 + \gamma \frac{k_1 [R_1]^{n_1}}{K^{n_1} + [R_1]^{n_1}} - d_1 \cdot [R_1] \tag{13.12}$$

where the parameter γ is a measure of the feedback control strength. Because at this point we are interested in using the model to reveal general trends, it is useful for the analysis to introduce new dependent variables to reduce the number of unknown parameters. For equation 13.12, a useful normalization is to use the dimensionless concentration r_1 defined by $r_1 = [R_1]/K_1$ and dimensionless time τ defined by $\tau = d_1 \cdot t$. This corresponds to expressing the protein concentration relative to that yielding 50% response and time relative to the protein lifetime, regardless of the actual value of these parameters. Using the chain rule, the normalized form of equation 13.12 is obtained as

$$\frac{dr_1}{d\tau} = a \cdot \kappa_1 + \gamma \frac{\kappa_1 \cdot r_1^{n_1}}{1 + r_1^{n_1}} - r_1 \tag{13.13}$$

where κ_1 is defined by $\kappa_1 = k_1/K_1/d_1$. Similarly normalized equations will be used in the remaining sections of this chapter.

   Figure 13.9B shows a bifurcation diagram obtained by plotting the steady state solutions of equation (13.13) as a function of the feedback control strength. The steady state curve has the "S"-shape characteristic of bistable systems. At low feedback strength, there is little or no activation, and expression occurs essentially at basal levels. At high feedback strength, the promoter is more or less fully activated

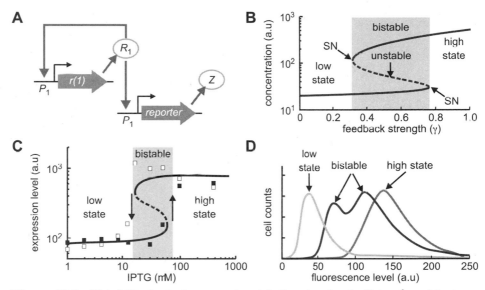

Figure 13.9 Bistability in single-gene autocatalytic networks. **A**. Network architecture. **B**. Bifurcation diagrams for the network model. Full and broken curves indicate stable and unstable steady states, respectively. The saddle-node (SN) bifurcations are located where the stable and unstable steady states collide. Parameter values are: $a = 0.1$, $\kappa = 5$, and $n = 3$. **C**. Bistability and hysteresis observed in a NtrC positive feedback network when the strength of NtrC-activated transcription is varied with IPTG. Closed and open squares correspond to cells initially in the low and high states, respectively. Based on data from (Atkinson et al., 2003). **D**. Transitions between uni- and bimodal population distributions observed in a CI/P_{RM} feedback network corresponding to different strengths of the feedback loop. Based on data from (Isaacs et al., 2003).

and expression takes place at a rate close to maximal. These two states co-exist at intermediate values of the feedback control strength parameter with the region of bistability demarcated by two saddle-node bifurcations located at a value of γ slightly above 0.3 and just just shy of 0.8.

Several synthetic single-gene autocatalytic gene networks have been constructed with the purpose of generating bistability (Becskei et al., 2001; Isaacs et al., 2003; Atkinson et al., 2003). One system (Atkinson et al., 2003) is constructed such that the transcription factor NtrC activates its own expression from a modified NtrC-responsive P_{glnA} promoter and that of a reporter gene from the promoter of the $glnK$ gene (P_{glnK}). The modified P_{glnA} promoter is engineered such that the ability of NtrC to activate transcription is attenuated by LacI. This is achieved by inserting LacI binding sites such that the repressor competes with the activator for promoter access. This allows for an indirect means of modulating the feedback control strength using IPTG. In cells that express LacI at high levels, increasing the IPTG concentration enables more efficient activation by NtrC of transcription from the modified P_{glnA} promoter.

Figure 13.9C shows the experimentally observed effect of varying the feedback control strength on the population-averaged expression of cells harboring the IPTG-sensitive, NtrC positive feedback system. They are in excellent agreement with the model predictions. When grown in the absence of IPTG (that is, low feedback strength), cells express the reporter protein at low levels because of efficient repression by LacI. When these cells are exposed to increased inducer concentrations (closed squares in figure 13.9C), the measured signal-response curve is fairly flat below a critical IPTG concentration where the reporter expression changes sharply from low to high levels. On the other hand, when cells initially grown with high IPTG concentrations and fully activated (open squares) are exposed to decreased concentrations of IPTG, expression levels remain high until a critical concentration where a sharp transition to low expression is observed. At identical intermediate IPTG concentrations, corresponding to intermediate strength of the feedback control, cell populations adopt a high or a low expression state depending on the initial conditions. Hence, the network endows cells with the ability to support bistability and hysteresis.

Another single-gene positive feedback system engineered to display bistability (Isaacs et al., 2003) employs the CI-activated P_{RM} promoter to control the expression of a mutated *cI* gene (designated *cI857*) encoding a temperature-sensitive variant of the CI protein. The CI protein also activates the transcription of a GFP-encoding gene allowing the measurement of gene expression at the level of single cells. The temperature-dependent activity of the *cI857*-encoded CI protein enables modulation of the feedback strength through temperature variation. The activity of the CI variant decreases with increased temperature. Hence, a low temperature corresponds to a high feedback strength and high temperature corresponds to a low value of this parameter. The model thus predicts a low expression state at high temperature, a high expression state at low temperature, and bistability at intermediate temperatures. Figure 13.9D illustrates the population-distribution of fluorescence from cells harboring the CI/P_{RM} feedback network at three different temperatures. For low and high temperatures, the population distributions contain a single peak and cells are in a high state when they are grown at low temperature and in a low expression state when they are grown at high temperature, respectively. At the intermediate temperature, the population-distribution is bimodal, indicating that cells transition frequently between the low and the high expression states due to noise-induced transitions. A detailed model of the circuit where the deterministic equations are augmented with stochastic terms accounts well for the observed distributions (Isaacs et al., 2003).

The toggle switch network, which is illustrated in figure 13.10A, is an example of a two-gene system designed and implemented to allow *E. coli* cells to be switched between two distinct expression states in response to external stimuli. This system, which represents a multi-component motif (Lee et al., 2002) with indirect positive feedback, is composed of two genes encoding transcription factor proteins, R_1 and R_2, that inhibit each other's expression. Because of this mutual repression, the network can be either in a state with high R_1 expression and repressed R_2 transcription, or in a state with high R_2 expression and repressed R_1 transcription.

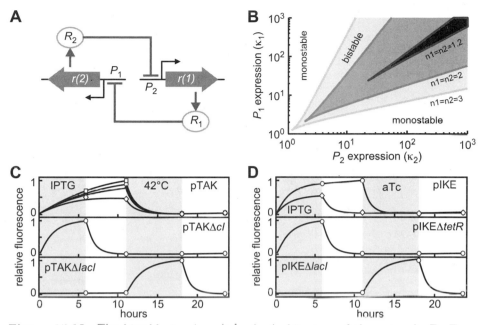

Figure 13.10 The bistable toggle switch. **A**. Architecture of the network. **B**. Two-dimensional bifurcation diagrams indicating the boundaries between bistable and monostable regions in the κ_1, κ_2 parameter space at different values of the Hill coefficients. **C−D**. Transitions between high and low expression states in the pTAK and the pIKE toggle switch networks, respectively. The networks, pTAKΔcI, pTAK$\Delta lacI$, pIKE$\Delta tetR$, and pTAK$\Delta lacI$, are controls in which one of the repressor genes is eliminated. Based on data from (Gardner et al., 2000).

The experimental implementation of the toggle switch network is guided by the analysis of the dimensionless ODEs (Gardner et al., 2000):

$$\frac{dr_1}{d\tau} = \frac{\kappa_2}{1 + r_2^{n_2}} - r_1, \quad \frac{dr_2}{d\tau} = \frac{\kappa_1}{1 + r_1^{n_1}} - r_2 \qquad (13.14)$$

where the repressor concentrations are expressed relative to the appropriate Hill constant and time relative to the protein lifetime (which is assumed to be the same for the two repressors). Conditions that make bistability more likely are high maximal expression levels (that is, high values of κ_1 and κ_2) and high Hill coefficients. This can be seen from the bifurcation diagrams in figure 13.10B, which show the location of saddle-node bifurcations, that is, the boundaries between mono- and bistability, in the κ_1, κ_2 parameter plane for different values of the Hill coefficients. Increased values of the Hill coefficients enlarge the bistable region in the κ_1, κ_2 parameter space, and an increased value of one of the maximal expression rates allows for bistability in a wider range of values of the other.

The toggle switch network is implemented experimentally using different transcriptional regulatory modules. Gardner et al. (2000) constructed two versions; one employs the LacI/P_{trc} and the TetR/P_{LtetO} modules and is designated as pIKE.

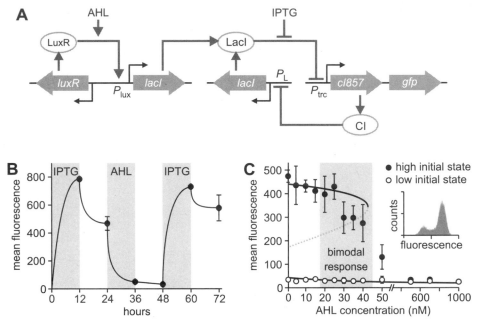

Figure 13.11 Extending the toggle switch network. **A**. The pTAK toggle switch is augmented with the AHL-activated LuxR/P_{lux} to convert the network into a AHL sensor. **B**. Flipping between stable expression state by application of AHL to activate LacI expression or IPTG to activate CI expression. **C**. Hysteresis is observed when cells initially in the low (open circles) or high (closed circles) expression states are exposed to AHL at varying concentrations. Based on data from (Kobayashi et al., 2004).

The other employs the LacI/P_{trc} and the temperature-sensitive CI/P_L modules and is designated as pTAK. The expression state is monitored by co-expressing GFP with the *cI857* gene (pTAK) or the *tetR* gene (pIKE). Two variants of the pIKE network with strong and weak *lacI*-RBS sequences demonstrate that the maximal rate of expression, in agreement with model predictions, is an important parameter for the emergence of bistability. Figure 13.10B shows the effect of treating cells in the high LacI state with IPTG to activate TetR expression. Both variants respond by expressing the reporter protein. When IPTG is removed, cells harboring the network with the weaker *lacI*-RBS maintain the high expression state while those harboring the variant with the stronger *lacI*-RBS revert to the low expression state. The cells that remain in the high TetR state require addition of aTc to reactivate LacI expression. Hence, only the network with the weak *lacI*-RBS supports bistability. Four variants of the pTAK network, also with different RBS sequences, all exhibit bistability. This is shown in figure 13.10C. Addition of IPTG to inhibit LacI induces a transition to a high expression state (low LacI/high CI), and the latter is maintained when IPTG inducer is removed. A subsequently applied transient temperature increase (to deactivate CI) induces a transition back to the low expression state (high LacI/low CI).

In an extension to the toggle switch (Kobayashi et al., 2004), mathematical modeling is used to guide the experimental implementation of "programmable" cells in which the pTAK toggle responds to signals from other gene regulatory networks. In one of the implementations, the LuxR/P_{lux} module is used to drive additional synthesis of LacI. The resultant five-gene network is depicted in figure 13.11A. In this system, the toggle switch can be flipped back and forth by adding AHL to increase the LacI synthesis rate and IPTG to increase the CI synthesis rate. This is illustrated in figure 13.11B, which shows population-averaged expression levels following induction with AHL and IPTG. Cells initially switch to the high expression state (high CI) following IPTG treatment and remain in this state when IPTG is removed. A transition to the low expression state (high LacI) occurs when these cells subsequently are treated with AHL, and the low state is maintained when AHL is removed. These cells are still responsive, and a second treatment with IPTG induces a transition to the high expression state.

The system in figure 13.11A also supports hysteresis. Figure 13.11D shows the result of an experiment where cells that were initially prepared in the high CI state (high fluorescence) or the high LacI state (low fluorescence), respectively, are exposed to AHL at varying concentrations. The high LacI state is, as expected, unaffected by AHL, and cells remain in the low fluorescence state regardless of the AHL concentration. However, the high CI state is sensitive to AHL. At inducer concentrations less than 20 nM, cells remain in the high expression state. On the other hand, at inducer concentrations higher than 40 nM, all the cells have switched to the low expression state. At intermediate inducer concentrations, the cell population contains a mixture of cells in high and low expression states. This bimodal response presumably arises from a combination of differences in induction threshold and noise-induced transitions, which are more likely to occur when the system is closer to the saddle-node bifurcation.

13.6.2   Oscillatory Networks

The discussion in the previous section provides examples of two complex properties, bistability and hysteresis, supported by genetic networks incorporating feedback regulation. Other complex behaviors that arise in feedback control systems are dampened and sustained oscillations. An example of a synthetic gene regulatory system capable of generating oscillations is obtained by adding a negative feedback to a three-step linear repressor cascade. The resultant system, which is referred to as the Repressilator (Elowitz and Leibler, 2000), is illustrated schematically in figure 13.12A. In the network, repressor R_1 inhibits the expression of repressor R_2, repressor R_2 inhibits the expression of repressor R_3, and repressor R_3 inhibits the expression of repressor R_1.

The implementation of the Repressilator network is based on the analysis of a model describing the dynamics of repressor mRNA and protein concentrations. The concentration of mRNA is included explicitly because the separation of transcription and translation contributes to a response delay that is important for the emergence of oscillations. Hence, equation 13.3 is used as the basis of the network model rather than equation 13.1, which, as described in section 13.3, assumes that the mRNA concentration is always in a steady state. To ease the analysis, it is assumed that each transcriptional module is characterized by the same set of parameters and that the rate constant associated with translation is equal to that associated with repressor decay. With these assumptions, the network is described by the following dimensionless equations (Elowitz and Leibler, 2000):

$$\frac{dm_i}{d\tau} = a\kappa + \frac{\kappa}{1 + r_j^n} - m_i, \quad \frac{dr_i}{d\tau} = \varepsilon(m_i - r_i) \tag{13.15}$$

where m_i and r_i represents the concentration of repressor R_i mRNA and protein, respectively, and r_j the concentration of repressor R_j regulating the expression of repressor R_i. The parameter ε is proportional to the ratio of the mRNA and the protein lifetimes.

What makes sustained oscillations possible in the Repressilator network? The answer to this question can be obtained by considering the dynamics of the linear three-step repressor network in section 13.4.4. Recall that in the linear network, the transcription of the R_1 repressor leads to increased expression from the P_3 promoter after a time delay. In the Repressilator, the R_1 repressor is expressed from the P_3 promoter. Hence, the network can be viewed as a negative feedback system with time delay (transcription of R_1 from the P_3 promoter eventually causes down-regulation of its own expression). The time delay in repression allows for the accumulation of protein product beyond the steady state level and, when the repression kicks in, for the subsequent decay in protein concentration. The mechanism causing oscillations in the Repressilator network is thus somewhat analogous to that leading to circadian clock oscillations as discussed in chapter 2 and chapter 12.

The conditions that make oscillations more likely to occur are high Hill coefficients, low basal expression levels, and short protein lifetimes. This is illustrated by the bifurcation diagrams in figure 13.12B, which show the location of the Hopf-bifurcations in the κ, ε parameter space that separate regions of oscillatory and steady state dynamics for different sets of parameter values. When the Hill coefficients are low and basal expression high, oscillations occur at intermediate values of κ when ε is greater than a critical value. In this case, the protein lifetime needs to be comparable to that of mRNA in order for oscillations to occur. Decreasing the basal expressions rate and increasing the Hill coefficients relax this requirement and allow oscillations to occur for a broader range of κ and ε values.

The Repressilator network is implemented experimentally by interconnecting the LacI/P_{LlacO}, the CI/P_{R}, and the TetR/P_{LtetO} modules such that LacI (R_1) is expressed from P_{R} (P_3), TetR (R_2) is expressed from P_{LlacO} (P_1), and CI (R_3) is expressed from P_{LtetO} (P_2). To decrease the lifetime of the repressor proteins, the repressor genes are "tagged" with a DNA sequence that targets the expressed protein for degradation. Figure 13.12C illustrates the temporal oscillations in fluorescence of a single cell measured when a fluorescent protein is expressed from a second

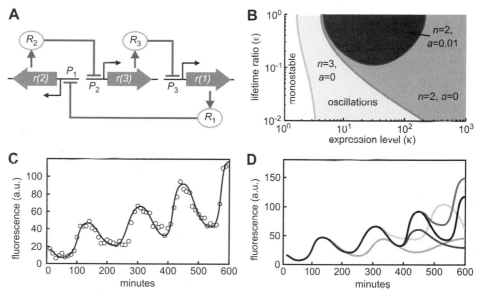

Figure 13.12 The Repressilator. **A.** Architecture of the three-gene network. **B.** Bifurcation diagram of the Repressilator model showing the regions of monostability and oscillations in the κ, ϵ parameter plane for different values of a and n. **C.** Time series of fluorescence emitted by a single cell harboring the Repressilator network. **D.** Comparison of the single-cell time series in **C** with those generated by its siblings. Based on experimental data from (Elowitz and Leibler, 2000).

P_{LtetO} promoter. At least 40% of cells oscillate with a period of 160 ± 40 minutes. This period is significantly longer than the average cell division time. Hence, an oscillation initiated in a mother cell is completed in a daughter cell, and the oscillation phase is passed down from one generation to the next. However, as shown in figure 13.12D, siblings display marked differences in their progression through the oscillation cycle, and the network fails to support coherent oscillations at the population level.

The Repressilator network fails to support coherent oscillations partly because of the significant differences in the rate at which the regulatory signal propagates through the three regulatory steps. Recall from section 13.4.4 that cells harboring the three-step linear cascade show marked variability in the onset of expression from the P_3 promoter. This is expected to translate directly into significant differences in the oscillation period as the differences in response times in the linear cascade are equivalent to differences in delay times in the feedback network. Hence, the number of steps in the network makes it especially susceptible to stochastic effects.

There are network architectures that generate oscillations with increased robustness against stochastic effects (Vilar et al., 2002). One example is a design, illustrated in figure 13.13A, where a transcriptional activator R_1 enhances its own expression from the multi-input promoter P_{12} and that of a repressor R_2 from the P_1 promoter. The R_2 repressor in turn attenuates the transcription of the activator

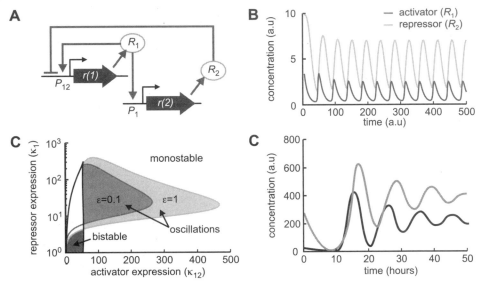

Figure 13.13   Mixed positive and negative feedback oscillator. **A**. Architecture of the network. **B**. Example of oscillations generated in simulations of the network model with parameter values $\kappa_1 = 10$, $\kappa_{12} = 100$, $n = 3$, $a = 10^{-3}$, and $\varepsilon = 0.1$. **C**. Bifurcation diagram showing the regions of monostability, bistability and oscillations in the κ_{12}, κ_1 plane for different values of ε for $a = 10^{-3}$ and $n = 3$. **D**. Coherent dampened oscillations observed in cell populations carrying two variants of the network differing approximately four-fold in the level of expression from the P_{12} promoter. Based on experimental data from (Atkinson et al., 2003).

by binding to the P_{12} promoter. Using the same assumptions as in the model of the Repressilator, the network can be modeled by the following dimensionless ODEs:

$$
\frac{dm_1}{dt} = a \cdot \kappa_{12} + \frac{\kappa_{12}}{1 + r_2^n} \cdot \frac{r_1^n}{1 + r_1^n} - m_1 \tag{13.16}
$$

$$
\frac{dm_2}{dt} = a \cdot \kappa_1 + \frac{\kappa_1 \cdot r_1^n}{1 + r_1^n} - m_2
$$

$$
\frac{dr_i}{d\tau} = \varepsilon(m_i - r_i) \quad i = 1, 2
$$

A simulation of the system for parameters yielding sustained oscillations is given in figure 13.13B. They arise because of a time delay between the activation and the repression of transcription from the P_{12} promoter. When the repressor concentration is initially low, the positive feedback causes an increase in both activator and repressor expression. However, because the negative feedback involves an additional step, the expression of the repressor is delayed. As a result, and akin to the pulse-generating network discussed in section 13.5.1, the activator accumulates to high levels before the repressor can reach a concentration that is sufficiently high to shut down the expression from the P_{12} promoter. Once this occurs, activator expression ceases and the activator concentration declines. This in turn causes the rate of repressor expression to decrease. The system subsequently returns to the low repressor state and a new cycle is initiated. Figure 13.13C shows the bifurcation diagram indicating the regions of monostability, bistability, and oscillations in the κ_1, κ_{12} parameter space. The bistability is associated with the positive feedback and is more likely to occur in the absence of the negative feedback (that is, for $\kappa_1 = 0$).

The network illustrated in figure 13.13A is implemented experimentally by augmenting the bistable NtrC single-gene positive feedback network discussed in section 13.6.1 with a negative feedback (Atkinson et al., 2003). Recall that in this network the strength of the positive NtrC feedback is dependent on LacI as the modified NtrC-activated P_{glnA} promoter (P_{12} in figure 13.13A) contains lacO operators. Hence, a negative feedback is readily added to the network by expressing LacI from a second NtrC-activated promoter (P_1 in figure 13.13A). Figure 13.13D shows two time series of the population-averaged expression level in cells harboring two different variants of the oscillator network that have an approximately four-fold difference in the expression rate from the modified P_{glnA} promoter. In both cases, the population-averaged expression level exhibit dampened oscillations with a period of 10–12 hours. The dampening of the oscillations is not due to cells becoming desynchronized. Measurements of the expression of a fluorescence reporter protein in individual cells (data not shown, see (Atkinson et al., 2003)) indicate that the dampening also occurs at the level of single cells. Given that the cells divide about once per hour, the coherence of the oscillations, which appears to be maintained for the duration of the experiments, that is, 40–50 generations, is quite remarkable. It confirms the prediction that a network architecture combining positive and negative feedback should be more robust against noise.

13.7 Conclusions

In this chapter, we have discussed selected synthetic gene regulatory systems implemented experimentally in *E. coli* to investigate the dynamics of transcriptional regulatory networks *in vivo* and to create strains with novel characteristics. These systems support a range of non-trivial behaviors such as cellular memory, pulse generation, spatial pattern formation, and oscillatory gene expression. In all the examples, the networks are designed with the aid of mathematical models based on fairly simple, phenomenological descriptions of relationships between input and output signals. These models are used to predict systems properties and how changes in DNA sequence affect performance and dynamics. In all cases, an excellent agreement between model predictions and experimental results is obtained. This demonstrates the close link between the current modeling methodologies and biological reality. Consequently, there is good reason to believe these methodologies are also useful for the analysis of the more complex regulatory systems found in nature. Indeed, the systems presented and analyzed elsewhere in this book strongly indicate that this is the case.

14 Multilevel Modeling in Systems Biology: From Cells to Whole Organs

Denis Noble

Successful physiological systems analysis requires that we understand the functional interactions between the key components of cells, organs, and physiological systems, and how these interactions change in disease states. This information resides neither in the genome nor even in the individual proteins that genes code for since no genes code for interactions as such. It lies at the level of protein network interactions within the context of sub-cellular, cellular, tissue, organ, and system structures. There is therefore no alternative to copying nature and computing these interactions to determine the logic of healthy and diseased states. The rapid growth in biological databases; models of cells, tissues, and organs; and the development of powerful computing hardware and algorithms have made it possible to explore functionality in a quantitative manner all the way from the level of genes to whole organs and systems. This chapter discusses the philosophy of multilevel modeling and illustrates this development in the case of the heart. Systems physiology of the 21st century is set to become highly quantitative, and therefore one of the most computationally-intensive disciplines.

14.1 Introduction: The Philosophy of Multilevel Simulation

The emphasis in recent decades of biological research has been on breaking cells, organs, and systems down into their smallest components: the genes, proteins, and other molecules whose interactions are essential to life. We have succeeded so well that the amount of molecular data generated by the new technologies has completely overwhelmed our ability to understand it. Genomics has provided us with a massive "parts catalog" for the human body, the 25,000 or so genes, while proteomics seeks to define these individual "parts" and the structures they form in detail. The parts catalog still needs a lot of annotating (gene ontology), and the proteomics side is still in its infancy, being much more challenging than sequencing genomes (see chapter 10). But, from the viewpoint of those interested in understanding cells, tissues, organs, and systems, there is as yet no "user's guide" describing how these parts are put together to allow those interactions that sustain life or cause disease. The project to model at multiple levels between cells and organs, which is the Human Physiome Project (Crampin et al., 2004; Hunter et al., 2002), can be seen to aim precisely to achieve this.

We have a long way to go because, in many cases, the cellular, organ, and system functions of genes and proteins are unknown, though clues sometimes come from homology in the gene sequences and other patterns being investigated by bioinformatics. Moreover, even when we understand function at the protein level, successful intervention, for example in drug therapy, depends on knowing how a protein behaves in context, as it interacts with the rest of the relevant cellular machinery to generate function at a higher level. Without this integrative knowledge, we may not even know in which disease states a receptor, enzyme, or transporter is relevant, and we will certainly encounter side effects that are unpredictable from molecular information alone. This is a major problem for the drug industry. My field of cardiac simulation is central to this problem since nearly half the compounds developed by the industry interact with the heart, sometimes with fatal effect (Muzikant and Penl, 2000).

Inspecting genome databases alone will not get us very far in addressing these problems. The reason is simple. Genes code for protein sequences. They do not explicitly code for the interactions between proteins and other cell molecules and organelles that generate function. As the geneticist Gabriel Dover (2000) remarks "We don't have a theory of interactions and until we do we cannot have a theory of development or a theory of evolution." The challenge of developing a theory of interactions, which must be one of the major goals of systems biology, therefore also has implications for biology as a whole. We need to lead the way towards biology maturing as a science to join the physical sciences as a fully quantitative science, with fully-fledged theories within which computational biology can be embedded. Otherwise, what we do will be piecemeal, not integrated together.

A major part of the difficulty is that much of the logic of the interactions in living systems is implicit. Wherever possible, nature leaves the interactions to the chemical properties of the molecules themselves and to the highly serendipitous way in which these properties have been exploited during evolution as nature has plundered its treasure chest of old genes to recruit new functions. It is as though the function of the genetic code is to build the components of a computer, which then self-assembles to run programs about which the genetic code knows nothing. The genetic code alone is not a program (Coen, 1999). Sydney Brenner (1998) expressed this very effectively when he wrote: "Genes can only specify the properties of the proteins they code for, and any integrative properties of the system must be 'computed' by their interactions." Brenner meant not only that biological systems themselves "compute" these interactions but also that in order to understand them *we* need to compute them, and he concluded "this provides a framework for analysis by simulation."

Brenner also coined the term that is being used to describe multilevel modeling, when he referred to it as "middle-out" (Novartis Foundation, 2001). An exhaustive "bottom-up" reconstruction is impossible ("I know one approach that will fail, which is to start with genes, make proteins from them and to try to build things bottom-up" (Novartis Foundation, 2005)). The approach that can work is to start modeling at any of the levels at which the data is sufficient to generate a model and then to reach out to lower and higher levels. In this way we can avoid the problems of information overload and combinatorial explosion (Feytmans et al., 2005).

In this chapter I will show how far we have advanced in using simulation to understand these interactions between the levels of genes, proteins, cells, and organs. I will refer mostly to the case of the heart since this is the organ in which such simulation is currently most advanced.

14.2   Cellular Models of the Heart

Many of the characteristic functions of the heart reside in the properties of the cells. Cells generate electrical signals that initiate a cascade of events leading to muscular contraction. Some of them also generate repetitive activity and so act as pacemakers. They also contain receptors that respond to neural and hormonal control to speed up or slow down the rhythm and to increase or decrease the force of contraction. Finally some, but not all, arrhythmic mechanisms can be found at the cellular level. Not surprisingly, therefore, modeling work in heart systems physiology has nearly always started at the cellular level.

The first cardiac cell models (Noble, 1960, 1962) sought insight into the most obvious difference between electrical activity in heart and nerve: the duration of the action potential. A nerve action potential may last only 1 msec. Its function is to encode information as rapidly as possible. A human ventricular action potential may last 400 msec, during which time many events are triggered that initiate and control mechanical contraction.

Weidmann's (1951) pioneering work showed that the conductance during the action potential is very low. The experimental reason for this became clear with the discovery of the inward-rectifier potassium channel current, I_{K1} (Carmeliet, 1961; Hall et al., 1963; Hutter and Noble, 1960) (see figure 14.1, top). The permeability of this channel falls almost to zero during strong depolarization. These experiments were also the first to show that there are at least two K^+ channels in the heart, I_{K1} and I_K (referred to as I_{K2} in early work, but now known to consist of I_{Kr} and I_{Ks} (Noble and Tsien, 1969; Sanguinetti and Jurkiewicz, 1990)). The 1962 model (Noble, 1960, 1962), (figure 14.1, bottom) was constructed to determine whether this combination of K^+ channels, together with a Hodgkin-Huxley type sodium channel (a channel protein showing voltage-dependent activation and inactivation processes) could explain all the classical Weidmann experiments on conductance changes. The model not only succeeded in doing this, it also demonstrated that an energy-conserving plateau mechanism was an automatic consequence of the inward-rectifying properties of I_{K1}. This has featured in all subsequent models, and it is a very important insight. The main advantage of a low conductance is minimizing energy expenditure.

Unfortunately, however, nature achieved a low conductance plateau at the cost of making the recovery (repolarization) process fragile. Pharmaceutical companies today are struggling to deal with evolution's answer to this problem, which was to entrust repolarization to the potassium channel i_{Kr}. This channel protein, hERG (Novartis Foundation, 2005), is one of the most promiscuous receptors known: large ranges of drugs can enter the channel mouth to block it, and even more interact with the G-protein coupled receptors that control it. The consequence can be failed repolarization, and the triggering of potentially fatal disorders of cardiac rhythm, called arrhythmias (see http://georgetowncert.org/qtdrugs_torsades.asp). Computer simulation is now playing a role in attempting to find a way around this difficult and seemingly intractable problem (Bottino et al., 2005; Fink et al., 2005; Muzikant and Penl, 2000).

The main defect of the 1962 model was that it included only one voltage-gated inward channel current, I_{Na}. There was a good reason for this. Calcium channels had not then been discovered. There was, nevertheless, a clue in the model that something important was missing. The only way in which the model could be made to work was to *greatly* extend the voltage range of the sodium current by reducing the voltage dependence of the sodium activation process. In effect, the sodium current was made to serve the function of both the sodium and calcium channels so far as the plateau is concerned. There was a clear prediction here: either sodium channels in the heart are quantitatively different from those in nerves, or other inward current-carrying channels must exist. Both predictions are correct.

The first successful measurements of ion channel activity under controlled membrane potential conditions (using the technique known as the voltage clamp) came in 1964 (Deck and Trautwein, 1964) and they rapidly led to the discovery of the cardiac calcium current (Reuter, 1967). By the end of the 1960s, therefore, it was already clear that the 1962 model needed replacing. However, the insights it gave on the behavior of the potassium currents are still valid. Systems biology can proceed in a stepwise fashion, in which different parts of an integrative analysis get clarified at different stages of the iteration between simulation and experiment.

In addition to the discovery of the calcium current, the early voltage clamp experiments also revealed multiple components of I_K (Noble and Tsien, 1969), now referred to as I_{Kr} and I_{Ks} (Sanguinetti and Jurkiewicz, 1990). We also showed that these slow gated currents in the plateau range of potentials were quite distinct from those near the resting potential, that is, that there were two separate voltage ranges in which very slow conductance changes could be observed (Noble and Tsien, 1969). These experiments formed the basis of the McAllister, Noble, and Tsien (MNT) model (McAllister et al., 1975).

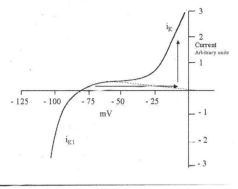

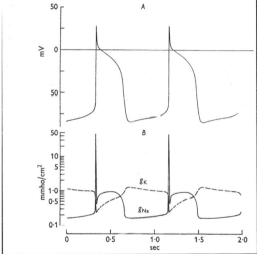

Figure 14.1   *Top*: experimental basis of the first analysis of the integrative role of potassium channels. Redrawn (Noble, 2002b) from (Hall et al., 1963). The solid line shows the total membrane current recorded in cardiac cells (from the conducting system of the heart called Purkinje fibres) in a sodium-depleted solution. The inward-rectifying current was identified as i_{K1}, which is extrapolated here as nearly zero at positive potentials. The outward-rectifying current, I_K, is now known to be mostly formed by the component I_{Kr}. The horizontal arrow indicates the trajectory at the beginning of the action potential, while the vertical arrow indicates the time-dependent activation of I_K, which initiates repolarization. *Bottom*: Sodium and potassium conductance changes computed from the 1962 model of the Purkinje fibre. Two cycles of activity are shown. The conductances are plotted on a logarithmic scale to accommodate the large changes in sodium conductance. Note the persistent level of sodium conductance during the plateau of the action potential, which is about 2% of the peak conductance. Note also the rapid fall in potassium conductance at the beginning of the action potential. This is attributable to the properties of the inward rectifier channel i_{K1}.

The MNT model reconstructed a much wider range of experimental results, and it did so with great accuracy in some cases. A good example of this was the reconstruction of the paradoxical effect of small current pulses on the pacemaker depolarization in Purkinje fibres—paradoxical because brief depolarizations (positive voltage deflections) *slow* the process and brief hyperpolarizations (negative voltage deflections) greatly *accelerate* it. This is paradoxical since the pacemaker potential itself is a positive deflection so that one would expect positive deflections to accelerate it. Reconstructing paradoxical or counterintuitive results is, of course, a major function of modeling work. This is one of the roles of modeling in unraveling complexity in biological systems.

But the MNT model also contained the seeds of a spectacular failure. Following the experimental evidence (Noble and Tsien, 1968) it attributed the slow conductance changes near the resting potential to a slow gated potassium current, I_{K2}. In fact, what became the "pacemaker current," or I_f, is an *inward* current activated by hyperpolarization (DiFrancesco, 1981), not an *outward* current activated by depolarisation. At the time it seemed hard to imagine a more serious failure than getting *both* the current direction *and* the gating by voltage completely wrong. There cannot be much doubt therefore that this stage in the iterative interaction between experiment and simulation created a major problem of credibility. Perhaps cardiac electrophysiology was not really ready for modeling cellular systems to be successful?

This is the point at which to emphasize one of the important points about the philosophy of simulation: it is one of the functions of models to be wrong. Of course, there are many ways of being wrong, and I am not talking here of failing in arbitrary or purely contingent ways, but in ways that advance our understanding by exploring the possible logics of complex systems and determining which are most accurate. Again, this situation is familiar to those working in simulation studies in engineering or cosmology or in many other physical sciences. And, in fact, the failure of the MNT model is one of the most instructive examples of experiment-simulation interaction in physiology, and of subsequent successful model development (see Noble (1984)).

The MNT model was also the point of departure for the ground-breaking work of Beeler and Reuter (1977) who developed the first ventricular cell model. (Ventricular cells are the real workhorse of the heart; they form the mass of muscle that does all the pumping into the arterial blood system.) As they wrote of their model: "in a sense, it forms a companion presentation to the recent publication of McAllister et al. (1975) on a numerical reconstruction of the cardiac Purkinje fibre action potential. There are sufficiently many and important differences between these two types of cardiac tissue, both functionally and experimentally, that a more or less complete picture of membrane ionic currents in the myocardium must include both simulations." For a recent assessment of this model and the subsequent Luo-Rudy models see Noble and Rudy (2001).

14.3   Connecting to Ion Pumps and Calcium Cycling

New ground in modeling cardiac cells was broken with the DiFrancesco-Noble model (DiFrancesco and Noble, 1985). The incorporation not only of ion channels (following the Hodgkin-Huxley paradigm of voltage-dependent gated channel proteins) but also of ion exchangers, such as Na-K exchange (the sodium pump), Na-Ca exchange, the SR calcium pump, and, more recently, the transporters involved in controlling cellular pH (Ch'en et al., 1998), was a fundamental advance since these are essential to the study of some disease states such as weak contraction (congestive heart failure) (Winslow et al., 1999) and impaired blood supply (ischaemic heart disease).

The greatly increased complexity of the DiFrancesco-Noble model, which for the first time also represented intracellular events by incorporating a model of calcium release from the sarcoplasmic reticulum, increased both the range of predictions and the opportunities for failure. Here I will limit myself to one example of each.

The most influential prediction was that relating to the sodium-calcium exchanger. In the early 1980s it was still widely thought that the electrically neutral stoichiometry (Na:Ca = 2:1) derived from early flux measurements was correct. The DiFrancesco-Noble model achieved two important conclusions. The first was that, with the experimentally known Na^+ gradient, there simply was not enough energy in a neutral exchanger to keep resting intracellular calcium levels below 1 μM, that is, at a level low enough to permit relaxation to occur. Switching to a stoichiometry of 3:1 readily allowed resting calcium to be maintained below 100 nM. This automatically led to the prediction that there must be a current carried by the Na-Ca exchanger and that, if this exchanger was activated by intracellular calcium, it must also be strongly time-dependent since intracellular calcium varies by an order of magnitude during each action potential. Even as the model was being published, experiments demonstrating the current I_{NaCa} were being performed (Kimura et al., 1986), and the variation of this current during activity was being revealed either as a late component of inward current or as a current tail on repolarization (Egan et al., 1989).

This prediction has turned out to have very important consequences for the elucidation of some of the mechanisms of cardiac arrhythmia in disease states in which cells accumulate sodium and calcium, either through loss of energy supply, as in ischaemia, or as a consequence of reduced activity of the Na-K ATPase (Na pump) as in treatment with cardiac glycosides. At a critical level of sodium and calcium accumulation, calcium release occurs spontaneously and becomes repetitive between sodium concentrations around 13 and 22 mM (Ch'en et al., 1998; Varghese and Winslow, 1994; Winslow et al., 1999)—see figure 14.2—a phenomenon also seen experimentally. Each calcium release activates inward current carried by sodium-calcium exchange which, if large enough, can trigger additional (ectopic) action potentials (Noble, 2002a) as shown in figure 14.6.

The main defect of the DiFrancesco-Noble model was that the intracellular calcium transient was far too large, mainly because the model did not represent

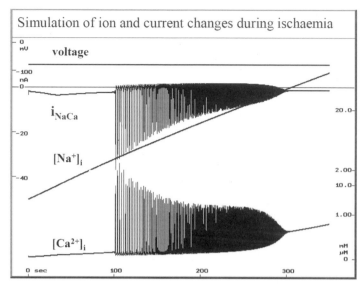

Figure 14.2   Calcium oscillations computed in a model of the Purkinje fibre during the rise in intracellular sodium [Na]$_i$ following blockage of the sodium-potassium ATPase. Each oscillation of intracellular calcium [Ca]$_i$ triggers inward sodium-calcium exchange current i$_{NaCa}$ (Ch'en et al., 1998). These computations were done under voltage clamp conditions.

the attachment of calcium to intracellular proteins. This signaled the need to incorporate intracellular calcium buffering.

This deficiency was tackled in the Hilgemann-Noble (Hilgemann and Noble, 1987) modeling of the atrial action potential. Although this was directed towards atrial cells, it also provided a basis for modeling ventricular cells in species (rat, mouse) with short ventricular action potentials, and many of its features were adopted in later ventricular cell models of species with high plateaus (Luo and Rudy, 1994, 1991; Noble et al., 1991, 1998).

The Hilgemann-Noble model addressed a number of integrative systems questions concerning calcium balance:

1. When does the calcium that enters during each action potential return to the extracellular space? Does it do this during the rest period between contractions (as most people had presumed) or during the contraction itself, that is, during, not after, the action potential? Hilgemann (Hilgemann, 1986) showed that the recovery of extracellular calcium (in intercellular clefts) occurs remarkably quickly (see figure 14.3, inset). In fact, net calcium efflux is established as soon as 20 msec after the beginning of the action potential, which at that time was considered to be surprisingly soon. Calcium activation of efflux via the Na-Ca exchanger achieved this in the model (see figure 14.3 – compare the computed trace [Ca]$_o$ with the experimental trace labeled [Ca]$_o$).

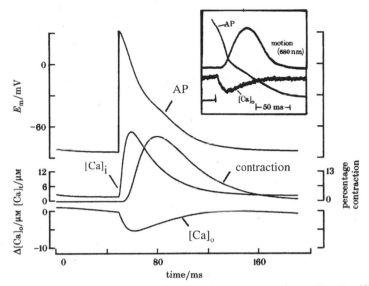

Figure 14.3   The first reconstruction of calcium balance in cardiac cells. The Hilgemann-Noble model (Hilgemann and Noble, 1987) incorporated complete calcium cycling, such that intracellular and extracellular calcium levels returned to their original state after each cycle and that the effects of sudden changes in frequency could be reproduced. Left: computed action potential (AP), intracellular calcium transient, contraction (represented by cross-bridge formation), and extracellular calcium transient. Inset: experimental recording of action potential (AP), cell motion, and extracellular calcium transient.

2. Where was the current that this would generate and did it correspond to the quantity of calcium that the exchanger needed to pump? Mitchell et al. (1984) had shown that replacement of sodium with lithium removes the late plateau. This was the first experimental evidence that the late plateau in action potentials with this shape might be maintained by sodium-calcium exchange current. The Hilgemann-Noble model showed that this is precisely what one would expect.

3. Could a model of the SR that reproduces at least the major features of Fabiato's experiments (Fabiato, 1983) showing calcium-induced calcium release (CICR) be incorporated into the cell models and integrate with whatever were the answers to questions 1–2? This was a major challenge. The model followed as much of the Fabiato data as possible, but the conclusions were that the modeling, while broadly consistent with the Fabiato work, could not be based on that alone. It is an important function of simulation to reveal when experimental data needs extending.

4. Were the quantities of calcium, free and bound, at each stage of the cycle consistent with the properties of the cytosol buffers? The answer here was a very satisfactory yes. The great majority of the cytosol calcium is bound so that, although much more calcium movement was involved, the free calcium transients were much smaller, within the experimental range.

There were, however, some gross inadequacies in the calcium dynamics. An additional voltage-dependence of Ca release was inserted to obtain a fast calcium transient. This was a compromise that requires more detailed modeling of the spaces immediately between the cell membrane and the intracellular machinery—a space where calcium channels and the ryanodine receptors interact—a problem later tackled by Jafri, Rice and Winslow (1998) and by Noble et al. (1998). Another problem was how the conclusions would apply to action potentials with high plateaus. This was tackled both experimentally (LeGuennec and Noble, 1994) and computationally (Noble et al., 1998). The answer is that the high plateau in ventricular cells of guinea pig, dog, human, and so forth greatly delays the reversal of the sodium calcium exchanger so that net calcium entry continues for a longer fraction of the action potential. This property is important in determining the way in which the force of contraction varies with the frequency of the heart beat.

Intracellular calcium dynamics have now become a major focus of simulation work (Coombes et al., 2004; Eisner et al., 2000; Hinch, 2004; Jafri et al., 1998; Puglisi et al., 2004; Soeller and Cannell, 2004). So also has the modeling of active transport and cardiac energetics (Matsuoka et al., 2004; Smith and Crampin, 2004), and the regulation by cell signaling networks (Saucerman and McCulloch, 2004). These developments are opening up the way for major developments in the use of cardiac models in understanding disease states, where calcium dynamics, active transport, and cell signaling are often affected.

14.4   Linking to the Genetic Level

An important strength of models based on reconstructing the functional properties of proteins in cellular structural contexts is that it is possible for the models to reach down to the genetic level, for example by reconstructing the effects of particular mutations when these are characterized by changes in protein function (Noble, 2002d).

An example of this approach is the use of state-specific Markov models of the cardiac sodium channel (Clancy and Rudy, 1999) simulating the behavior of the wild-type and of a mutant sodium channel. The simulated mutation was the Δ KPQ mutation, a three-amino-acid deletion that affects the channel inactivation and is associated with a congenital form of the long-QT syndrome, known as LQT3. The simulations showed that mutant channel reopenings from the inactivated state and channel bursting due to a transient failure of inactivation generate a persistent inward sodium current during the action potential plateau in the mutant cell. This causes major prolongation of repolarization and the development of arrhythmogenic early-after-depolarizations at slow pacing rates, a behavior that is consistent with the clinical presentation of bradycardia-related arrhythmogenic episodes during sleep or relaxation in LQT3 patients.

Another sodium channel mutation that has been, at least partially, reconstructed is a mis-sense mutation that affects the voltage dependence of sodium channel inactivation and which is responsible for one form of idiopathic ventricular fibrillation. In this case, small shifts of the voltage dependence of inactivation generate early-after-depolarizations that may underlie fatal arrhythmia (Noble and Noble, 2000). Simulation can also unravel the way in which the effects of these genetic mutations interact with drugs to explain why some people are particularly prone to arrhythmic side effects of many drugs (Noble, 2003b).

Early after-depolarizations are also responsible for the arrhythmias of congestive heart failure. Winslow et al. (2001) have modeled this process based on experimentally determined changes in gene expression levels for several of the transporter proteins involved.

These examples highlight the ability of cellular models to predict the arrhythmogenic consequences of genetic and ion channel abnormalities either of behavior or of expression levels. Given the present explosion of genetic information, such studies will continue to be at the forefront of modeling efforts in the next decade. Connecting the genome to physiology is one of the exciting prospects for computational systems biology.

14.5 Linking to Biochemistry: Counterintuitive Predictions

Complex systems are characterized by the fact that the results of modeling them are frequently counterintuitive. Beyond a certain degree of complexity, armchair (qualitative) thinking is not only inadequate, it can even be misleading. A good example of this comes from the extension of cellular models to include some of the biochemical changes that occur during ischaemia (Ch'en et al., 1998). This work succeeds in reconstructing arrhythmias attributable to delayed after-depolarizations that arise as a consequence of intracellular calcium oscillations in conditions of sodium-calcium overload. These oscillations generate an inward current carried by the sodium-calcium exchanger which can lead to premature excitation. This work has led to some interesting counterintuitive predictions concerning up- and down-regulation of sodium-calcium exchange in disease states (Noble, 2002c). This transporter is currently a focus of anti-arrhythmia drug therapy. Simulation is playing an important role in clarifying and assessing the mechanism of action of such drugs.

Another area in which modeling has been rich in counterintuitive results is that of mechano-electric feedback. Kohl and Sachs (2001) describe the extent to which this feedback mechanism has been unraveled in elegant experimental and computational work. Some of the results, particularly on the actions of changes in cell volume (which are important in many disease states) are unexpected and have been responsible for determining the next stage in experimental work. Indeed, it is hard to see how such unraveling of complex physiological processes can occur without the iterative interaction between experiment and simulation.

14.6   Linking to Pharmacology: Assessing and Predicting Drug Actions

Most drugs act on proteins such as receptors, channels, transporters, and enzymes. Models that reach down to the protein level are therefore relevant to assessing and predicting drug actions. Cardiac simulations have already been used in assessing drug action by the Food and Drug Administration in the United States, and we can expect this kind of use of biological models to increase as their complexity and power grows. An example of the detailed use of cardiac cell models in drug development can be found in Bottino et al. (2005) who used reverse engineering to determine the profile of action of drugs on ion channels from information on their effects on the action potential. I have reviewed some of these developments in more detail elsewhere (Noble and Colatsky, 2000; Noble et al., 1999). One obvious use in the case of the heart is in assessing the cardiac safety of drugs. Around half the drug withdrawals that have occurred in the United States post-launch since 1998 have been attributable to cardiac side effects, often in the form of effects on the electrocardiogram and consequent arrhythmias. This is a large and very expensive form of attrition. Since virtually all the ion transporters involved in cardiac repolarization are now modeled and very realistic simulations of the T wave of the electrocardiogram can be obtained, when these models are incorporated into 3-dimensional cardiac tissue models it is possible to use *in silico* screens for drug development. One of the reasons that this is necessary is that the electrocardiogram is, unfortunately, an unreliable indicator of potential arrhythmogenicity. Similar changes in form of the electrocardiogram can be induced by very different molecular and cellular effects, some benign, others dangerous. We need to understand and predict the mechanisms all the way from individual channel properties through to the electrocardiogram. This goal is within reach, particularly as we acquire more experience of the incorporation of accurate cellular models into anatomically detailed organ models (see below).

Another use of simulation in drug discovery will be in screening drugs for multiple actions. Very few drugs that act on the heart bind to just one receptor. It is much more common for 2, 3 or, even more receptors or channels to be affected. This is particularly true for drugs that act on the sodium-calcium exchanger (Watanabe and Kimura, 2000). An important point to realize here is that multisite action may actually be beneficial. The reviews referred to above give examples of multireceptor drug actions that would be expected to be beneficial. I predict that this will in fact be one of the ways in which more rational discovery of anti-arrhythmic drugs may occur. In regulating cardiac function, nature has developed many multiple-action processes, particularly those regulated by G-protein coupled receptors. In seeking for more "natural" ways of intervening in disease states, we should also be seeking to play the orchestra of proteins in more subtle ways. We need simulation to guide us through the complexity and to understand multiple action functionality. Examples of this approach to combinatorial drug action in computational biology of the heart now exist (Noble, 2003b; Noble and Colatsky, 2000).

14.7   Linking to Tissues and Organs

In the case of the heart, in addition to the data-rich cellular level, there is also data-rich modeling of the 3-dimensional geometry of the whole organ (Costa et al., 1996; LeGrice et al., 2001). Connecting this level to that of cell modeling has been an exciting venture (Crampin et al., 2004; Kohl et al., 2000; Smith et al., 2001). Anatomically detailed models of the ventricles, including fiber orientations and sheet structure, have been used to incorporate the cellular models in an attempt to reconstruct the electrical and mechanical behavior of the whole organ.

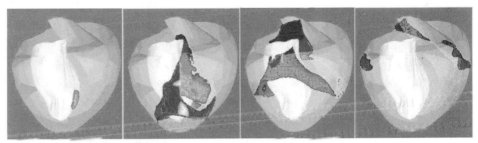

Figure 14.4   Spread of the electrical activation wavefront in an anatomically detailed cardiac model. Earliest activation occurs at the left ventricular endocardial surface near the apex (left). Activation then spreads in endocardial-to-epicardial direction (outwards) and from the apex towards the base of the heart (upwards, middle frames). The activation sequence is strongly influenced by the fibrous-sheet architecture of the myocardium, as illustrated by the non-uniform transmission of excitation. Black = activation wavefront; white = endocardial surface.

Figure 14.4 shows stills from a simulation in which the spread of the activation wavefront is reconstructed. This is heavily influenced by cardiac ultra-structure, with preferential conduction along the fiber-sheet axes, and the result corresponds well with that obtained from multi-electrode recording from dog hearts *in situ*. Accurate reconstruction of the depolarization wavefront promises to provide reconstruction of the largest phases of the electrocardiogram. Other parts of the organ, including the pacemaker region (sinus node), the atrium (the chambers receiving venous blood), and the specialized conducting system are now being incorporated into the model heart so that we can look forward to the first example of reconstruction of a complete physiological process from the level of protein function right up to routine clinical observation. Work is in progress in a number of laboratories on simulation of the sinus node (Boyett et al., 2003, 1999; Dobrzynski et al., 2003; Garny et al., 2003) and atrium (Blanc et al., 2001; Garny et al., 2000; Harrild and Henriquez, 2000). The whole ventricular model has already been incorporated into a virtual torso (Bradley et al., 1997), including the electrical conducting properties of the different tissues, to extend the external field computations to reconstruction of multiple-lead chest and limb recording.

14.8 Coronary Circulation

Ischaemic heart disease is a major cause of serious incapacity and mortality. It is also a good example of the fact that most disease states are multifactorial. Very few diseases are attributable to single gene or protein malfunction. As noted above, cellular reconstructions of the metabolic and electrophysiological processes that occur following deprivation of the energy supply to cardiac cells have already advanced to the point at which some arrhythmic mechanisms can be reproduced. The initiating process in such energy deprivation is restriction or blockage of coronary arteries. This is another example where modeling at different data-rich levels is holding out the prospect of very exciting integration of function. Figure 14.5 shows some of the spectacular modeling of the coronary circulation (Smith et al., 2000, 2001). These are stills from a simulation in which the blood flow through an anatomically-detailed model of the coronary circulation is computed while the ventricles are beating. The simulation therefore also included the deformation that occurs as mechanical events influence blood flow.

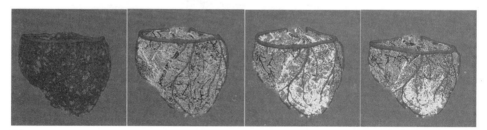

Figure 14.5 Flow calculations coupled to the deforming myocardium. The color coding represents transmural pressure acting on the coronary vessels from the myocardial stress (dark gray = zero pressure, light gray = peak pressure). The deformation states are (from left to right) zero pressure, end-diastole, early systole, and late systole.

This model has already been used to investigate the changes in blood flow that occur following constriction or block of one of the main arterial branches, and work is in progress to connect this to the modeling of ischaemia at the cell and tissue level (see figure 14.6). If we can also connect the cellular mechanisms of arrhythmia to the processes by which regular excitation breaks down into the multiple wavelets of ventricular fibrillation (Panfilov and Kerkhof, 2004) then yet another "grand challenge" for integrative physiological computation will come within range: the full-scale reconstruction of a coronary heart attack.

This is a suitable point at which to note that I chose the term *grand challenge* deliberately. This kind of work requires massive computer power. The whole organ simulations described here require many hours of computation using supercomputers. (By contrast, the single cell models can be run faster on a PC or laptop than in real time.) Future progress will be determined partly by the availability of computing capacity.

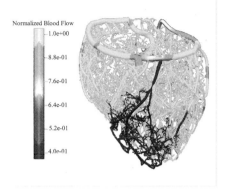

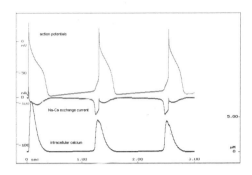

Figure 14.6   Left: the coronary circulation model shown in figure 14.5 has been subjected to a constriction of one of the main branches leading to blocked blood flow in the regions colored black. (Figure kindly provided by Nic Smith.) Right: simulation of ectopic beats using the DiFrancesco-Noble 1985 Purkinje fiber model (Noble, 2002a) in conditions of calcium overload of the kind that occurs in ischaemic tissue. To simulate sodium/calcium overload, $[Na]_i$ was increased from 8 to 12 mM (see figure 14.2). Oscillatory calcium changes (bottom) induce inward sodium-calcium exchange current (middle) leading to initiation of action potentials (above). The first action potential is evoked by a current pulse. The second two are initiated by calcium oscillations. Note that the rise in $[Ca]_i$ and the flow of inward Na-Ca exchange current occur *before* the depolarization. Linking these two levels of modeling to create a complete model of coronary heart attack is one of the "grand challenges" requiring massive computer power.

Blood flow within the chambers of the heart, including the movement of valves, has been elegantly modeled by Peskin and McQueen (1993) and this has been extended to the study of diastolic mechanical function (Kovacs et al., 2001).

14.9   The Future: From Genome to Proteome to Physiome

Integrative multilevel modeling of biological systems is an important technique for organizing and integrating vast amounts of biological information. Although this article has focused on modeling of the heart, it is important to note that multilevel biological simulation is now done for a wide range of pathways, cells, and systems. The role of *in silico* biology in medical and pharmaceutical research is likely to become increasingly prominent as we seek to exploit the data generated through rapid gene sequencing and proteomic mapping through to creating the physiome.

However, progress will be significantly enhanced by enabling ever greater numbers of researchers to use and verify models in the course of their everyday experimental work. It has been extremely difficult to transfer models between research centers or to extend existing models so that more complex models can be constructed in an object-oriented or modular fashion. This process will be enhanced by the development of uniform standards for representing and communicating the content of models and by the wide distribution of software tools that permit even non-modelers to access, execute, and improve existing models. Increasingly, publication of models is accompanied by their availability on Web sites. And the process of establishing standards of communication and languages is developing (Lloyd et al., 2004).

Once this is achieved, we can confidently predict an explosion in the development of integrated model cells, organs, and systems. In a few years we shall all wonder how we ever managed to do without them in biological research. So far as drug development is concerned, there will certainly be a major change as these tools come on line and rapidly increase in their power. This will grow in a nonlinear way with the degree of biological detail that is incorporated. The number of interactions modeled increases much faster than the number of components. Biology is set to become highly quantitative in the 21st century. It will become a computer-intensive discipline.

Acknowledgments

Work in the author's laboratory is funded by the BHF, EPSRC, MRC, and Wellcome Trust.

IV COMPUTATIONAL MODELING

15 Computational Constraints on Modeling in Systems Biology

Vipul Periwal

Computational power has followed Moore's Law quite well, and modeling in biology has taken full advantage of this. Nevertheless, as models get bigger and more aspects of the models need to be inferred, many techniques that are applicable for small scale models become inapplicable. A review of basic aspects of algorithms and data structures is provided, along with a summary of the computational aspects of sorting, searching, dynamic programming, graph theory, dynamical systems, and noise reduction algorithms.

15.1 Introduction

How many interacting species of molecules can one realistically mathematically model? Quite apart from the question of comprehending the interactions of several hundred thousand molecular species, there are limitations on systems biology that arise from simple combinatorics coupled with the lack of precise biological knowledge of the properties in vivo of most molecules.

The only way to overcome such limitations is to incorporate as much biological knowledge as possible. The task of modeling without using biological knowledge is, frankly, computationally impossible. As a simple example, consider a set of ten independent hypothetical interactions. In a eukaryotic system, a single protein may well have about ten splice sites or interactions. Our task is to ascertain which of these hypotheses is present or absent, given some experimental data. We are immediately faced with the task of generating $2^{10} = 1,028$ independent models, to fit each individual model with the data, and then determine the correct model by looking at some goodness-of-fit criterion. If this is the case for a single protein, consider the situation for modeling the complete proteome. This is an example of the combinatorial explosion of modeling in the face of partial knowledge and limited observability. Of course, there are sampling strategies that one can attempt to use to lessen the burden in this particular example, but in general, in biology, one must use as much biological information as possible to reduce the number of models that need to be evaluated.

Many properties, such as adaptability and robustness (chapter 2), are apparent only in dynamical simulations with limited changes in inputs and/or environment leading to no qualitative change in behavior in the case of adaptability, or limited changes in reaction rates or deletions of a few species producing no qualitative change in the case of robustness. Even if we know in advance which qualitative feature of the dynamical behaviour we wish to preserve (and this is rarely the case), there is no simple way to infer which changes will or will not preserve the qualitative feature from the static properties of a mathematical model, unless the model is specifically constructed in accord with standard techniques in control theory (chapter 12) and robust design. Is a model so designed mimicking biology at the biochemical level? Not necessarily. This underscores the fact that a mathematical model is, first and foremost, a model. It is useful as an attempted abstraction and simplification (chapter 3) of the essential features of the phenomenon being modeled and as a hypothesis-generating tool for further experiments, but it is not computationally feasible to make models that are *in silico* exact replicas of biochemistry in vivo for anything beyond a trivial scale.

Much has been made of the availability of large-scale data sets in systems biology. While it is certainly true that these cell- or organ-level complete coverage data sets allow screening of most of the relevant factors in any given biological phenomenon, none of the results of such screening can be translated directly into mathematical models, for several reasons:

1. Cellular localization information is absent (chapter 11).

2. Due to resource and experimental limitations, the time scales of measurement are rarely fine enough to allow observation of both very quick transient initial or priming responses and enough detail of following longer term behavior (chapter 6).

3. Determinations of the interactions of proteins oftentimes have large false-positive and false-negative rates (chapter 10).

4. The models are too large and have too much missing information (reaction rates, localization, concentrations) to be computationally tractable, even on massively parallel supercomputers.

While new and improving technologies will alleviate these problems in the fullness of time, predictive modeling at the present juncture requires cognizance of this reality. This chapter is concerned with an overview of various computational algorithms that are relevant for systems biology, with the specific aim of fleshing out (4) above.

15.2 Algorithms

Algorithms for solving computational problems, either discrete (such as finding cliques in graphs) or continuous (parameter optimization for a dynamical model), are judged on correctness, efficiency, and ease of implementation. A heuristic, on the other hand, is probably intuitively clear in the way it approaches a problem, but does not come with any guarantee of correctness (Rawlins, 1991; Knuth, 1997).

There are two main ways to compare algorithms. One corresponds to the random-access-machine (RAM) model of computation, and the other is an asympototic analysis of worst case complexity. The RAM model counts computational and memory access steps, but is somewhat simplistic in that it assumes that all basic and arithmetic operations take equal amounts of time. This is not true for real processors, of course, but anything more detailed would be processor-dependent, and therefore not of as much utility in comparing algorithms. The RAM model tells us how an algorithm might work on a given input, but it does not give an indication of how the algorithm might fare on a typical input, or the worst or best case scenarios for different inputs. These are important issues, especially if we do not know too well the kinds of data we might obtain in biological experiments.

The growth rate of an algorithm gives us an idea of how big an input we can realistically compute with it. The growth rate is computed by counting all the steps needed to carry out the algorithm, with each computation or memory access step counting as one step, for a given input size n, which might, for example, be the number of protein nodes in a graph of yeast-2-hybrid predictions of protein-protein interactions.

1. An algorithm whose growth rate is $n!$ becomes useless before $n = 20$.

2. An algorithm whose growth rate is 2^n becomes useless before $n = 40$.

3. An algorithm whose growth rate is n^2 will work reasonably up to $n = 100$, but will rapidly become impractical beyond this. For $n = 10^5$, such an algorithm requires 10^{10} steps.

4. An algorithm whose growth rate is n will likely be useful up to $n = 10^9$, and this holds even for algorithms with logarithmic corrections such as $n \log n$.

5. An algorithm whose growth rate is $\log n$ will be useful forever. There *are* such algorithms, for example, binary search.

There are usually constant multiplicative factors associated with these growth rates, and they make some difference to assessments of the viability of the algorithms for small values of n, but for asymptotically large values of n, these constants are usually irrelevant. Algorithms of types 1 and 2 are not relevant for the scale of problems usually of interest in systems biology. For problems of type 3, bigger and faster computers will make a difference.

To understand which algorithm to use, we must describe the modeling problem in a manner such that we are able to look in repositories of algorithms and find appropriate results. The common terms used in formal descriptions of algorithms are abstract structures such as graphs, permutations, trees, and sets. *Permutations* are relevant if the problem requires different arrangements, orderings, or sequences. An example in systems biology is system identification, when we do not know the order in which molecules appear in a pathway, or in what temporal order the molecules interact with each other. *Sets* are relevant when the problem seeks a cluster, a collection, or any selection from a set of items. *Trees* appear in problems with hierarchical relationships. Submodules embedded in modules (chapter 3) might be one example in hierarchical dynamical models. *Graphs* appear in problems involving networks, circuits, or webs. *Points* representing locations in some space appear in problems like protein conformations or protein localization in a cell. *Strings* appear in problems involving patterns or labels. Finding consensus sequences for transcription factor binding sites upstream of a set of genes is a prototypical string matching problem.

Data structures are the flip side of the computational cost coin. The organization of the data impacts algorithm performance greatly. A basic understanding of the types of common data types used in computer science is helpful in deciding the storage of experimental data. While biologists typically will not need the detailed implementations of the data structures introduced below, an acquaintance may facilitate communications with collaborators in other sciences.

A *container* is a data type which permits storage and retrieval of data irrespective of the content of the data. Such a data type has access only through insertion or retrieval operations, and might be implemented as a *stack* (permitting last-in-first-out (LIFO) retrieval)

$$\leftrightarrow [\text{Item}_n][\text{Item}_{n-1}][\text{Item}_{n-2}]\ldots[\text{Item}_1][\text{Item}_0] \tag{15.1}$$

as a *queue* (permitting first-in-first-out (FIFO) retrieval

$$\rightarrow [\text{Item}_n][\text{Item}_{n-1}][\text{Item}_{n-2}]\ldots[\text{Item}_1][\text{Item}_0] \rightarrow \tag{15.2}$$

useful for algorithms where the order of the stored data is important) or as a *table* (permitting retrieval indexed by position, as in an array)

$$\begin{array}{ccccc} 0 & 1 & 2 & \ldots & n \\ \downarrow & \downarrow & \downarrow & \downarrow & \downarrow \\ [\text{Item}_0] & [\text{Item}_1] & [\text{Item}_2] & \ldots & [\text{Item}_n] \end{array} \tag{15.3}$$

A *dictionary* is a data type suited for accessing data by content. Thus a dictionary has *keys* and *data referred to by the keys*. Dictionaries permit search for data given a key, insertion, and deletion.

$$
\begin{aligned}
\text{Key}_0 &\rightarrow \text{Value}_0 \\
\text{Key}_1 &\rightarrow \text{Value}_1 \\
\text{Key}_2 &\rightarrow \text{Value}_2 \\
\vdots \quad &\quad \vdots \quad \quad \vdots \\
\text{Key}_n &\rightarrow \text{Value}_n
\end{aligned}
\tag{15.4}
$$

Structures that can implement the dictionary data type include *linked lists* (constant time insertion and deletion but more computational steps for search), *arrays* (constant time search but longer insertion and deletion times), *binary search trees*, or *hash tables*. Binary search trees store data in a tree structure with each node having 0, 1, or 2 offspring. If the keys in the dictionary have an ordering (in other words, for any two keys we can decide if $x < y$ or $y < x$; an example of an ordering is alphabetical ordering), when we search for the data labeled by a key x, we search from the root node up the tree, taking the left branch if the key at the root node is larger than x and the right branch otherwise, and proceeding onwards recursively.

$$
\begin{array}{ccccccc}
K_1 & & & K_5 & K_4 & & K_0 \\
 & \searrow \quad \nearrow & & & \searrow \quad \nearrow & \\
 & K_2 & & & K_3 & & \\
 & & \searrow \qquad \nearrow & & & & \\
 & & \text{root} & & & &
\end{array}
\tag{15.5}
$$

If the items were inserted more or less at random, the search will involve about $\log n$ steps. As an example of the difference between worst-case performance and average performance, notice that if the items were inserted in an ordered fashion, the search will take n steps instead.

While containers and dictionaries are the most common data types, more specialized data types are valuable when the data to be stored has more structure. Examples of such data structures are *suffix trees* to store strings, *kd-trees* to store geometric objects, *adjacency lists* (lists of pairs of connected vertices) or *adjacency matrices* (matrices indexed by the vertices in the graph, with non-zero entries corresponding to vertices linked by an edge) to store sparse or densely connected graphs respectively, and set data stored as *hypergraphs* or in the form of dictionaries associated with subsets.

15.3   Sorting

Sorting is a basic building block of algorithm design and takes about $n \log n$ steps for sorting a set of n elements. Thus, sorting is the type of building block in an algorithm that can be used for fairly large sets of items. One should use sorting as much as necessary without worrying about rendering the problem computationally intractable. Sorting is the basis for

A. Searching: After sorting the keys, one can test for the presence of an item in a dictionary in $\log n$ time.

B. Closest pair: To find the closest pair of numbers, sort the numbers and then do a linear time scan through the sorted list. Total time required (including sorting): $n \log n$.

C. Selection: What is the kth largest item in a set? Sort the set and look at the kth position.

Two general principles of algorithms are *divide and conquer* and *randomization.* Divide and conquer is the principle of dividing the original problem into several smaller ones, solving the smaller problems, then combining the solutions of the smaller problems into a solution of the original problem. This is typically possible in a recursive fashion. Randomization is the principle of randomizing the input data in order to ensure with high probability that a given algorithm's good average behavior is utilized, as opposed to (possibly) much worse worst-case behavior.

A sorting algorithm called *mergesort* is a good example of divide and conquer. The data to be sorted is split into two subpiles, each of which is then sorted.

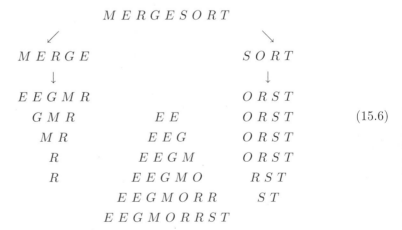

$$
\begin{array}{ccc}
& M\ E\ R\ G\ E\ S\ O\ R\ T & \\
\swarrow & & \searrow \\
M\ E\ R\ G\ E & & S\ O\ R\ T \\
\downarrow & & \downarrow \\
E\ E\ G\ M\ R & & O\ R\ S\ T \\
G\ M\ R & E\ E & O\ R\ S\ T \\
M\ R & E\ E\ G & O\ R\ S\ T \\
R & E\ E\ G\ M & O\ R\ S\ T \\
R & E\ E\ G\ M\ O & R\ S\ T \\
& E\ E\ G\ M\ O\ R\ R & S\ T \\
& E\ E\ G\ M\ O\ R\ R\ S\ T &
\end{array}
\tag{15.6}
$$

We then *merge* the two sorted subpiles by comparing the first (lowest) elements in each sorted subpile. The lowest of the two elements is removed, leaving the next lowest element as the lowest element in one of the two subpiles, and so on. This merging process takes on the order of n steps, and the recursion into smaller subpiles takes on the order of $\log n$, so the total running time of this algorithm is $n \log n$.

Quicksort is another sorting algorithm, in which we pick an arbitrary element x of the data set. The rest of the data is separated into the elements larger than x and the elements smaller than x, and each of these subsets is sorted. The complete sorted set is then obtained by merging the results with x inserted in between.

$$
\begin{array}{ccc}
& Q\,U\,I\,C\,K\,S\,O\,R\,T & \\
\swarrow & & \searrow \\
I\,C\,K\,O & Q & U\,S\,R\,T \\
\downarrow & & \downarrow \\
C\,I\,K\,O & Q & R\,S\,T\,U \\
& C\,I\,K\,O\,Q\,R\,S\,T\,U &
\end{array}
\tag{15.7}
$$

The total cost is on the order of $n \log n$ steps *on average* since x is more likely to be closer to the center of the sorted set than to the edges. However, if x happens to be at either end, the number of steps will be more like n^2. To ensure the average good behavior is obtained with high probability, we use *randomization* along with quicksort.

If we know more about the distribution of the data, we can apply more specialized algorithms like *distribution sort*. The key point to note here is that the specialized algorithms may perform much worse if our hypothesis about the distribution happens to be incorrect.

15.4 Dynamic Programming

Dynamic algorithms (Denardo, 2003) are algorithms which solve problems by solving and storing the solutions to small problems, and then combining these solutions into a solution of the larger problem. There is, of course, a trade-off: memory is traded for speed. The memory requirements must be kept in mind for such algorithms. An important feature of dynamic programming algorithms is optimal sub-structure: the sub-problems which are the solution to the problem posed are themselves optimal solutions. In other words, all future steps depend only on which state the algorithm is in, not on how the algorithm got to that state. Dynamic programming is, in a sense, the opposite of recursion.

Thus, the entire focus in looking for a dynamic programming solution is to establish what are the appropriate steps in the solution, what are the decisions at any step, *and* what are the states that are associated with each step. The decision at any step must determine the next state, *given* the state you are in. Dynamic programming is particularly useful in cases where there is a natural ordering to each input such as the left-to-right ordering of bases in a DNA sequence. The reason is that the number of partial solutions found must stay bounded. If the order of the input did not matter, there would be an exponential explosion in possible states which we would not be able to store in memory. The traveling salesman problem, for example, has no inherent order in the vertices, and results in an exponential sized set of states.

For example: Fibonacci numbers are defined by a simple two-term relation: $f_n = f_{n-1} + f_{n-2}$, with $f_0 = 0$, $f_1 = 1$. They can be computed recursively, but in exponential time because each recursive step branches into more recursive steps: $f_n = f_{n-2} + 2f_{n-3} + f_{n-4}$ and so on. This requires no storage. A dynamic programming algorithm, directly iterating over the definition, runs in linear time, but stores the last two values computed as it runs. As another example, binomial coefficients can be computed recursively

$$C_m^n = C_m^{n-1} + C_{m-1}^{n-1} \tag{15.8}$$

(with $C_m^n = 1$ if $m = 0$ or $n = m$) in exponential time, or dynamically from the last computed row of Pascal's triangle

$$
\begin{array}{ccccccccccc}
 & & & & & 1 & & & & & \\
 & & & & 1 & & 1 & & & & \\
 & & & 1 & & 2 & & 1 & & & \\
 & & 1 & & 3 & & 3 & & 1 & & \\
 & 1 & & 4 & & 6 & & 4 & & 1 & \\
1 & & 5 & & 10 & & 10 & & 5 & & 1 \\
\end{array}
$$

in n^2 time, but using n integers to store the last computed row.

Dynamic programming is a standard technique in sequence analysis, for example: given two sequences of symbols, find the longest subsequence of symbols that appears in both the given sequences. This problem takes about mn steps, where m, n are the lengths of the two sequences and uses two $m \times n$ arrays to store the partial results. From the perspective of biological modeling, a more interesting application of dynamic programming is for *stochastic control*. Model-free control theory may be particularly interesting as a way to make progress in predicting the response of a biological system in the absence of a complete model of its biochemistry. A recent review of applications of dynamic programming to stochastic control is Lee and Lee (2004).

15.5 Graphs

A graph is, simply put, a set of relationships between objects. Networks of protein interactions are graphs, with the relationship being the evidence for interaction between two proteins, for example from a yeast-2-hybrid screeen. Many complex systems can be described in terms of the relationships between their parts. Hence graph theory appears prominently in many attempts to find common structures in large-scale data sets, for example expression array measurements, and is likely to become even more prominent as dynamical models are expressed as graphs of interactions and molecular species.

Several common graph problems are directly relevant to biology. For example, one might want to know if a graph of interactions remains connected if one deletes a certain set of interactions (edges in the graph) (Jeong et al., 2000). One might want to find the number of paths of length less than 5 connecting two given proteins. One might want to find the set of shortest paths that still connect all the proteins—this corresponds to a *minimal spanning tree*. In computer science, a graph with costs associated with the edges is termed a *network*. In biological examples, one might consider the $-\log(\text{probability})$ of a given interaction being a true positive as the cost associated with the interaction. Then one might want to consider the cheapest paths from protein A to protein B.

The solutions to many graph problems require the use of a class of algorithms termed *greedy*, in which a solution can be found by using only knowledge possessed at the time the next choice is made. A characteristic of these algorithms is that they are short-sighted, in other words, they take the step that seems intuitively to be the best one for the next step, *but* eventually their steps converge to the best solution. From a heuristic perspective this makes greedy algorithms intuitive, but it is often difficult to prove that they actually will lead to a correct solution without getting trapped in a sub-optimal solution. For example, finding the shortest path that goes through all the nodes in a graph (the traveling salesman problem) might intuitively require adding the nearest node to the path at any given step, but this will not usually lead to a solution of the problem. Greedy algorithms are particularly useful in problems with an exponentially or factorially growing search space, for example the number of possible interaction graphs for n proteins, or the graph of models obtained by elaborating or simplifying reactions in a dynamical model (chapter 4 and chapter 11). Searching in the latter graph is an important problem in biological system identification.

Examples of greedy algorithms are the algorithms that find the solution to the minimum spanning tree problem, the problem of finding the subgraph in a graph such that every node in the graph is a node in the subgraph, and there are no cycles (closed loops) in the subgraph. For sparse graphs, which are graphs with the number of nodes roughly equal to the number of edges, the best algorithm runs in about $n \log \log n$ steps. Are graphs of biological interest sparse? Scale-free graphs have the number of edges roughly proportional to the number of nodes, for example.

Many uses for information present in a graph require a traversal of the graph. Such traversals are usually depth-first, that is, visit all nodes attached to a node attached to a starting node, before visiting all the nodes attached to a second node attached to the starting node and so on (as numbered below),

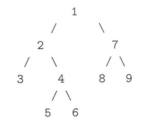

or breadth-first (as numbered below),

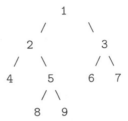

that is, visit all the nodes attached to the starting node first, then proceed outwards to the next-nearest neighbors and so on. Both these traversals of graphs take on the order of (number of nodes + number of edges) time steps, but different applications require different traversals. For example, traversing the graph of possible models mentioned above should probably be a breadth-first search, at least initially.

15.6   Search

Search algorithms depend on the search problem (Russell and Norvig, 2003). Searching amounts to locating an item in a set by probing elements in the set. In detail, it is a protean problem: the probe may be inaccurate (experimental uncertainty); probes may have unequal costs (for example, protein mass spectroscopy versus expression measurements); the search space may be infinite or very large (for example, parameter optimization for a large system of differential equations); the item to be located may not be uniquely identifiable (for example, there may be two dynamical models consistent with the data); resources for the search may be limited (for example, there may be a limited amount of extract available for expression measurements); or some combination of all these characteristics. Accordingly, there are many different types of search algorithms, most of which are variants or combinations of a few basic strategies.

The simplest strategy might be termed generate-and-test. It is usually simple to implement, and it will clearly find the solution. It may, however, take a long time to find the solution in a problem with even a modest amount of complexity. Improvements on this algorithm include *hill-climbing* (adding a heuristic distance function to guide the generation of solutions that minimize the distance from the desired state) and *adding stochasticity* (avoiding entrapment in a local minimum of the distance function by adding some stochasticity to the distance function, and reducing the amplitude of the stochasticity as the search proceeds).

An improved combination of depth-first search and breadth-first search is called best-first search. The search problem is formulated as a graph. Every node in the graph represents a model. A queue of nodes is set up. The current node is at the head of the queue. If the current node is not the solution of the search problem, all its nearest neighbor nodes are added to the queue, and the current node is removed from the queue. The queue is re-ordered according to a (possibly heuristic) distance function. The lowest node in the queue is made the current node and the algorithm repeats. Thus the search always tries the best move available from its list of possible moves, regardless of whether the move is a horizontal or vertical move. Only horizontal moves (breadth-first search) avoid getting trapped in dead ends but may end up searching all the nodes, while only vertical moves (depth-first search) avoid searching all the nodes but can get trapped in dead ends. Best-first search avoids the pitfalls of both breadth-first and depth-first searches by hopping around in the search graph between areas more likely to contain the desired model.

Perhaps the most popular search algorithm is $A*$ search. A difference between best-first and $A*$ search is that the heuristic function is the sum of two contributions. One is an underestimate of the distance from the current node to the goal of the search, and the other measures the distance from the current node to the putative next node. The heuristic behind this summation is that this sum (by the triangle inequality) is an estimate of the distance from the putative next node to the goal, the underestimation in the heuristic function compensating for the triangle inequality approximately. Having chosen a new current node, if the node chosen is not the goal, then this node and all its nearest neighbors are removed from the queue, another difference from best-first search. All these search algorithms have a worst-case performance proportional to the number of nodes in the graph, but $A*$ search will out-perform the others by about an order of magnitude on typical problems.

For the iterative modeling that is needed in systems biology, another search heuristic is useful: means-ends analysis. This allows both backwards and forwards searching, so it is possible to iteratively refine models from gross features to more detailed features. It should be obvious that any application of this algorithm to modeling requires detailed biological input, so we give here only a rough sketch of the search strategy. The strategy examines the current state (the present dynamical model), the desired state (the present data), and the differences between them. The difference is used to iterate over adding additional interactions to the model that may bring the model closer to the desired predictions. If the data is organized in a hierarchical fashion, the resulting model will have a hierarchical structure. The model remains, of course, a model, in that the actual organization of biochemical interactions in the cell may not mirror the mathematical interactions incorporated into the model.

In search problems with uncertainty, it is often interesting to pose the problem as a *constraint satisfaction* problem (chapter 5). Multiple alignments of DNA sequences are an example of constraint satisfaction problems, with biological knowledge guiding which alignments are good. Another example would be to apply constraint satisfaction heuristics to modeling the response of various related cell-lines to different stimuli—we do not want totally different models for each cell-line. We would like the models to be roughly "aligned," just as in the sequence alignment problem, with biology dictating the meaning of a good alignment—for example, the timescales for corresponding parts of a particular response may be different between cell-lines so models which reproduce this difference should be regarded as "aligned." The constraints in this example might be modeled as statements off the form: "The time at which transcription factor X concentration rises precedes the time at which genes with upstream binding sites for X are expressed."

15.7   Identification, Simulation and Optimization of Dynamical Models

Dynamical models of biological systems usually have unknown parameters such as reaction rates and initial concentrations for molecules not observed. The experimental data is used to constrain such unknown parameters. This process typically involves searching through different values of the unknown parameters, solving the dynamical system with a given set of values of these parameters, then adjusting the values of these parameters trying to improve the fit to experimental data.

Dynamical systems come in many varieties, ranging from entirely stochastic descriptions of molecular dynamics (chapter 8 and chapter 16) to systems of ordinary differential equations expressing smooth variation (chapter 6 and chapter 12). If we have a system of m differential equations governing the dynamics of m molecular concentrations that we need to fit to T data points, with a certain time resolution, the number of computational steps required is roughly proportional to mT, with a proportionality constant depending on the accuracy required and the computational complexity of the system of equations. If every species interacts with every other species, the computational complexity may be of the order of m or higher, but if only a few molecules interact with any given molecule with simple Michaelis-Menten dynamics, the complexity will be independent of m.

These two extremes reflect a major consideration in modeling. One has a choice: one can model a large number of molecular species with a few interactions per species, or one can model only a few marker species with complicated interactions between possibly all the other marker species. Given a good understanding of the fundamental interactions, it would appear from this discussion that the crossover point between the two approaches occurs at about (number of markers)$^2 \approx$ (number of molecules). These are, however, not the only considerations. One almost never has detailed and accurate knowledge of the fundamental interactions of molecules in vivo so available data is used for finding likely values of reaction rates, which adds greatly to the computational burden. In addition, one usually cannot obtain many time-points for a large number of molecules due to resource constraints, which implies that the data used to constrain the reaction rates has low dynamical resolution. Another problem is that incomplete biochemical information may lead to finding a model that matches the data but only because important reactions are unknown—predictions for the effects of novel perturbations from such a model may be wrong. On the other hand, if one opts for using more accurately and more frequently sampled marker concentrations, one has to infer the whole complex of effective interactions, since many fundamental interactions are subsumed into the set of signals propagating from one marker molecule to another. This means that many more effective reaction rates need to be determined per marker molecule. Furthermore, one has to have enough biological knowledge to select the marker species, or factor in the cost of exploratory experiments to determine markers.

From a larger perspective, modelers are interested in deducing dynamics from data (chapter 11). Experimental time series data is often noisy, reflecting both underlying stochasticity and experimental systematic variation. Noise reduction in time series is an extensively studied topic (Kostelich and Schreiber, 1993). A considerable part of the literature deals with systems characterized as low-dimensional chaos which is not generally the case for biological systems. Rarely does a biologist have the resources to make measurements over many periods of the asymptotic behavior of a given system. Nevertheless, some techniques have broader applicability.

One of the most flexible and robust techniques is a variant on the Takens time-delay-embedding method (Kostelich and Schreiber, 1993), usually called singular spectrum analysis (Golyandina et al., 2001). In this procedure, the available time series $x_i, i = 1, \ldots, N$ is used to construct a set of vectors $\mathbf{v}_i, i = 1, \ldots, N - L + 1$ defined as

$$\mathbf{v}_i = (x_i, x_{i+1} \ldots x_{i+L-1}) \tag{15.9}$$

The $\mathbf{v}$ vectors are used to construct a matrix M of size $(N - L + 1) \times L$ with the vectors serving as the columns of M. A Hankel matrix is a matrix with entries that are constant along anti-diagonals, so M is an example of a Hankel matrix. A singular value decomposition of M results in $M = Y_1 + Y_2 + \ldots$ with each X_i corresponding to a particular singular value. While this is an exact decomposition, the noise reduction is achieved by using only the Y_i that correspond to the largest few singular values and add up to an approximately Hankel matrix. The final step is to add the chosen Y_i and derive a noise-reduced time series by averaging over the anti-diagonal elements, thereby inverting the process by which M was initially formed from the experimental time series x_i. The value of this technique is that it works without assuming anything about the underlying signal, unlike Fourier analysis or other noise filters based on specific special functions. The only assumption in singular spectrum analysis is the choice of the window length L. A rough rule of thumb is to pick a value of L in a range where small changes in L do not affect the noise-reduced time series, while keeping $N - L$ considerably larger than L.

Time series data has an advantage over a permutable data set, in that there is a definite time ordering to the data. How does one exploit the time ordering to help in the noise reduction? A simple approach, adapted from a procedure applied in chaotic systems (Kostelich and Schreiber, 1993), is applicable if the time series experiment has been repeated three times or more: We can use an interpolation of the form $x_n = a_{n,n-1}x_{n-1} + a_{n,n+1}x_{n+1} + b_n$ to find coefficients $a_{n,n-1}, a_{n,n+1}, b_n$ (for example, by a least-squares fit to the replicates) and use these coefficients to compute a noise-reduced value $\hat{x}_n$ for the time series. The procedure can even be carried out with only two samples in conjunction with a Bayesian marginalization (chapter 4) over the coefficients. This procedure reduces the noise in the time series for all save the first and last points. The caveats are that the interpolation may not be an accurate description of the dynamics, and there may be some sensitivity to outliers in the data if fewer repetitions of the data are available.

Optimization of parameters is a search problem dealing with continuous families of parameters. As such, more methods are available to guide the search procedure. Optimization algorithms are broadly divided into searches that take derivatives of an objective cost function to find paths in the parameter space that minimize the cost, and searches that do not require derivatives of the cost function. Often in biological modeling, one does not know the character of the landscape associated with the cost function. If the cost varies smoothly with the parameter values simple ideas like going down the path of steepest descent may find the minimum. More likely, the cost function may have multiple local minima, and getting trapped in the vicinity of a local minimum will trap local optimization strategies such as steepest descent. A more global approach to finding a minimum of the cost function uses simulated annealing, where the size of a step in parameter space that the algorithm takes as it searches for a global minimum is gradually reduced so that the algorithm avoids getting trapped in a wrong local minimum. Simulated annealing is much slower than gradient descent type methods but is much more likely to avoid false minima. Deterministic global optimization methods like branch and bound are computationally too demanding for most interesting biological modeling problems. Evolutionary strategies are search strategies that apply the biological example of evolution to searching by evolving a population consisting of different points in the parameter space according to how the points lower the cost function. A detailed description of the algorithms is beyond the scope of this chapter, but results by Moles et al. (2003) suggest that such strategies are the only ones capable of finding the true minimum in biological modeling. Since evolutionary strategies are generally the best suited to multimodal optimization problems, a possible implication is that cost function landscapes in biological systems may be multimodal. However, it is important to note that von Dassow et al. (2000) found that within the *Drosophila* stripe formation module, the parameter optimization problem was remarkably easy to solve, suggesting a quite different picture of the cost function landscape, provided that one has the right model. This suggests that model search should include ease of optimization as a criterion. A speculative hypothesis is that preferring a model that is easy to optimize is a computational analog of one kind of evolutionary pressure.

15.8 Summary

It may not be entirely obvious that a given problem of biological interest has a solution expressed as an algorithm studied before in an abstracted setting. Nevertheless, it is extremely important to make sure that there is no relevant abstract problem in computer science that has been studied prior to embarking on one's own reinvention of the wheel. As the examples given in this chapter show, brute force large-scale computation to solve problems in systems biology is a still-born endeavor. Even worse, ideas with conceptual merit may be dismissed as computationally impractical if one does not look for efficient algorithms suited to the task.

An important consideration for modeling is the idea of relaxing the requirements in order to find approximate solutions in a reasonable amount of time. Given that a lot of biological information is uncertain or unavailable, *probabilistic* algorithms are a useful tool. These algorithms will find the correct answer usually, but they will always give an answer quickly. A simple example of a probabilistic approach in system inference, for example, is to generate a variety of different models constrained by available knowledge, and without parameter optimization, just check the qualitative agreement between the data and the models. One can then do the computationally expensive step of parameter optimization for the best of the generated models, if needed. The point here is that we made no attempt to exhaustively enumerate and test all the possible models, given a set of hypotheses. We figuratively threw a bunch of models all at once at the experimental data and picked the model that came closest for further evaluation. We could be wrong, and this would be something we could check by repeatedly running this algorithm and comparing the common features of the selected models. A probabilistic model is not always right, while a proven algorithm is not always fast.

For the scale of problems of interest in systems biology, for the foreseeable future there will be no such thing as "unlimited computational power." This does not imply that systems biology is impossible. It does imply that the results obtained by computer scientists in the past must be utilized just as much as known biology in order to make effective use of the large-scale data sets now becoming available.

16 Numerical Simulation for Biochemical Kinetics

Daniel T. Gillespie and Linda R. Petzold

In chemical systems formed by living cells, the small numbers of molecules of a few reactant species can result in dynamical behavior that is discrete and stochastic, rather than continuous and deterministic (McAdams and Arkin, 1999, 1997; Arkin et al., 1998; Elowitz et al., 2002; Fedoroff and Fontana, 2002). By "discrete," we mean the integer-valued nature of small molecular populations, which makes their representation by real-valued (continuous) variables inappropriate. By "stochastic," we mean the random behavior that arises from the lack of total predictability in molecular dynamics. In this chapter we introduce some concepts and techniques that have been developed for mathematically describing and numerically simulating chemical systems that take proper account of discreteness and stochasticity. Throughout, we shall make the simplifying assumption that the system is well-stirred or spatially homogeneous. In practice this assumption is often justified, and it allows us to specify the state of the system simply by giving the molecular populations of the various chemical species. But in some circumstances the well-stirred assumption will not be justified; then the locations of the molecules and the dynamics of their movement must also be considered. Some approaches to this more computationally challenging situation are described in chapter 8.

16.1 Chapter Overview

We begin in section 16.2 by outlining the foundations of "stochastic chemical kinetics" and deriving the *chemical master equation* (CME), the time-evolution equation for the probability function of the system's state. Unfortunately, the CME cannot be solved, either analytically or numerically, for any but the simplest of systems. But we can generate numerical realizations (sample trajectories in state space) of the stochastic process defined by the CME by using a Monte Carlo strategy called the *stochastic simulation algorithm* (SSA). The SSA is derived and discussed in section 16.3. Although the SSA is an ideal algorithm in the sense that it provides exact realizations of the CME, there is a computational price for this: Because the SSA simulates every reaction event, it will be painfully slow for systems that involve enormous numbers of such events, which most real chemical systems do. This has motivated a search for algorithms that give up some of the exactness of the SSA in return for greater simulation speed.

One such approximate accelerated algorithm is known as *tau-leaping*, and it is described in section 16.4. In tau-leaping, instead of advancing the system to the time of the next reaction event, the system is advanced by a pre-selected time τ, which typically encompasses more than one reaction event. The number of times each reaction fires in time τ is approximated by a Poisson random variable, and we explain why that can be done in section 16.4. In section 16.5 we show how, under certain conditions, tau-leaping further approximates to a stochastic differential equation called the *chemical Langevin equation* (CLE), and then how the CLE can in turn sometimes be approximated by an ordinary differential equation called the *reaction rate equation* (RRE). Tau-leaping, the CLE, and the RRE are successively coarser-grained approximations which usually become appropriate as the molecular populations of the reacting species are made larger and larger.

In the past, virtually all chemically reacting systems were analyzed using the deterministic RRE, even though that equation is accurate only in the limit of infinitely large molecular populations. Near that limit though, the RRE practically always provides the most efficient description. One reason for this is the extensive theory that has been developed over the years for efficiently solving ordinary differential equations, especially those that are *stiff*. A stiff system of ordinary differential equations is one that involves processes occurring on vastly different time scales, the fastest of which is stable. Stiff RREs arise for chemical systems that contain a mixture of fast and slow reactions, and many if not most cellular systems are of this type. The practical consequence of stiffness is that, even though the system itself is stable, naive simulation techniques will be unstable unless they proceed in extremely small time steps. In section 16.6 we describe the problem of stiffness in a deterministic (RRE) context, along with its standard numerical resolution: implicit methods.

Given the connections described above between tau-leaping, the CLE, and the RRE, it should not be surprising that stiffness is also an issue for tau-leaping and the CLE. In section 16.7 we describe an *implicit tau-leaping algorithm* for stochastically simulating stiff chemical systems. We conclude in section 16.8 by describing and illustrating yet another promising algorithm for dealing with stiff stochastic chemical systems, which we call the *slow-scale SSA*.

16.2 Foundations of Stochastic Chemical Kinetics and the Chemical Master Equation

We consider a well-stirred system of molecules of N chemical species $\{S_1, \ldots, S_N\}$ interacting through M chemical reaction channels $\{R_1, \ldots, R_M\}$. The system is assumed to be confined to a constant volume Ω, and to be in thermal (but not necessarily chemical) equilibrium at some constant temperature. With $X_i(t)$ denoting the number of molecules of species S_i in the system at time t, we want to study the evolution of the state vector $\mathbf{X}(t) = (X_1(t), \ldots, X_N(t))$, given that the system was initially in some state $\mathbf{X}(t_0) = \mathbf{x}_0$.

Each reaction channel R_j is assumed to be "elemental" in the sense that it describes a distinct physical event which happens essentially instantaneously. Elemental reactions are either unimolecular or bimolecular; more complicated chemical reactions (including trimolecular reactions) are actually coupled sequences of two or more elemental reactions.

Reaction channel R_j is characterized mathematically by two quantities. The first is its *state change vector* $\boldsymbol{\nu}_j = (\nu_{1j}, \ldots, \nu_{Nj})$, where ν_{ij} is defined to be the change in the S_i molecular population caused by one R_j reaction; thus, if the system is in state $\mathbf{x}$ and an R_j reaction occurs, the system immediately jumps to state $\mathbf{x} + \boldsymbol{\nu}_j$. The array $\{\nu_{ij}\}$ is commonly known as the stoichiometric matrix.

The other characterizing quantity for reaction channel R_j is its *propensity function* a_j. It is defined so that $a_j(\mathbf{x})\, dt$ gives the probability, given $\mathbf{X}(t) = \mathbf{x}$, that one R_j reaction will occur somewhere inside Ω in the next infinitesimal time interval $[t, t + dt)$. This probabilistic definition of the propensity function finds its justification in physical theory (Gillespie, 1992b,a). If R_j is the unimolecular reaction $S_i \rightarrow$ products, the underlying physics is quantum mechanical, and implies the existence of some constant c_j such that $a_j(\mathbf{x}) = c_j x_i$. If R_j is the bimolecular reaction $S_i + S_{i'} \rightarrow$ products, the underlying physics implies a different constant c_j, and a propensity function $a_j(\mathbf{x})$ of the form $c_j x_i x_{i'}$ if $i \neq i'$, or $c_j \frac{1}{2} x_i (x_i - 1)$ if $i = i'$. The stochasticity of a bimolecular reaction stems from the fact that we do not know the precise positions and velocities of all the molecules in the system, so we can predict only the probability that an S_i molecule and an $S_{i'}$ molecule will collide in the next dt and then react according to R_j.

It turns out that c_j for a unimolecular reaction is numerically equal to the reaction rate constant k_j of conventional deterministic chemical kinetics, while c_j for a bimolecular reaction is equal to k_j/Ω if the reactants are different species, or $2k_j/\Omega$ if they are the same (Gillespie, 1976, 1992b,a). But it would be wrong to infer from this that the propensity functions are simple heuristic extrapolations of the rates used in deterministic chemical kinetics; in fact, the inference flow actually goes the other way. The existence and forms of the propensity functions follow directly from molecular physics and kinetic theory, and not from deterministic chemical kinetics.

The probabilistic nature of the dynamics described above implies that the most we can hope to compute is the probability $P(\mathbf{x}, t \,|\, \mathbf{x}_0, t_0)$ that $\mathbf{X}(t)$ will equal $\mathbf{x}$, given that $\mathbf{X}(t_0) = \mathbf{x}_0$. We can deduce a time-evolution equation for this function by using the laws of probability to write $P(\mathbf{x}, t + dt \,|\, \mathbf{x}_0, t_0)$ as:

$$P(\mathbf{x}, t + dt \,|\, \mathbf{x}_0, t_0) = P(\mathbf{x}, t \,|\, \mathbf{x}_0, t_0) \times [1 - \sum_{j=1}^{M} a_j(\mathbf{x}) dt]$$

$$+ \sum_{j=1}^{M} P(\mathbf{x} - \boldsymbol{\nu}_j, t \,|\, \mathbf{x}_0, t_0) \times a_j(\mathbf{x} - \boldsymbol{\nu}_j) dt$$

The first term on the right is the probability that the system is already in state $\mathbf{x}$ at time t, and no reaction of any kind occurs in $[t, t + dt)$. The generic second term is the probability that the system is one R_j reaction removed from state $\mathbf{x}$ at time t, and one R_j reaction occurs in $[t, t + dt)$. That these $M + 1$ routes from time t to state $\mathbf{x}$ at time $t + dt$ are mutually exclusive and collectively exhaustive is ensured by taking dt so small that no more than one reaction of any kind can occur in $[t, t + dt)$. Subtracting $P(\mathbf{x}, t \,|\, \mathbf{x}_0, t_0)$ from both sides, dividing through by dt, and taking the limit $dt \to 0$, we obtain (McQuarrie, 1967; Gillespie, 1992b)

$$\frac{\partial P(\mathbf{x}, t \,|\, \mathbf{x}_0, t_0)}{\partial t}$$

$$= \sum_{j=1}^{M} [a_j(\mathbf{x} - \boldsymbol{\nu}_j) P(\mathbf{x} - \boldsymbol{\nu}_j, t \,|\, \mathbf{x}_0, t_0) - a_j(\mathbf{x}) P(\mathbf{x}, t \,|\, \mathbf{x}_0, t_0)] \quad (16.1)$$

This is the *chemical master equation* (CME). In principle, it completely determines the function $P(\mathbf{x}, t \,|\, \mathbf{x}_0, t_0)$. But the CME is really a set of nearly as many coupled ordinary differential equations as there are combinations of molecules that can exist in the system. So it is not surprising that the CME can be solved analytically for only a very few very simple systems, and numerical solutions are usually prohibitively difficult.

One might hope to learn something from the CME about the behavior of averages like $\langle f(\mathbf{X}(t)) \rangle \equiv \sum_{\mathbf{x}} f(\mathbf{x}) P(\mathbf{x}, t \,|\, \mathbf{x}_0, t_0)$, but this too turns out to pose difficulties if any of the reaction channels are bimolecular. For example, it can be proved from equation 16.1 that

$$\frac{d \langle X_i(t) \rangle}{dt} = \sum_{j=1}^{M} \nu_{ij} \langle a_j(\mathbf{X}(t)) \rangle \quad (i = 1, \dots, N)$$

If all the reactions were unimolecular, the propensity functions would all be linear in the state variables, and we would have $\langle a_j(\mathbf{X}(t)) \rangle = a_j(\langle \mathbf{X}(t) \rangle)$. The above equation would then become a closed set of ordinary differential equations for the first moments, $\langle X_i(t) \rangle$. But if any reaction is bimolecular, the right hand side will contain at least one quadratic moment of the form $\langle X_i(t) X_{i'}(t) \rangle$, and the equation then becomes merely the first of an infinite, open-ended set of coupled quations for all the moments.

In the hypothetical case that there are *no fluctuations*, we would have $\langle f\left(\mathbf{X}(t)\right)\rangle = f\left(\mathbf{X}(t)\right)$ for all functions f. The above equation for $\langle X_i(t)\rangle$ would then reduce to

$$\frac{dX_i(t)}{dt} = \sum_{j=1}^{M} \nu_{ij}\, a_j(\mathbf{X}(t)) \quad (i = 1, \ldots, N) \tag{16.2}$$

This is the *reaction rate equation* (RRE) of traditional deterministic chemical kinetics—a set of N coupled first-order ordinary differential equations for the $X_i(t)$, which are now continuous (real) variables. The RRE is more commonly written in terms of the concentration variables $X_i(t)/\Omega$, but that scalar transformation is inconsequential for our purposes here. Examples of RREs in a biological context abound in Chapter 6.

Although the deterministic RRE would evidently be valid in the absence of fluctuations, it is not clear what the justification and penalty might be for ignoring fluctuations. We shall later see how the RRE follows more deductively from a series of physically transparent approximating assumptions to the stochastic theory.

16.3   The Stochastic Simulation Algorithm

Since the CME (eq. 16.1) is rarely of much use in computing the probability density function $P(\mathbf{x}, t \,|\, \mathbf{x}_0, t_0)$ of $\mathbf{X}(t)$, we need another computational approach. One approach that has proven fruitful is to construct *numerical realizations* of $\mathbf{X}(t)$, that is, simulated trajectories of $\mathbf{X}(t)$-versus-t . This is *not* the same as solving the CME numerically, as that would give us the probability density function of $\mathbf{X}(t)$ instead of samplings of that random variable. However, much the same effect can be achieved by either histogramming or averaging the results of many realizations. The key to generating simulated trajectories of $\mathbf{X}(t)$ is not the CME or even the function $P(\mathbf{x}, t \,|\, \mathbf{x}_0, t_0)$, but rather a new function, $p(\tau, j \,|\, \mathbf{x}, t)$ (Gillespie, 1976). It is defined so that $p(\tau, j \,|\, \mathbf{x}, t)\, d\tau$ is the probability, given $\mathbf{X}(t) = \mathbf{x}$, that the *next* reaction in the system will occur in the infinitesimal time interval $[t + \tau, t + \tau + d\tau)$, *and* will be an R_j reaction. Formally, this function is the joint probability density function of the two random variables "time to the next reaction" (τ) and "index of the next reaction" (j).

To derive an analytical expression for $p(\tau, j \,|\, \mathbf{x}, t)$, we begin by noting that if $P_0(\tau \,|\, \mathbf{x}, t)$ is the probability, given $\mathbf{X}(t) = \mathbf{x}$, that no reaction of any kind occurs in the time interval $[t, t + \tau)$, then the laws of probability imply the relation

$$p(\tau, j \,|\, \mathbf{x}, t)\, d\tau = P_0(\tau \,|\, \mathbf{x}, t) \times a_j(\mathbf{x})d\tau$$

The laws of probability also imply

$$P_0(\tau + d\tau \,|\, \mathbf{x}, t) = P_0(\tau \,|\, \mathbf{x}, t) \times [1 - \sum_{j'=1}^{M} a_{j'}(\mathbf{x})d\tau]$$

An algebraic rearrangement of this last equation and passage to the limit $d\tau \to 0$ results in a differential equation whose solution is easily found to be $P_0(\tau \mid \mathbf{x}, t) = \exp(-a_0(\mathbf{x})\tau)$, where

$$a_0(\mathbf{x}) \equiv \sum_{j'=1}^{M} a_{j'}(\mathbf{x}) \tag{16.3}$$

When we insert this result into the equation for p, we get

$$p(\tau, j \mid \mathbf{x}, t) = a_j(\mathbf{x}) \exp(-a_0(\mathbf{x})\tau) \tag{16.4}$$

Equation 16.4 is the mathematical basis for the stochastic simulation approach. It implies that the joint density function of τ and j can be written as the product of the τ-density function, $a_0(\mathbf{x}) \exp(-a_0(\mathbf{x})\tau)$, and the j-density function, $a_j(\mathbf{x})/a_0(\mathbf{x})$. We can generate random samples from these two density functions by using the inversion method of Monte Carlo theory (Gillespie, 1992a). Draw two random numbers r_1 and r_2 from the uniform distribution in the unit-interval, and select τ and j according to

$$\tau = \frac{1}{a_0(\mathbf{x})} \ln\left(\frac{1}{r_1}\right) \tag{16.5a}$$

$$j = \text{ the smallest integer satisfying } \sum_{j'=1}^{j} a_{j'}(\mathbf{x}) > r_2\, a_0(\mathbf{x}) \tag{16.5b}$$

Thus we arrive at the following version of the *stochastic simulation algorithm* (SSA) (Gillespie, 1976, 1977):

1. Initialize the time $t = t_0$ and the system's state $\mathbf{x} = \mathbf{x}_0$.

2. With the system in state $\mathbf{x}$ at time t, evaluate all the $a_j(\mathbf{x})$ and their sum $a_0(\mathbf{x})$.

3. Generate values for τ and j according to equations 16.5a and b.

4. Effect the next reaction by replacing $t \leftarrow t + \tau$ and $\mathbf{x} \leftarrow \mathbf{x} + \boldsymbol{\nu}_j$.

5. Record $(\mathbf{x}, t)$ as desired. Return to step 2, or else end the simulation.

The $\mathbf{X}(t)$ trajectory that is produced by the SSA might be thought of as a "stochastic version" of the trajectory that would be obtained by solving the RRE. But note that the time step τ in the SSA is *exact* and is not a finite approximation to some infinitesimal dt, as is the time step in most numerical solvers for the RRE. If it is found that every SSA-generated trajectory is practically indistinguishable from the RRE trajectory, then we may conclude that microscale fluctuations are ignorable. But if the SSA trajectories deviate noticeably from the RRE trajectory, then we must conclude that microscale fluctuations are not ignorable, and the deterministic RRE does not provide an accurate description of the system's real behavior.

The SSA and the CME are logically equivalent to each other; yet even when the CME is completely intractable, the SSA is quite straightforward to implement. The problem with the SSA is that it is often very slow. The source of this slowness can be traced to the factor $1/a_0(\mathbf{x})$ in the τ equation 16.5a: $a_0(\mathbf{x})$ can be very large if the population of one or more reactant species is large, and that is often the case in practice.

There are variations on the above method for implementing the SSA that make it more computationally efficiency (Gibson and Bruck, 2000b; Cao et al., 2004a). But *any* procedure that simulates every reaction event one at a time will inevitably be too slow for most practical applications. This prompts us to look for ways of giving up some of the exactness of the SSA in return for greater simulation speed.

16.4   Tau-Leaping

One approximate accelerated simulation strategy is tau-leaping (Gillespie, 2001). It advances the system by a *pre selected time* τ which encompasses more than one reaction event. In its simplest form, tau-leaping requires that τ be chosen small enough that the following *leap condition* is satisfied: The expected state change induced by the leap must be sufficiently small that no propensity function changes its value by a significant amount.

We recall that the Poisson random variable $\mathcal{P}(a, \tau)$ is by definition the number of events that will occur in time τ given that $a\, dt$ is the probability that an event will occur in any infinitesimal time dt, where a can be any positive constant. Therefore, if $\mathbf{X}(t) = \mathbf{x}$, and if we choose τ small enough to satisfy the leap condition, so that the propensity functions stay approximately constant during the leap, then reaction R_j should fire approximately $\mathcal{P}_j\left(a_j(\mathbf{x}), \tau\right)$ times in $[t, t + \tau)$. Thus, to the degree that the leap condition is satisfied, we can leap by a time τ simply by taking

$$\mathbf{X}(t + \tau) \doteq \mathbf{x} + \sum_{j=1}^{M} \boldsymbol{\nu}_j\, \mathcal{P}_j\left(a_j(\mathbf{x}), \tau\right) \tag{16.6}$$

Doing this evidently requires generating M Poisson random numbers for each leap (Press et al., 1986). It will result in a faster simulation than the SSA to the degree that the total number of reactions leapt over, $\sum_{j=1}^{M} \mathcal{P}_j\left(a_j(\mathbf{x}), \tau\right)$, is large compared to M.

In order to use this simulation technique efficiently, we obviously need a way to estimate the largest value of τ that is compatible with the leap condition. One possible way of doing that (Gillespie and Petzold, 2003) is to estimate the largest value of τ for which no propensity function is likely to change its value during τ by more than $\varepsilon a_0(\mathbf{x})$, where ε $(0 < \varepsilon \ll 1)$ is some pre-chosen accuracy-control parameter. Whatever the method of selecting τ, the (explicit) tau-leaping simulation procedure goes as follows (Gillespie, 2001; Gillespie and Petzold, 2003):

1. In state $\mathbf{x}$ at time t, choose a value for τ that satisfies the leap condition.

2. For each $j = 1, \ldots, M$, generate the number of firings k_j of reaction R_j in time τ as a sample of the Poisson random variable $\mathcal{P}(a_j(\mathbf{x}), \tau)$.

3. Leap, by replacing $t \leftarrow t + \tau$ and $\mathbf{x} \leftarrow \mathbf{x} + \sum_{j=1}^{M} k_j \boldsymbol{\nu}_j$.

In the limit that $\tau \to 0$, tau-leaping becomes mathematically equivalent to the SSA. But tau-leaping also becomes very inefficient in that limit, because all the k_j's will approach zero, giving a very small time step with usually no reactions firing. As a practical matter, tau-leaping should not be used if the largest value of τ that satisfies the leap condition is less than a few multiples of $1/a_0(\mathbf{x})$, the expected time to the next reaction in the SSA, since it would then be more efficient to use the SSA.

Tau-leaping has been shown to significantly speed up the simulation of some systems (Gillespie, 2001; Gillespie and Petzold, 2003). But it is not as foolproof as the SSA. If one takes leaps that are too large, bad things can happen; for example, some species populations might be driven negative. If the system is stiff, meaning that it has widely varying dynamical modes with the fastest mode being stable, the leap condition will generally limit the size of τ to the time scale of the fastest mode, with the result that large leaps cannot be taken. Stiffness is very common in cellular chemical systems and will be considered in more detail later.

It is tempting to try to formulate a "higher-order" tau-leaping formula by extending higher-order ODE methods in a straightforward manner for discrete stochastic simulation. However, doing this correctly is much harder than it might at first appear. Most such extensions are not even first order accurate for the stochastic part of the system. An analysis of the consistency, order, and convergence of tau-leaping methods is given by Rathinam et al. (2005), where it is shown that the tau-leaping method defined above, and the "implicit" tau-leaping method to be described in section 16.7, are both first-order accurate as $\tau \to 0$.

16.5   Transitioning to the Macroscale: The Chemical Langevin Equation and the Reaction Rate Equation

Suppose we can choose τ small enough to satisfy the leap condition, so that approximation 16.6 is good, but nevertheless large enough that

$$a_j(\mathbf{x})\, \tau \gg 1 \quad \text{for all } j = 1, \ldots, M \tag{16.7}$$

Since $a_j(\mathbf{x})\tau$ is the mean of the random variable $\mathcal{P}_j(a_j(\mathbf{x}), \tau)$, the physical significance of condition 16.7 is that each reaction channel is expected to fire many more times than once in the next τ. It will not always be possible to find a τ that satisfies both the leap condition and condition 16.7, but it usually will be if the populations of all the reactant species are sufficiently large.

When condition 16.7 does hold, we can make a useful approximation to the tau-leaping formula 16.6. This approximation stems from the purely mathematical fact that the Poisson random variable $\mathcal{P}(a, \tau)$, which has mean and variance $a\tau$, can be well approximated when $a\tau \gg 1$ by a normal random variable with the same mean and variance. Denoting the normal random variable with mean m and variance σ^2 by $\mathcal{N}(m, \sigma^2)$, it thus follows that when condition 16.7 holds,

$$\mathcal{P}_j\left(a_j(\mathbf{x}), \tau\right) \doteq \mathcal{N}_j\left(a_j(\mathbf{x})\tau, a_j(\mathbf{x})\tau\right) = a_j(\mathbf{x})\tau + (a_j(\mathbf{x})\tau)^{1/2}\,\mathcal{N}_j(0, 1)$$

the last step following from the fact that $\mathcal{N}(m, \sigma^2) = m + \sigma\mathcal{N}(0, 1)$. Inserting this approximation into equation 16.6 gives (Gillespie, 2000, 2002)

$$\mathbf{X}(t+\tau) \doteq \mathbf{x} + \sum_{j=1}^{M} \boldsymbol{\nu}_j a_j(\mathbf{x})\tau + \sum_{j=1}^{M} \boldsymbol{\nu}_j \sqrt{a_j(\mathbf{x})}\mathcal{N}_j(0, 1)\sqrt{\tau} \qquad (16.8)$$

where the $\mathcal{N}_j(0, 1)$ are statistically independent normal random variables with means 0 and variances 1. Equation 16.8 is called the Langevin leaping formula. It evidently expresses the state increment $\mathbf{X}(t+\tau) - \mathbf{x}$ as the sum of two terms: a deterministic *drift term* proportional to τ, and a fluctuating *diffusion term* proportional to $\sqrt{\tau}$. It is important to keep in mind that equation 16.8 is an approximation, which is valid only to the extent that τ is (i) small enough that no propensity function changes its value significantly during τ, yet (ii) large enough that every reaction fires many more times than once during τ. The approximate nature of equation 16.8 is underscored by the fact that $\mathbf{X}(t)$ therein is now a continuous (real-valued) random variable instead of a discrete (integer-valued) random variable; we lost discreteness when we replaced the integer-valued Poisson random variable with a real-valued normal random variable. The Langevin leaping formula 16.8 gives faster simulations than the tau-leaping formula 16.6 not only because condition 16.7 implies that very many reactions get leapt over at each step, but also because the normal random numbers that are required by equation 16.8 can be generated much more easily than the Poisson random numbers that are required by equation 16.6 (Press et al., 1986).

The "small-but-large" character of τ in equation 16.8 marks that variable as a "macroscopic infinitesimal." If we subtract $\mathbf{x}$ from both sides and then divide through by τ, the result can be shown to be the following (approximate) stochastic differential equation, which is called the *chemical Langevin equation* (CLE) (Gillespie, 2000, 2002):

$$\frac{d\mathbf{X}(t)}{dt} \doteq \sum_{j=1}^{M} \boldsymbol{\nu}_j\, a_j\left(\mathbf{X}(t)\right) + \sum_{j=1}^{M} \boldsymbol{\nu}_j \sqrt{a_j\left(\mathbf{X}(t)\right)}\,\Gamma_j(t) \qquad (16.9)$$

The $\Gamma_j(t)$ here are statistically independent "Gaussian white noise" processes satisfying $\langle\Gamma_j(t)\,\Gamma_{j'}(t')\rangle = \delta_{jj'}\,\delta(t - t')$, where the first delta function is Kronecker's and the second is Dirac's. The CLE (equation 16.9) is mathematically equivalent to the Langevin leaping formula (equation 16.8), and is subject to the same conditions for validity. Stochastic differential equations arise in many areas of physics, but the usual way of obtaining them is to start with a macroscopically inspired drift term (the first term on the right side of the CLE) and then *assume* a form for the diffusion term (the second term on the right side of the CLE) with an eye to obtaining some pre-conceived outcome. So it is noteworthy that our derivation here of the CLE did not proceed in that ad hoc manner; instead, we used careful mathematical approximations to infer the forms of both the drift and diffusion terms from the premises underlying the CME/SSA.

Molecular systems become "macroscopic" in what physicists and chemists call the *thermodynamic limit*. This limit is formally defined as follows: the system volume Ω and the species populations X_i all approach ∞, but in such a way that the species concentrations X_i/Ω all remain constant. The large molecular populations in chemical systems near the thermodynamic limit generally mean that such systems will be well described by the Langevin formulas 16.8 and 16.9. To discern the implications of those formulas in the thermodynamic limit, we evidently need to know the behavior of the propensity functions in that limit. It turns out that all propensity functions grow *linearly* with the system size as the thermodynamic limit is approached. For a unimolecular propensity function of the form $c_j x_i$ this behavior is obvious, since c_j will be independent of the system size. For a bimolecular propensity function of the form $c_j x_i x_{i'}$ this behavior is a consequence of the fact that bimolecular c_j's are always inversely proportional to Ω, reflecting the fact that two reactant molecules have a harder time finding each other in larger volumes.

It follows that, as the thermodynamic limit is approached, the deterministic drift term in equation 16.8 grows like the size of the system, while the fluctuating diffusion term grows like the square root of the size of the system, and likewise for the CLE. This establishes the well known rule-of-thumb in chemical kinetics that relative fluctuation effects in chemical systems typically scale as the inverse square root of the size of the system.

In the full thermodynamic limit, the size of the second term on the right side of equation 16.9 will usually be negligibly small compared to the size of the first term, in which case the CLE reduces to the RRE. Thus we have derived the RRE as a series of limiting approximations to the stochastic theory that underlies the CME and the SSA. The tau-leaping and Langevin-leaping formulas evidently provide a conceptual bridge between stochastic chemical kinetics (the CME and SSA) and conventional deterministic chemical kinetics (the RRE), enabling us to see how the latter emerges as a limiting approximation of the former.

16.6   Stiffness in Deterministic Reaction Rate Equations

Stiffness can be defined roughly as the presence of widely varying time-scales in a dynamical system, the fastest of which is stable. It poses special problems for the numerical solution of both deterministic ordinary differential equations (ODEs) and stochastic differential equations (SDEs), particularly in the context of chemical kinetics. Stiffness also impacts both the SSA and the tau-leaping algorithm equation 16.6. In this section we will describe the phenomenon of stiffness for deterministic systems of ODEs, and show how it restricts the timestep size for all "explicit" methods. Then we will show how the use of "implicit" methods overcomes this restriction.

Consider the deterministic ODE system

$$\frac{d\mathbf{x}}{dt} = \mathbf{f}(t, \mathbf{x}) \tag{16.10}$$

In simplest terms, this system is said to be "stiff" if it has a strongly damped, or "superstable" mode. To get a feeling for this concept, consider the solutions $\mathbf{x}(t)$ of an ODE system starting from various initial conditions. For a typical nonstiff system, if we plot a given component of the vector $\mathbf{x}$-versus-t we might get a family of curves resembling those shown in figure 16.1a: The curves either remain roughly the same distance apart as t increases, as in the figure, or they might show a tendency to merge very slowly. But when such a family of curves is plotted for a typical stiff system, the result looks more like what is shown in figure 16.1b: The curves merge rapidly to one or more smoother curves, with the deviation from the smoother curves becoming very small as t increases.

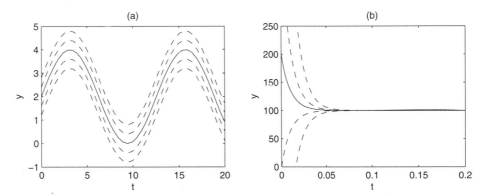

Figure 16.1   A system of ODEs is said to be "stiff" if its solutions show strongly damped behavior as a function of the initial conditions. The family of curves shown in (a) represents the behavior of solutions to a nonstiff system for various initial conditions. In contrast, solutions to the stiff system shown in (b) tend to merge quickly.

Stiffness in a system of ODEs corresponds to a strongly stable behavior of the physical system being modeled. At any given time the system will be in a sort of equilibrium, although not necessarily a static one, and if some state variable is perturbed slightly, the system will respond rapidly to restore the equilibrium. Typically, the true solution $\mathbf{x}(t)$ of the ODE system does not show any rapid variation, except possibly during an initial transient phase. But the potential for rapid response is always present and will manifest itself if we perturb $\mathbf{x}$ out of equilibrium. A stiff system has (at least) two time scales. There is a long (slow) timescale for the quasi-equilibrium phase, and a short (fast) timescale for the transient phase following a perturbation. The more different these two time scales are, the stiffer the system is said to be.

The smallest (fastest) timescale in a stiff system manifests itself in another way when we try to carry out a numerical solution of the system. Solution by an explicit time stepping method, such as the simple *explicit Euler method*

$$\mathbf{x}_n = \mathbf{x}_{n-1} + \tau \mathbf{f}(t_{n-1}, \mathbf{x}_{n-1}) \tag{16.11}$$

where $t_n = t_{n-1} + \tau$ and $\mathbf{x}_n$ is the numerical approximation to $\mathbf{x}(t_n)$, will produce very inaccurate results unless the time stepsize τ is kept smaller than the smallest time scale in the system.

To see why this is so, let us consider a simple example: the reversible isomerization reaction, $S_1 \underset{c_2}{\overset{c_1}{\rightleftharpoons}} S_2$. Let x_T denote the (constant) total number of molecules of the two isomeric species, and $x(t)$ the time-varying number of S_1 molecules. The deterministic RRE for this system is the ODE

$$\frac{dx}{dt} = -c_1 x + c_2(x_T - x) = -(c_1 + c_2)x + c_2 x_T \tag{16.12}$$

The solution to this ODE for the initial condition $x(0) = x_0$, is given by

$$x(t) = \frac{c_2 x_T}{c_1 + c_2} + \left(x_0 - \frac{c_2 x_T}{c_1 + c_2} \right) e^{-(c_1 + c_2)t}$$

From the form of this solution, we can see that if the initial value x_0 differs from the asymptotic value $\frac{c_2 x_T}{c_1 + c_2}$, the solution will relax to that asymptotic value in a time of order $(c_1 + c_2)^{-1}$; therefore, if $(c_1 + c_2)$ is very large, this system will be stiff. In figure 16.2 we show the exact solution of the reversible isomerization reaction 16.12 along with numerical solutions obtained using the explicit Euler method (equation 16.11) with two different stepsizes τ. Note that the smaller stepsize Euler solution is accurate, but the larger stepsize solution is unstable.

To see why this instability arises, let us write down the explicit Euler method (equation 16.11) with stepsize τ for the ODE (equation 16.12):

$$x_n = x_{n-1} - \tau(c_1 + c_2)x_{n-1} + \tau c_2 x_T \tag{16.13}$$

If we expand the true solution $x(t)$ in a Taylor series about t_{n-1}, we get

$$x(t_n) = x(t_{n-1}) - \tau(c_1 + c_2)x(t_{n-1}) + \tau c_2 x_T + O(\tau^2) \tag{16.14}$$

Subtracting 16.14 from 16.13, and defining the error $e_n = x_n - x(t_n)$, we obtain

$$e_n = e_{n-1} - \tau(c_1 + c_2)e_{n-1} + O(\tau^2) \tag{16.15}$$

Thus, e_n is given by the recurrence formula

$$e_n = (1 - \tau(c_1 + c_2)) e_{n-1} + O(\tau^2) \tag{16.16}$$

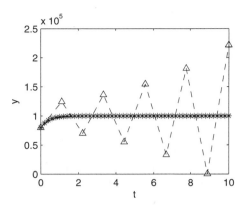

Figure 16.2 Exact solution of equation 16.12 (solid line) and its explicit Euler approximations for stepsizes 0.2 (asterisks) and 1.1 (triangles) with $c_1 = c_2 = 1$ and $x_T = 2 \times 10^5$. The fast time constant for this problem is $(c_1 + c_2)^{-1} = 0.5$.

If $\tau > 2(c_1 + c_2)^{-1}$, then $|1 - \tau(c_1 + c_2)|$ will be greater than 1, and we will have $|e_n| > |e_{n-1}|$. The recurrence will then be unstable. In general, to ensure the stability of an explicit method, we must restrict the stepsize to the timescale of the fastest mode, even though much larger stepsizes might seem perfectly acceptable for getting an adequate resolution of the solution curve.

The restriction of the explicit Euler method to timesteps τ that are on the order of the short (fast) timescale makes the method very slow for stiff systems. So it is natural to ask if there are other solution methods for which the timesteps are not restricted by stability, but only by the need to resolve the solution curve. It is now widely recognized that a general way of doing this is provided by implicit methods (Ascher and Petzold, 1998), the simplest of which is the *implicit Euler method*. For the ODE (equation 16.10), it reads

$$\mathbf{x}_n = \mathbf{x}_{n-1} + \tau \mathbf{f}(t_n, \mathbf{x}_n) \tag{16.17}$$

In contrast to the explicit Euler formula (equation 16.11), this method is implicit because $\mathbf{x}_n$ is not defined entirely in terms of past values of the solution; instead, it is defined implicitly as the solution of the (possibly nonlinear) equation 16.17. We can write this system abstractly as

$$\mathbf{F}(\mathbf{u}) = 0 \tag{16.18}$$

where $\mathbf{u} = \mathbf{x}_n$ and $\mathbf{F}(\mathbf{u}) = \mathbf{u} - \mathbf{x}_{n-1} - \tau \mathbf{f}(t_n, \mathbf{u})$. Usually, the most efficient way to numerically solve equation 16.18 is by *Newton iteration*: One iterates the formula

$$\left(\frac{\partial \mathbf{F}}{\partial \mathbf{u}}\right)[\mathbf{u}^{(m+1)} - \mathbf{u}^{(m)}] = -\mathbf{F}(\mathbf{u}^{(m)}) \tag{16.19}$$

over m, where $\mathbf{u}^{(m)}$ is the mth iterated approximation to the exact root of F, and the Jacobian matrix $\partial \mathbf{F}/\partial \mathbf{u}$ is evaluated at $\mathbf{u}^{(m)}$. This is a linear system of equations, which is to be solved at each iteration for $\mathbf{u}^{m+1}$. Newton's method converges in one iteration for linear systems, and the convergence is quite rapid for most nonlinear systems given a good initial guess. The initial guess is usually obtained by evaluating a polynomial that coincides with recent past solution values at t_n. In practice, the Jacobian matrix is usually not reevaluated at each iteration; also, it is often approximated by numerical difference quotients rather than evaluated exactly. The use of an approximate Jacobian matrix that is fixed throughout the iteration is called *modified Newton* iteration. On first glance, it might seem that the expense of solving the nonlinear system at each time step would outweigh the advantage of increased stability; however, this is usually not so. For stiff systems, implicit methods are usually able to take timesteps that are so much larger than those of explicit methods that the implicit methods wind up being much more efficient.

To see why the implicit Euler method does not need to restrict the step size to maintain stability for stiff systems, let us consider again the reversible isomerization reaction (equation 16.12). For it, the implicit Euler method reads (cf. equation 16.13)

$$x_n = x_{n-1} - \tau(c_1 + c_2)x_n + \tau c_2 x_{\mathrm{T}} \tag{16.20}$$

Expanding the true solution in a Taylor series about t_n, we get (cf. equation 16.14)

$$x(t_n) = x(t_{n-1}) - \tau(c_1 + c_2)x(t_n) + \tau c_2 x_{\mathrm{T}} + O(\tau^2) \tag{16.21}$$

Subtracting 16.21 from 16.20, we find that the error $e_n = x_n - x(t_n)$ now satisfies (cf. equation 16.15)

$$e_n = e_{n-1} - \tau(c_1 + c_2)e_n + O(\tau^2) \tag{16.22}$$

Solving this for e_n, we get

$$e_n = \frac{e_{n-1}}{1 + \tau(c_1 + c_2)} + O(\tau^2) \tag{16.23}$$

In contrast to the error eq. 16.16 for the *explicit* Euler method, the error for the implicit Euler method remains small for arbitrarily large values of $\tau(c_1 + c_2)$, as seen in figure 16.3.

For the general ODE system (eq. 16.10), the negative eigenvalues of the matrix $J = \partial \mathbf{f}/\partial \mathbf{x}$ play the role of $(c_1 + c_2)$. For stiff systems, the eigenvalues λ of J will include at least one with a relatively large negative real part, corresponding in the case of an RRE to the fastest reactions. The set of complex numbers $\tau\lambda$ satisfying $|1 + \tau\lambda| < 1$ is called the *region of absolute stability* of the explicit Euler method. The corresponding region for the implicit Euler method is given by $1/|1 - \tau\lambda| < 1$, and it will be much larger.

A great deal of work has gone into the numerical solution of stiff systems of ODEs (and of ODEs coupled with nonlinear constraints, called differential algebraic equations (DAEs)). There is extensive theory and highly efficient and reliable

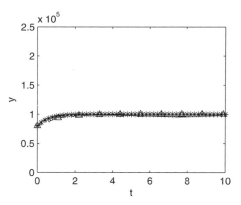

Figure 16.3   The implicit Euler method overcomes a weakness of the explicit Euler method in that it does not need to restrict the step size to provide stable solutions for stiff systems. The figure shows the true solution of the deterministic reversible isomerization reaction 16.12 (solid line), and the numerical solution by the implicit Euler method for stepsizes 0.2 (asterisks) and 1.1 (triangles) with $c_1 = c_2 = 1$ and $x_T = 2 \times 10^5$. Note the excellent agreement, in contrast to the case with the explicit Euler method shown in figure 16.2.

software which adapts both the method order and the timestep to the given problem. See Ascher and Petzold (1998), for more details.

16.7   Stiffness in Stochastic Chemical Kinetics: The Implicit Tau-Leaping Method

When stochasticity is introduced into a chemical system that has fast and slow time scales, with the fast mode being stable as before, we may still expect there to be a slow manifold corresponding to the equilibrium of the fast reactions. But stochasticity changes the picture in a fundamental way: once the system reaches the slow manifold, naturally occurring fluctuations will drive it back off, leading to persistent random fluctuations transverse to the slow manifold. If these fluctuations are negligibly small, then an implicit scheme which takes large steps (on the time scale of the slow mode) will do just fine. But if the fluctuations off the slow manifold are noticeable, then an implicit scheme that takes steps much larger than the time scale of the fast dynamics will dampen the fluctuations, and thus fail to reproduce them correctly. Fortunately, this failing can usually be corrected by using a procedure called *down-shifting*, which we will describe shortly.

The original tau-leaping method (equation 16.6) is explicit because the propensity functions a_j are evaluated at the current (known) state, so the future (unknown) random state $\mathbf{X}(t + \tau)$ is given as an explicit function of $\mathbf{X}(t)$. It is this explicit nature of equation 16.6 that leads to stability problems when stiffness is present, just as with ordinary differential equations. One way of making the explicit tau-leaping formula 16.6 implicit is to modify it as follows (Rathinam et al., 2003):

$$\mathbf{X}(t+\tau) \doteq \mathbf{X}(t) + \sum_{j=1}^{M} \boldsymbol{\nu}_j a_j(\mathbf{X}(t+\tau))\,\tau$$

$$+ \sum_{j=1}^{M} \boldsymbol{\nu}_j \left[\mathcal{P}_j(a_j(\mathbf{X}(t)), \tau) - a_j(\mathbf{X}(t))\,\tau \right] \quad (16.24)$$

Since the random variables $\mathcal{P}_j(a_j(\mathbf{X}(t), \tau)$ here can be generated without knowing $\mathbf{X}(t+\tau)$, then once values for those random variables are set, equation 16.24 becomes an ordinary implicit equation for the unknown state $\mathbf{X}(t+\tau)$, and $\mathbf{X}(t+\tau)$ can then be found by applying Newton iteration to equation 16.24.

Just as the explicit tau method segues to the explicit Euler methods for SDEs and ODEs, the implicit tau method segues to the implicit Euler methods for SDEs and ODEs. In the SDE regime we get, approximating Poissons random variables by normal random variables, the implicit version of the Langevin leaping formula :

$$\mathbf{X}(t+\tau) \doteq \mathbf{X}(t) + \tau \sum_{j=1}^{M} \boldsymbol{\nu}_j a_j(\mathbf{X}(t+\tau)) + \sum_{j=1}^{M} \boldsymbol{\nu}_j \sqrt{a_j(\mathbf{X}(t))} \mathcal{N}_j(0,1)\sqrt{\tau} \quad (16.25)$$

Here, the $\mathcal{N}_j(0,1)$ are, as in eq. 16.8, independent normal random variables with mean zero and variance 1. And in the thermodynamic limit, where the random terms in eq. 16.25 may be ignored, it reduces to the implicit Euler method

$$\mathbf{X}(t+\tau) \doteq \mathbf{X}(t) + \tau \sum_{j=1}^{M} \boldsymbol{\nu}_j a_j(\mathbf{X}(t+\tau))\,. \quad (16.26)$$

for the deterministic RRE .

We noted earlier that the implicit tau method, when used with a relatively large timestep, will dampen the natural fluctuations of the fast variables. Thus, although the implicit tau-leaping method computes the slow variables with their correct distributions, it computes the fast variables with the correct means but with spreads about those means that are too narrow. Fortunately, a time-stepping strategy called *down-shifting* can restore the overly-damped fluctuations in the fast variables. The idea is to interlace the implicit tau-leaps, each of which is on the order of the time scale of the slow variables and hence "large," with a sequence of much smaller time steps on the time scale of the fast variables, these being taken using either the explicit tau method or the SSA. This sequence of smaller steps "regenerates" the correct statistical distributions of the fast variables. Further details on implicit tau-leaping and down-shifting can be found in Rathinam et al. (2003).

16.8 Stiffness in Stochastic Chemical Kinetics: The Slow-Scale SSA

Another way to deal with stiffness in stochastic systems is to use the recently developed (Cao et al., 2005) *slow-scale* SSA (ssSSA). The first step in setting up the ssSSA is to divide (and reindex) the M reaction channels $\{R_1, \ldots, R_M\}$ into fast and slow subsets, $\{R_1^{\mathrm{f}}, \ldots, R_{M_{\mathrm{f}}}^{\mathrm{f}}\}$ and $\{R_1^{\mathrm{s}}, \ldots, R_{M_{\mathrm{s}}}^{\mathrm{s}}\}$, where $M_{\mathrm{f}} + M_{\mathrm{s}} = M$. We initially do this provisionally (subject to possible later change) according to the following criterion: the propensity functions of the fast reactions, $a_1^{\mathrm{f}}, \ldots, a_{M_{\mathrm{f}}}^{\mathrm{f}}$, should usually be very much larger than the propensity functions of the slow reactions, $a_1^{\mathrm{s}}, \ldots, a_{M_{\mathrm{s}}}^{\mathrm{s}}$. The broad result of this partitioning will be that the time to the occurrence of the next fast reaction will usually be very much smaller than the time to the occurrence of the next slow reaction.

Next we divide (and reindex) the N species $\{S_1, \ldots, S_N\}$ into fast and slow subsets, $\{S_1^{\mathrm{f}}, \ldots, S_{N_{\mathrm{f}}}^{\mathrm{f}}\}$ and $\{S_1^{\mathrm{s}}, \ldots, S_{N_{\mathrm{s}}}^{\mathrm{s}}\}$, where $N_{\mathrm{f}} + N_{\mathrm{s}} = N$. This gives rise to a like partitioning of the state vector $\mathbf{X}(t) = \big(\mathbf{X}^{\mathrm{f}}(t), \mathbf{X}^{\mathrm{s}}(t)\big)$, and also the generic state space variable $\mathbf{x} = (\mathbf{x}^{\mathrm{f}}, \mathbf{x}^{\mathrm{s}})$, into fast and slow components. The criterion for making this partitioning is simple: a fast species is any species whose population gets changed by some fast reaction; all the other species are called slow. Note the asymmetry in this definition: a slow species cannot get changed by a fast reaction, but a fast species can get changed by a slow reaction. Note also that a_j^{f} and a_j^{s} can both depend on both fast and slow variables. The state-change vectors can now be re-indexed

$$\boldsymbol{\nu}_j^{\mathrm{f}} \equiv \big(\nu_{1j}^{\mathrm{ff}}, \ldots, \nu_{N_{\mathrm{f}}j}^{\mathrm{ff}}\big), \quad j = 1, \ldots, M_{\mathrm{f}},$$

$$\boldsymbol{\nu}_j^{\mathrm{s}} \equiv \big(\nu_{1j}^{\mathrm{fs}}, \ldots, \nu_{N_{\mathrm{f}}j}^{\mathrm{fs}}, \nu_{1j}^{\mathrm{ss}}, \ldots, \nu_{N_{\mathrm{s}}j}^{\mathrm{ss}}\big), \quad j = 1, \ldots, M_{\mathrm{s}},$$

where $\nu_{ij}^{\sigma\rho}$ denotes the change in the number of molecules of species S_i^{σ} ($\sigma = \mathrm{f}, \mathrm{s}$) induced by one reaction R_j^{ρ} ($\rho = \mathrm{f}, \mathrm{s}$). We can regard $\boldsymbol{\nu}_j^{\mathrm{f}}$ as a vector with the same dimensionality (N_{f}) as $\mathbf{X}^{\mathrm{f}}$, because $\nu_{ij}^{\mathrm{sf}} \equiv 0$ (slow species do not get changed by fast reactions).

The next step in setting up the ssSSA is to introduce the *virtual fast process* $\hat{\mathbf{X}}^{\mathrm{f}}(t)$. It is composed of the same fast species state variables as the real fast process $\mathbf{X}^{\mathrm{f}}(t)$, but it evolves only through the fast reactions; that is, $\hat{\mathbf{X}}^{\mathrm{f}}(t)$ is $\mathbf{X}^{\mathrm{f}}(t)$ with all the slow reactions switched off. To the extent that the slow reactions don't occur very often, we may expect $\hat{\mathbf{X}}^{\mathrm{f}}(t)$ and $\mathbf{X}^{\mathrm{f}}(t)$ to be very similar to each other. But from a mathematical standpoint there is an profound difference: $\mathbf{X}^{\mathrm{f}}(t)$ by itself is not a Markov (past-forgetting) process, whereas $\hat{\mathbf{X}}^{\mathrm{f}}(t)$ is. Since the evolution of $\mathbf{X}^{\mathrm{f}}(t)$ depends on the evolving slow process $\mathbf{X}^{\mathrm{s}}(t)$, $\mathbf{X}^{\mathrm{f}}(t)$ is not governed by a master equation of the simple Markovian form (equation 16.1); indeed, the easiest way to find $\mathbf{X}^{\mathrm{f}}(t)$ would be to solve the Markovian master equation for the full process $\mathbf{X}(t) \equiv \big(\mathbf{X}^{\mathrm{f}}(t), \mathbf{X}^{\mathrm{s}}(t)\big)$, which is something we have tacitly assumed cannot be done. But for the *virtual* fast process $\hat{\mathbf{X}}^{\mathrm{f}}(t)$, the slow process $\mathbf{X}^{\mathrm{s}}(t)$ stays fixed at some constant initial value $\mathbf{x}_0^{\mathrm{s}}$; therefore, $\hat{\mathbf{X}}^{\mathrm{f}}(t)$ evolves according to the Markovian master equation,

$$
\frac{\partial \hat{P}(\mathbf{x}^{\mathrm{f}}, t \,|\, \mathbf{x}_0, t_0)}{\partial t}
$$
$$
= \sum_{j=1}^{M_{\mathrm{f}}} \left[a_j^{\mathrm{f}}(\mathbf{x}^{\mathrm{f}} - \boldsymbol{\nu}_j^{\mathrm{f}}, \mathbf{x}_0^{\mathrm{s}}) \hat{P}(\mathbf{x}^{\mathrm{f}} - \boldsymbol{\nu}_j^{\mathrm{f}}, t \,|\, \mathbf{x}_0, t_0) - a_j^{\mathrm{f}}(\mathbf{x}^{\mathrm{f}}, \mathbf{x}_0^{\mathrm{s}}) \hat{P}(\mathbf{x}^{\mathrm{f}}, t \,|\, \mathbf{x}_0, t_0) \right]
$$

wherein $\hat{P}(\mathbf{x}^{\mathrm{f}}, t \,|\, \mathbf{x}_0, t_0)$ is the probability that $\hat{\mathbf{X}}^{\mathrm{f}}(t) = \mathbf{x}^{\mathrm{f}}$, given that $\mathbf{X}(t_0) = \mathbf{x}_0$. This master equation for $\hat{\mathbf{X}}^{\mathrm{f}}(t)$ will be simpler than the master equation for $\mathbf{X}(t)$ because it has fewer reactions and fewer species.

Finally, in order to apply the ssSSA, we require that two conditions be satisfied. The first condition is that the virtual fast process $\hat{\mathbf{X}}^{\mathrm{f}}(t)$ be stable, in the sense that it approaches a well defined, time-independent random variable $\hat{\mathbf{X}}^{\mathrm{f}}(\infty)$ as $t \to \infty$; thus, we require the limit

$$
\lim_{t \to \infty} \hat{P}(\mathbf{x}^{\mathrm{f}}, t \,|\, \mathbf{x}_0, t_0) \equiv \hat{P}(\mathbf{x}^{\mathrm{f}}, \infty \,|\, \mathbf{x}_0)
$$

to exist. $\hat{P}(\mathbf{x}^{\mathrm{f}}, \infty \,|\, \mathbf{x}_0)$ can be calculated from the stationary form of the time-dependent master equation,

$$
0 = \sum_{j=1}^{M_{\mathrm{f}}} \left[a_j^{\mathrm{f}}(\mathbf{x}^{\mathrm{f}} - \boldsymbol{\nu}_j^{\mathrm{f}}, \mathbf{x}_0^{\mathrm{s}}) \hat{P}(\mathbf{x}^{\mathrm{f}} - \boldsymbol{\nu}_j^{\mathrm{f}}, \infty \,|\, \mathbf{x}_0) - a_j^{\mathrm{f}}(\mathbf{x}^{\mathrm{f}}, \mathbf{x}_0^{\mathrm{s}}) \hat{P}(\mathbf{x}^{\mathrm{f}}, \infty \,|\, \mathbf{x}_0) \right]
$$

which will be easier to solve since it is purely algebraic. The second condition we impose is that the relaxation of $\hat{\mathbf{X}}^{\mathrm{f}}(t)$ to its stationary asymptotic form $\hat{\mathbf{X}}^{\mathrm{f}}(\infty)$ happen very quickly on the time scale of the slow reactions. More precisely, we require that the relaxation time of the virtual fast process be very much less than the expected time to the next slow reaction.

These two conditions will generally be satisfied if the system is stiff. If satisfying them can be accomplished only by making some changes in the way we originally partitioned the reactions into fast and slow subsets, then we do that, regardless of propensity function values. But if these conditions cannot be satisfied, we must conclude that the ssSSA is not applicable.

Given the forgoing definitions and conditions, it is possible to prove the *slow-scale approximation*(Cao et al., 2005): if the system is in state $(\mathbf{x}^{\mathrm{f}}, \mathbf{x}^{\mathrm{s}})$ at time t, and if Δ_{s} is a time increment that is very large compared to the relaxation time of $\hat{\mathbf{X}}^{\mathrm{f}}(t)$ but very small compared to the expected time to the next slow reaction, then the probability that one R_j^{s} reaction will occur in the time interval $[t, t + \Delta_{\mathrm{s}})$ can be well approximated by $\bar{a}_j^{\mathrm{s}}(\mathbf{x}^{\mathrm{s}}; \mathbf{x}^{\mathrm{f}}) \, \Delta_{\mathrm{s}}$, where

$$
\bar{a}_j^{\mathrm{s}}(\mathbf{x}^{\mathrm{s}}; \mathbf{x}^{\mathrm{f}}) \triangleq \sum_{\mathbf{x}^{\mathrm{f}'}} \hat{P}(\mathbf{x}^{\mathrm{f}'}, \infty \,|\, \mathbf{x}^{\mathrm{f}}, \mathbf{x}^{\mathrm{s}}) \, a_j^{\mathrm{s}}(\mathbf{x}^{\mathrm{f}'}, \mathbf{x}^{\mathrm{s}}) \tag{16.28}
$$

We call $\bar{a}_j^s(\mathbf{x}^s; \mathbf{x}^f)$ the *slow-scale propensity* function for reaction channel R_j^s because it serves as a propensity function for R_j on the timescale of the slow reactions. Mathematically, it is the average of the regular R_j^s propensity function over the fast variables, treated as though they were distributed according to the asymptotic virtual fast process $\hat{\mathbf{X}}^f(\infty)$.

The slow-scale SSA is an immediate consequence of this slow-scale Approximation. The idea is to move the system forward in time in the manner of the SSA one slow reaction at a time, updating the fast variables after each step by randomly sampling $\hat{\mathbf{X}}^f(\infty)$ (Cao et al., 2005).

To illustrate how the ssSSA works, consider the simple reaction set

$$S_1 \underset{c_2}{\overset{c_1}{\rightleftharpoons}} S_2 \overset{c_3}{\longrightarrow} S_3 \tag{16.29}$$

under the condition

$$c_2 \gg c_3 \tag{16.30}$$

Here, an S_2 molecule is most likely to change into an S_1 molecule, a change that is relatively unimportant since it will eventually be reversed. On rare occasions, though, an S_2 molecule will instead change into an S_3 molecule, a potentially more important change since it is irreversible. This simple model has been used to help understand certain features of the heat shock response mechanism in *E. Coli* (El-Samad and Khammash, 2006). Roughly, S_2 can be thought of as the active form of an enzyme which either gets deactivated via reaction R_2 (and subsequently reactivated via reaction R_1), or gets bound to a DNA promoter site via reaction R_3 to allow the transcription of an important gene. In the heat shock model, we are particularly interested in the case in which the average number of S_2 molecules is very small, even less than 1.

We shall take the fast reactions to be R_1 and R_2, and the slow reaction to be R_3. Then the fast species will be S_1 and S_2, and the slow species S_3. The virtual fast process $\hat{\mathbf{X}}^f(t)$ will be the S_1 and S_2 populations undergoing only the fast reactions R_1 and R_2. Unlike the real fast process, which gets affected whenever R_3 fires, the virtual fast process obeys the conservation relation

$$\hat{X}_1(t) + \hat{X}_2(t) = x_T \quad \text{(constant)} \tag{16.31}$$

This relation greatly simplifies the analysis of the virtual fast process, since it reduces the problem to a single independent state variable.

Eliminating $\hat{X}_2(t)$ in favor of $\hat{X}_1(t)$ by means of equation 16.31, we see that given $\hat{X}_1(t) = x_1'$, $\hat{X}_1(t + dt)$ will equal $x_1' - 1$ with probability $c_1 x_1' dt$, and $x_1' + 1$ with probability $c_2(x_T - x_1')dt$. $\hat{X}_1(t)$ is therefore what is known mathematically as a "bounded birth-death" Markov process. It can be shown (Gillespie, 2002) that this process has, for any initial value $x_1 \in [0, x_T]$, the asymptotic stationary distribution

$$\hat{P}(x_1', \infty \mid x_T) = \frac{x_T!}{x_1!\,(x_T - x_1')!} q^{x_1'}(1 - q)^{x_T - x_1'}, \quad (x_1' = 0, 1, \ldots, x_T) \tag{16.32}$$

where $q \equiv c_2/(c_1 + c_2)$. This tells us that $\hat{X}_1(\infty)$ is the binomial random variable $\mathcal{B}(q, x_\mathrm{T})$, whose mean and variance are given by

$$\left\langle \hat{X}_1(\infty) \right\rangle = x_\mathrm{T} q = \frac{c_2 x_\mathrm{T}}{c_1 + c_2} \tag{16.33a}$$

$$\mathrm{var}\left\{ \hat{X}_1(\infty) \right\} = x_\mathrm{T} q(1 - q) = \frac{c_1 c_2 x_\mathrm{T}}{(c_1 + c_2)^2} \tag{16.33b}$$

It can also be shown (Cao et al., 2005) that $\hat{X}_1(t)$ relaxes to $\hat{X}_1(\infty)$ in a time of order $(c_1 + c_2)^{-1}$.

The slow scale propensity function for the slow reaction R_3 is, according to equation 16.28, the average of $a_3(\mathbf{x}) = c_3 x_2$ with respect to $\hat{\mathbf{X}}^\mathrm{f}(\infty)$. Therefore, using equations 16.31 and 16.33a,

$$\bar{a}_3(x_3; x_1, x_2) = c_3 \left\langle \hat{X}_2(\infty) \right\rangle = \frac{c_3 c_1 (x_1 + x_2)}{c_1 + c_2} \tag{16.34}$$

Since the reciprocal of $\bar{a}_3(x_3; x_1, x_2)$ estimates the average time to the next R_3 reaction, the condition that the relaxation time of the virtual fast process be very much smaller than the mean time to the next slow reaction is

$$c_1 + c_2 \gg \frac{c_3 c_1 (x_1 + x_2)}{c_1 + c_2} \tag{16.35}$$

This condition will be satisfied if the inequality 16.30 is sufficiently strong. In that case, the slow-scale SSA for reactions 16.29 goes as follows:

1. Given $\mathbf{X}(t_0) = (x_{10}, x_{20}, x_{30})$, set $t \leftarrow t_0$ and $x_i \leftarrow x_{i0}$ $(i = 1, 2, 3)$.

2. In state (x_1, x_2, x_3) at time t, compute $\bar{a}_3(x_3; x_1, x_2)$ from equation 16.34.

3. Draw a unit-interval uniform random number r, and compute

$$\tau = \frac{1}{\bar{a}_3(x_3; x_1, x_2)} \ln\left(\frac{1}{r}\right)$$

4. Advance to the next R_3 reaction by replacing $t \leftarrow t + \tau$ and

$$x_3 \leftarrow x_3 + 1, \; x_2 \leftarrow x_2 - 1$$

$$\begin{cases} \text{With } x_\mathrm{T} = x_1 + x_2 \\[2mm] x_1 \leftarrow \text{ sample of } \mathcal{B}\left(\dfrac{c_2}{c_1 + c_2}, x_\mathrm{T}\right) \\[2mm] x_2 \leftarrow x_\mathrm{T} - x_1 \end{cases}$$

5. Record (t, x_1, x_2, x_3) if desired. Then return to step 2 or else stop.

In step 4, the x_3 update and the first x_2 update actualize the R_3 reaction. The bracketed procedure then "relaxes" the fast variables in a manner consistent with the

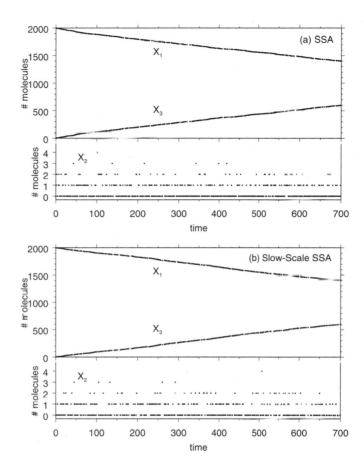

Figure 16.4   Two simulations of reactions 16.29 using the parameter values 16.36. Graph (a) shows an exact SSA run in which the populations are plotted essentially after each R_3 reaction (see text for details). Over 23 million reactions make up this run, the overwhelming majority of which are R_1 and R_2 reactions. Graph (b) shows an approximate ssSSA run in which only R_3 reactions, which totaled 587, were directly simulated, and the populations are plotted after each of those. The ssSSA simulation ran over 1,000 times faster than the SSA simulation.

stationary distribution in equation 16.32 and the new value of x_T. See Press et al. (1986), for a way to generate samples of the binomial random variable $\mathcal{B}(q, x_T)$.

Figure 16.4a shows the results of an exact SSA run of reactions 16.29 for the parameter values

$$c_1 = 10, \quad c_2 = 4 \times 10^4, \quad c_3 = 2; \quad x_{10} = 2000, \quad x_{20} = x_{30} = 0 \qquad (16.36)$$

The S_1 and S_3 populations here are plotted out immediately after each R_3 reaction. The S_2 population, which is shown on a separate scale, is plotted out at a like number of equally spaced time intervals; this gives a more typical picture of the S_2 population than plotting it immediately after each R_3 reaction because R_3 reactions are more likely to occur when the S_2 population is larger.

For the parameter values 16.36, condition 16.35 is satisfied by 4 orders of magnitude initially, and even more so as the total population of S_1 and S_2 declines; therefore, this reaction set should be amenable to simulation using the slow-scale SSA. Figure 16.4b shows the results of such a simulation, plotted after each R_3 reaction. We note that all the species trajectories in this approximate ssSSA run agree very well with those in the exact SSA run of figure 16.4a; even the behavior of the sparsely populated species S_2 is accurately replicated by the ssSSA. But whereas the SSA run in figure 16.4a had to simulate over 23 million reactions, the slow-scale SSA run in figure 16.4b simulated only 587 reactions, with commensurate differences in their computation times.

16.9   Concluding Remarks

In this chapter we have discussed two broad themes. The first is the "logical bridge" that connects the chemical master equation (CME) and stochastic simulation algorithm (SSA) on one side with the reaction rate equation (RRE) on the other side. Under the well-stirred (spatially homogeneous) assumption, the CME/SSA provides a mathematical description that is exact, discrete, and stochastic. If the system is such that the leap condition can be satisfied, the CME/SSA can be approximated by the Poissonian tau-leaping formula (equation 16.6) to obtain a description that is *approximate*, discrete, and stochastic. Further, if the reactant populations are large enough that the Poissonian tau-leaping formula can be approximated by the Gaussian tau-leaping formula (equation 16.8), which in turn is equivalent to the chemical Langevin equation (CLE) (equation 16.9), we obtain a description that is approximate, *continuous*, and stochastic. And finally, in the thermodynamic limit of an infinitely large system, the random terms in the CLE usually become negligibly small compared to the deterministic terms, and the CLE reduces to the RRE, which is approximate, continuous, and *deterministic*.

This progression—from the CME and SSA to tau-leaping to the CLE to the RRE—in which each successive level is an approximation of the preceding level, would, along with the corresponding numerical methods at each level, give us all the tools we need to efficiently simulate spatially homogeneous systems were it not for the multiscale nature of most biochemical systems: Both the species populations and the rates of the various chemical reactions typically span many orders of magnitude. As a consequence, in most cases the system as a whole does not fit efficiently into one level of description exclusive of the others. The second theme of our development in this chapter has been to describe two strategies for coping with multiscale problems:

implicit tau-leaping, and the slow-scale SSA. But much more remains to be done on the problem of multiscale.

17 Software Infrastructure for Effective Communication and Reuse of Computational Models

Andrew Finney, Michael Hucka, Benjamin J. Bornstein, Sarah M. Keating, Bruce E. Shapiro, Joanne Matthews, Ben L. Kovitz, Maria J. Schilstra, Akira Funahashi, John Doyle, and Hiroaki Kitano

Until recently, the majority of computational models in biology were implemented in custom programs and published as statements of the underlying mathematics. However, to be useful as formal embodiments of our understanding of biological systems, computational models must be put into a consistent form that can be communicated more directly between the software tools used to work with them. In this chapter, we describe the Systems Biology Markup Language (SBML), a format for representing models in a way that can be used by different software systems to communicate and exchange those models. By supporting SBML as an input and output format, different software tools can all operate on an identical representation of a model, removing opportunities for errors in translation and assuring a common starting point for analyses and simulations. We also take this opportunity to discuss some of the resources available for working with SBML as well as ongoing efforts in SBML's continuing evolution.

17.1 Introduction

The chapters of this book testify to the rising importance of computational modeling in biological research as a means of helping to better understand biological function. The increasing interest in this approach, coupled with our modern ability to generate ever-more complex models more rapidly than ever before, make it clear that practical computational modeling requires the use of software tools.

Until recently, the majority of models were implemented in custom programs and published only as statements of the underlying mathematics (that is, intended for human consumption). However, to be useful as formal embodiments of our understanding of biological systems (Bower and Bolouri, 2001), computational models must be put into a consistent form that can be communicated more directly between the software tools used to work with them. This format must help overcome a number of problems facing a systems biologist:

- Users often need to work with complementary resources from multiple software tools in the course of a project because different tools have different strengths and capabilities. For example, one tool may have a good model editing interface, another tool may provide novel facilities for analyzing system properties, yet another may implement an advanced simulation capability but lack a good graphical interface, etcetera. If the tools do not share a common model storage format, users are forced to re-encode their models in each tool separately, a time-consuming and error-prone practice.

- Models published in peer-reviewed journals are sometimes accompanied by instructions for obtaining the definitions in electronic form. However, because each author may use a different software environment (and associated model representation language), such definitions are often not straightforward to examine, test, and reuse. Researchers who wish to use a published model typically must transcribe it manually into a format compatible with their particular software.

- When simulation software packages are no longer supported, models developed in those systems can become stranded and unusable. This has already happened on a number of occasions, with a resulting loss of usable models to the research community. Continued innovation and development of new tools will only aggravate this problem unless the issue of standard formats is addressed.

- Reuse of existing models requires that those models can be clearly identified, easily retrieved, and related to their published descriptions in the scientific literature. Moreover, because of the increasing size and complexity of models continually being developed, the model structure should be documented to allow for efficient handling and sound modification.

We developed the Systems Biology Markup Language (SBML) in an effort to address these problems. SBML is a format for representing computational models in a way that can be used by different software systems to communicate and exchange those models (Finney and Hucka, 2003; Hucka et al., 2003, 2004). By supporting SBML as an input and output format, different software tools can all operate on an identical representation of a model, removing opportunities for errors in translation and assuring a common starting point for analyses and simulations. SBML is by no means a perfect format, but it has proven useful and achieved widespread acceptance within the domain of modeling at the level of biochemical reaction networks. Over 90 open-source and commercial software tools support SBML as of November 2005.

A gratifying by-product of the SBML project has been the way it has catalyzed a community of interested researchers, developers, and users who are now collaborating on evolving SBML and creating new resources around it. This is undoubtedly a reflection of an urgent need in the community for *any* format such as SBML to address issues of interoperability. At the same time, we suspect that the challenges faced by the SBML community and the solutions that are arising have underlying components that would be faced by any effort to define a similar standard exchange format. We discuss two examples in this chapter. One is the difficulty of balancing ease of language implementation against representational power. Today this is being answered by progress towards SBML Level 2 Version 2, which is expected to be ratified in 2005, and the modular SBML Level 3, which is expected in 2006. A second is the unexpected difficulty of ensuring correct interpretation of SBML by different software applications. We describe our current attempts to address this problem using a combination of (i) a carefully-designed software library, libSBML, which among other features provides rule-based model consistency testing, and (ii) a semantic validation suite for testing correct interpretation of SBML constructs by software applications.

17.2   Software Assistance for Biological Modeling

As an example of how software technologies such as SBML assist modelers today, consider the following hypothetical (but still quite plausible) sequence of events.

A computationally-savvy biologist named Albert is investigating one of the mitogen-activated protein kinase (MAPK) cascades. The MAPK pathways lead from growth factor receptors on cell membranes to effector molecules located in the cell cytoplasm and nucleus. This family of signaling pathways is one that has received much attention in both experimental (Seger and Krebs, 1995) and computational biology (Schoeberl et al., 2002).

Our hypothetical biologist might begin with a body of experimental data gathered by himself and other members of the laboratory in which he works. In order to understand his experiments in the context of other data and other published results, he decides to develop a computational model so that he can integrate different sources of existing knowledge and his own hypotheses into a common, formalized framework. Since the MAPK system is a popular topic of study, he has no trouble finding related work in the literature, including existing computational models. He chooses to begin with a relatively simple model by Kholodenko (Kholodenko, 2000). The original publication gives a complete listing of the mathematical equations that define Kholodenko's model, but no software implementation. (Even though that particular article is from this decade, it still predates the development of SBML and most of today's software tools.) The model is not complex, but he knows that recreating a model from a research paper will take time, so before starting, he visits the BioModels Database (BioModels Team, 2005) to check if the model is available in a machine-readable format. He searches the database and quickly finds an existing implementation (figure 17.1), which he can download in SBML format.

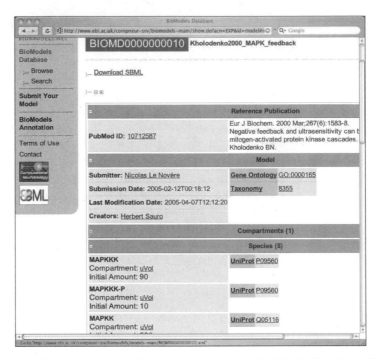

Figure 17.1   Screenshot of a model display page in the BioModels Database (BioModels Team, 2005).

Once he has the SBML file, Albert starts up his favorite Windows-based model editing package, JDesigner (Sauro et al., 2003). This package provides a friendly, graphical diagram view of a model (figure 17.2). He spends a significant amount of time experimenting with the model running time-course simulations to examine the behavior under different conditions, as well as making modifications and exploring the results. After becoming familiar with the Kholodenko model, he next begins to make modifications based on his own experimental work and that of his colleagues.

Eventually, Albert's model grows and becomes substantially different from the original. He reaches a point where he has to find values for parameters in the model that are not directly measurable, but he believes he has enough converging evidence from other data that he can search for plausible values by a process known as parameter estimation. This is a resource-intensive task requiring many repeated simulation runs and analyses—more than he can comfortably run on his laptop computer. Albert enlists the aid of a colleague, Bernadette, who works at another institution and who has access to clusters of computers on which she can quickly perform large computations. Bernadette is less a biologist and more a computational scientist, but she has had enough exposure to biological modeling that she can perform the parameter estimation tasks for Albert. Despite the geographical distance separating them and the fact that Bernadette is adamant about using Linux rather than Windows as her computer operating system of choice, Albert has no difficulty conveying an unambiguous model definition to Bernadette

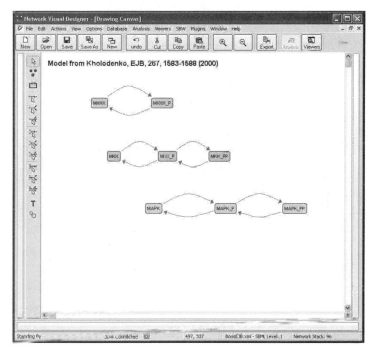

Figure 17.2   Screenshot of JDesigner (Sauro et al., 2003), a free computational modeling system for biochemical reaction networks. It runs on the Microsoft Windows operating system.

because JDesigner can produce SBML output and Bernadette has at her disposal several software tools that can read SBML.

Bernadette writes command scripts in Linux that take Albert's model and his experimental data (which he stored in ordinary comma-delimited tabular format) and perform parameter estimation using an optimization package written in MAT-LAB (The Mathworks, Inc., 2005). To convert the SBML model into appropriate MATLAB data structures, she uses one of the free MATLAB toolboxes available for this purpose (Keating, 2005). After some iterations back and forth with Albert to clarify his goals, and many computer runs, the pair eventually determine best-estimate values for the unknown parameters in Albert's model. Bernadette also performs a large number of additional simulation and analysis runs on her Linux computers using COPASI (Mendes, 2003) to explore the behaviors of the model. The results enable Albert to continue further with his research, comparing his predictions to experimental data and refining his model to incorporate new hypotheses. The model and its results are novel enough that Albert writes an article about them with Bernadette. They also submit the SBML model to the BioModels Database, where the curators annotate the model and enter it into the database for other researchers to use and build upon.

Some time after the article is published, a researcher working at a pharmaceutical company reads Albert and Bernadette's paper on MAPK signaling. It turns out

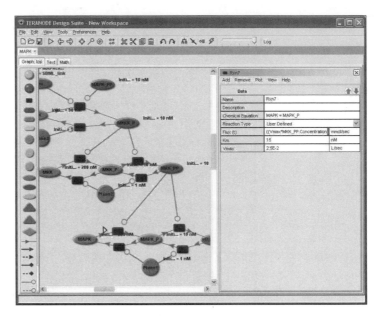

Figure 17.3 Screenshot of TERANODE Design Suite (TERANODE, Inc., 2005), an example of a modern commercial software package using SBML and integrating model editing, analysis, and simulation.

that this researcher, Carl, has been investigating novel therapeutic interventions on this same pathway. Thanks to the availability of the model in SBML form, Carl is able to quickly obtain and try out the model in his software tool of his choice (figure 17.3), a full-featured commercial package called TERANODE Design Suite (TERANODE, Inc., 2005). The model's structure and behavior are consistent with his own findings, and moreover, it provides new insights that could lead to an investigation of new pharmacological agents. Carl is interested in pursuing this further. The copyright on the model stipulates that commercial users must contact the authors, so he contacts Albert and Bernadette and begins a promising new collaboration.

17.3   The SBML Representation of Models

The SBML project is not an attempt to define a universal language for representing quantitative models; the rapidly evolving views of biological function, coupled with the vigorous rates at which new computational techniques and individual tools are being developed today, are incompatible with a one-size-fits-all idea of a universal language. A more realistic alternative is to acknowledge the diversity of approaches and methods being explored by different software tool developers, and seek a common intermediate format—a *lingua franca*—enabling communication of the most essential aspects of the models.

17.3.1 Brief Summary of the Form and Features of SBML

SBML is a machine-readable *model definition language* defined neutrally with respect to programming languages and software encoding. It is defined using a subset of UML, the Unified Modeling Language (Eriksson and Penker, 1998; Oestereich, 1999), and in turn, this definition is used to create an XML Schema (Biron and Malhotra, 2000; Fallside, 2000; Thompson et al., 2000) for SBML. The XML Schema specifies how SBML can be expressed using XML, the eXtensible Markup Language (Bosak and Bray, 1999; Bray et al., 2000). XML is a simple and portable text-based substrate that has been gaining widespread acceptance in computational biology and bioinformatics (Achard et al., 2001; Augen, 2001).

The main focus of SBML is encoding models consisting of biochemical entities (species) linked by reactions to form biochemical networks. An important principle in SBML is that models are decomposed into explicitly-labeled constituent elements, the set of which resembles a verbose rendition of chemical reaction equations. The representation deliberately does not cast the model directly into a set of differential equations or other specific mathematical frameworks. This explicit, modeling-framework-agnostic decomposition makes it easier for different software tools to interpret the model and translate the SBML form into whatever internal form each tool actually uses.

SBML is being developed in levels, with each higher SBML *level* adding richness to the model definitions that can be represented by the language. Level 2 is the highest level of SBML currently defined; it represents an incremental evolution of the language resulting from the practical experiences of many users and developers working with Level 1 since its introduction in the year 2001. A definition of a model in SBML Level 2 consists of lists of one or more of the following components:

- *compartment*: a container of finite dimensions where one or more chemical substances (well-mixed) are located;

- *species*: a pool of a chemical substance located in a specific compartment (a species represents the concentration or amount of a substance and not a single molecule);

- *reaction*: a statement describing some transformation, transport or binding process that can change one or more species (each reaction is characterized by the stoichiometry of its products and reactants and optionally by a rate equation);

- *parameter*: a quantity that has a symbolic name;

- *unit definition*: a name for a unit used in the expression of quantities in a model;

- *rule*: a mathematical expression that is added to the model equations constructed from the set of reactions (rules can be used to set parameter values, establish constraints between quantities, etcetera.);

- *function*: a named mathematical function that can be used in place of repeated expressions in rate equations and other formulas; and

- *event*: a set of mathematical formulas evaluated at a specified moment in the time evolution of the system.

```xml
<?xml version="1.0" encoding="UTF-8"?>
<sbml xmlns="http://www.sbml.org/sbml/level2" level="2" version="1">
  <model id="EnzymeKinetics">
    <listOfCompartments>
      <compartment id="Cell" size="1"/>
    </listOfCompartments>
    <listOfSpecies>
      <species id="S"  compartment="Cell" initialAmount="1" boundaryCondition="true"/>
      <species id="E"  compartment="Cell" initialAmount="1"/>
      <species id="ES" compartment="Cell" initialAmount="0.01"/>
      <species id="P"  compartment="Cell" initialAmount="0.01" boundaryCondition="true"/>
    </listOfSpecies>
    <listOfReactions>
      <reaction id="Reaction1">
        <listOfReactants>
          <speciesReference species="S"/>
          <speciesReference species="E"/>
        </listOfReactants>
        <listOfProducts>
          <speciesReference species="ES"/>
        </listOfProducts>
        <kineticLaw>
          <math xmlns="http://www.w3.org/1998/Math/MathML">
            <apply> <minus/>
              <apply> <times/>  <ci> k_1 </ci>  <ci> S </ci>  <ci> E </ci>  </apply>
              <apply> <times/>  <ci> k_r </ci>  <ci> ES </ci>  </apply>
            </apply>
          </math>
          <listOfParameters>
            <parameter id="k_1" value="3"/>
            <parameter id="k_r" value="6"/>
          </listOfParameters>
        </kineticLaw>
      </reaction>
      <reaction id="Reaction2" reversible="false">
        <listOfReactants>
          <speciesReference species="ES"/>
        </listOfReactants>
        <listOfProducts>
          <speciesReference species="E"/>
          <speciesReference species="P"/>
        </listOfProducts>
        <kineticLaw>
          <math xmlns="http://www.w3.org/1998/Math/MathML">
            <apply> <times/>  <ci> k_2 </ci>  <ci> ES </ci>  </apply>
          </math>
          <listOfParameters>
            <parameter id="k_2" value="9"/>
          </listOfParameters>
        </kineticLaw>
      </reaction>
    </listOfReactions>
  </model>
</sbml>
```

Figure 17.4 Simple SBML Level 2 model of a system of reactions involving enzyme kinetics.

Additional features in SBML Level 2 include support for a systematic way of including metadata, and support for delay functions. The latter are useful for representing biological processes having a delayed response, but where the details of the processes and the actual delay mechanism are not relevant to the operation of the model.

To make this discussion concrete, figure 17.4 gives the complete SBML Level 2 listing of a simple model of enzyme kinetics, $E + S \rightleftharpoons ES \rightarrow P$, where E, S, and P represent the enzyme, substrate, and product species, respectively, and ES is an intermediate complex formed during the reaction. In this particular SBML rendition, the system is represented as two reaction structures: the reversible reaction $E+S \rightleftharpoons ES$, here defined with a forward reaction rate of $k_1 * [S] * [E]$ and a reverse reaction rate of $k_r * ES$, and the irreversible reaction $ES \rightarrow P$, here defined with a forward reaction rate of $k_2 * [ES]$. The symbols E, S, and ES, when used in rate expressions (SBML's kineticLaw elements), stand for the concentrations of the different species, and the parameters k_1, k_r, and k_2 are set to values $k_1 = 3$, $k_r = 6$, and $k_2 = 9$. When specific units are omitted from quantities in an SBML model (as they are here), the model is assumed to use the default units for those quantities, which in SBML are moles for substance amounts and liters for volumes. Other formulations of this model might, for example, express this system explicitly as three irreversible reactions, change the units on quantities to be millimoles and microliters, and so on. This model is presented here only to give a sense for the structure of SBML and the relative simplicity, and we reiterate that people are not meant to edit models directly at this level; instead, software tools read and write this kind of representation on the user's behalf.

SBML's representational power extends far beyond the kind of simple enzyme kinetics model used here as an illustration. Its simple formalisms allow a wide range of biological phenomena to be modeled, including cell signaling, metabolism, gene regulation, and more. There is no assumption about the kinds of kinetics or interactions or network organizations that can be represented. Significant flexibility and power come from the ability to define arbitrary formulas for the rates of change of variables as well as the ability to express other constraints mathematically.

17.3.2   Relationships to Other Efforts

Many XML-based formats have been proposed for representing data and models in biology; however, we know of only two XML-based formats that are suitable for representing compartmental reaction network models with sufficient mathematical depth that the descriptions can be used as direct input to simulation software. The two are SBML and CellML (Hedley et al., 2001b,a; Lloyd et al., 2004).

CellML is built around an approach of composing systems of equations by linking together the variables in those equations; this is augmented by features for declaring biochemical reactions explicitly, as well as encapsulating arbitrary components into modules. Its focus is on a component-based architecture to facilitate reuse of models and parts of models, and the mathematical description of models. By contrast, SBML provides constructs that are more similar to the internal data objects used in many contemporary simulation/analysis software packages specialized for biochemical networks.

These differences notwithstanding, the SBML and CellML efforts share much in common and represent somewhat different approaches to solving the same general problems. They were initially developed independently, but the primary developers of both languages are actively engaged in exchanges of ideas and are seeking ways of making the languages more interoperable. SBML Level 2 borrows a number of approaches from CellML, making it that much easier to translate between the two formats.

17.4   The Continued Evolution of SBML

The need for a language like SBML was manifest during the first Workshop on Software Platforms for Systems Biology, held at the California Institute of Technology in early 2000. The two or three dozen attendees at the time represented less than a dozen software projects, yet even within this small group, it proved impossible to share models without having to re-encode them anew in each software tool. This needless impediment to collaboration directly inspired the SBML effort.

Defining a language such as SBML and encouraging its use by other groups has always involved balancing conflicting demands. For example, there is pressure to include a wide variety of features to support the various kinds of modeling and analysis capabilities being explored in different tools. But if the capabilities are too advanced or too specialized for most tools, then few if any software packages will implement support for the entire language specification, with the consequence that most tools will still not be able to exchange models in a meaningful way. On the other hand, if SBML does not expand quickly enough to support features satisfying more advanced research efforts, then SBML risks losing the groups' patience, potentially leading to the creation of incompatible dialects of the language.

In an attempt to help achieve this balance, we are proceeding with a staged approach to SBML development, embodied in the already-mentioned concept of SBML *levels*. Each higher SBML level adds richness to the model definitions that can be represented by the language. By delimiting sets of features at incremental stages, the SBML development process provides software authors with stable standards, and the community can gain experience with the language definitions before new features are introduced. Two levels have been defined so far (Finney et al., 2002; Hucka et al., 2001). Level 1 is simpler (but less powerful) than Level 2. The separate levels are intended to coexist; SBML Level 2 does not render Level 1 obsolete. Software tools that cannot support higher levels can go on using lower levels; tools that can read higher levels are assured of also being able to interpret models defined in the lower levels. The open-source software infrastructure we have been developing around SBML (see Section 17.5) allows developers to support both Levels 1 and 2 in their software with a minimum amount of effort.

17.4.1   Community Involvement

One component of SBML's success has been the community-oriented method of its continued evolution. SBML's popularity has led to the formation of an active international group of researchers and software developers who are now working together to push SBML in new directions. As is the case with many projects today, the primary mode of interaction between members is electronic mail, with discussions taking place on the community mailing list, sbml-discuss@caltech.edu. The list currently contains over 200 members coming from academic, commercial and private environments, from all continents. Besides discussions over the list, another important mode of interaction has been regular face-to-face meetings during the Workshops on Software Platforms for Systems Biology (also known informally as the SBML Forum meeting), held since mid-2000.

These meetings serve many vital functions. First, they provide a forum where proposals for potential new SBML features can be presented and where consensus decisions can be made about the development of SBML, with the aim of enabling SBML to support a wider range of model paradigms and modes of interoperability. Second, they ensure that systems biology software interoperability is maximized by discussing the correct use of SBML and (related to this) exposing software developers to issues in the correct interpretation and handling of SBML in all software. Third, they inform developers of the latest developments in software infrastructure for SBML. And finally, they educate the systems biology community about the range of modeling paradigms that are being used to understand biological phenomena. The ninth SBML Forum meeting was held on October 14–15, 2004, in Heidelberg, Germany, and was attended by 49 representatives of different international research groups. All presentation materials from SBML meetings are made publicly available on the project Web site (SBML Team, 2005b).

In 2003, a new type of meeting was instituted: *SBML Hackathons*, in which software developers gather together to work simultaneously on their software next to other developers, discovering and resolving interoperability problems as they go. The third SBML Hackathon was held on May 9–10, 2005, at the National Museum of Emerging Science and Innovation in Tokyo, Japan, and was attended by 45 delegates, nearly three times as many as attended the first SBML Hackathon in 2003.

17.4.2   SBML Level 2 Version 2

As a practical consequence of how SBML develops and evolves, it reflects how theoreticians and software developers conceptualize and structure their computational models of biochemical reaction networks. The exact form of the language matters less than the representational elements comprising the language. Though the incremental development path taken for SBML has led to a less-than-elegant structure, it is fair to say that SBML represents a consensus view of how computational models of reaction networks are understood today. The dedicated community of interested researchers has kept up the evolution of SBML and continues to result in improvements to meet increasingly sophisticated needs.

The next specification of SBML is expected to be an incremental update, Level 2 Version 2, to be followed closely by SBML Level 3, which has been in development for over a year. The following are illustrative of the enhancements likely to be introduced in SBML Level 2 Version 2 and the reasons for them.

■ *SpeciesType*. In SBML, the amount (concentration or molecular count) of every chemical species must be defined with respect to a location. Locations in SBML are represented as compartments, where a compartment can represent a physical structure such as "cytoplasm" or a purely theoretical location used solely for modeling expediency. If the same kind of species appears in more than one location (for example, both inside a cell's cytoplasm and outside the cell), this must be represented as two different species, each having separate identifiers in the model. The reason is that when SBML models are translated into typical computational forms, those species are represented as variables (again, either concentrations or molecular counts) whose values can change over time. Species located in different compartments are assumed to comprise different *pools* of the species—that is the logical point of having compartments in the first place. However, a number of software developers have expressed the need for specifying that two species variables in SBML refer to the same kind or type of chemical irrespective of compartmental location. Therefore, one of the changes planned for SBML Level 2 Version 2 is the introduction of a SpeciesType data structure for this purpose. This will make it possible for a model to define a list of SpeciesType structures. Each species definition will then be able to refer to a particular SpeciesType definition, stating, in effect, that it is "of this species type." For example, a model could contain a SpeciesType for aspartate, and could have multiple species definitions, one for aspartate located in the cytosol and others for mitochondrial matrix compartments. The species representing the different pools of aspartate would have different identifiers (for instance, "aspartate_cytosol" and "aspartate_mitochon"), but each would refer to the common aspartate SpeciesType.

▪ *Nested Unit Definitions.* Not all software tools provide a means of changing the units of measurement used for the numerical quantities in a user's model; they often assume specific units for different quantities and rely on users to adjust numerical values as necessary when encoding models in the software environment. Unfortunately, sometimes different tools make different unit assumptions; thus, *some* capability for redefining units in SBML is necessary in addition to specifying default units. We thought that a small, simple scheme would serve best, and this is what we attempted in the first definition of SBML (Level 1 Version 1). The scheme turned out to be too limited; for example, it did not allow for the definition of some types of units that are not in the SI unit system, and it was significantly less capable than the unit scheme in CellML, making it difficult to translate some models between CellML and SBML. The consensus in the SBML community was that more definitional power was warranted, so SBML Level 2 Version 1 introduced a fuller unit scheme. Arguably the one feature it lacked was a provision to allow unit definitions to be defined in terms of other defined units rather than solely in terms of the base units. The reason was our continuing attempts to make the unit scheme simple—after all, what use is it if many software tools don't support it? But in the end, the consensus of users was that it should not be up to the SBML language to arbitrarily limit the capabilities in this area because it impacts a researcher's ability to represent their intentions precisely. The SBML community felt that tools that lack adequate support for units should either be enhanced appropriately, or else that unit manipulation functionality could be encoded in separate software tools and libraries such as libSBML (Section 17.5).

- *ConstraintRule.* It is sometimes important to be able to express the idea that certain model conditions should hold true and if, during a simulation, the conditions are exceeded, then the user should be alerted that the model is operating outside the assumptions made by the model's author. Although SBML has always had facilities for expressing mathematical relationships between quantities, it lacked a provision for expressing these kinds of constraints. SBML Level 2 Version 2 will extend the types of SBML rules available to include ConstraintRule. This structure will allow the statement of mathematical expressions whose values evaluate to a Boolean value (true or false). If at some point in time during a time-course evaluation of the model, the expression evaluates to false, the constraint is not satisfied. The ConstraintRule will contain an optional note (in XHTML format) that can contain a message to be displayed to the user if the constraint expression evaluates to false. An example of the application of this rule would be to make explicit the assumptions of an Henri-Michaelis-Menten rate law about relative species concentrations between product and substrate as well as between enzyme and substrate.

17.4.3   SBML Level 3

As a language that is an intersection rather than a union of features needed by all tools, SBML currently cannot support all the representational capabilities that all software systems offer to users. Some tools offer features that have no explicit equivalent in SBML Level 2, and those tools currently can only store those features as annotations in an SBML model. But in many cases those features could potentially be used by more than one tool, and thus it would be appropriate to have some representation for them in SBML. Using Level 2 as a starting point, the SBML community has been developing proposals and prototype implementations of many new capabilities that will become part of SBML Level 3. The main current areas of interest are:

- Diagram layout: enabling the inclusion of diagrammatic renditions of a model of the sort visible in the screenshots of figures 17.2 and 17.3.
- Model composition: allowing construction of models from instances of submodels
- Multicomponent species: allowing species to be composed from instances of species types, enabling such things as the representation of complexes of phosphorylated proteins and generalized reactions acting on them
- Arrays: allowing models to contain indexed collections of objects of the same type
- Spatial features: allowing the representation of the geometric features of compartments, the diffusion rates of species and the spatial distribution of model parameters and boundary conditions
- Constraints: enabling the definition of constraints on model variables

It is unreasonable to expect a tool to support every feature planned for Level 3 in order to be called Level 3 compatible. One of the challenges for SBML Level 3 will be to design a modular feature set. The idea is to enable a model to contain explicit information about which capabilities are necessary to interpret it correctly, so that tools encountering the model may reject it gracefully if they do not possess the necessary facilities. For reasons of efficiency and correctness, an explicit indication is preferable to requiring tools to read and interpret the entire model and inferring the capabilities needed.

We anticipate that Level 3 will take the form of a core, consisting of minimal extensions to Level 2, and a set of Level 3 modules, each encapsulating the definition of one of the major features listed above. One of the extensions making up the Level 3 core will be explicit feature indicators, such that each of the modules has a corresponding feature tag which will appear in a list at the beginning of the model definition. The presence of a feature tag will signal to software tools reading the model that the model uses that particular feature. The software tool may then make a decision about whether it can handle the model or whether it should alert the user to a problem.

17.5   Enabling Efficient and Correct Interpretation of SBML Using a Dedicated Software Library

To make it easier for software developers and users to work with SBML, and more generally to promote the language's use as a common exchange format, our group has released and continues to develop a number of open-source SBML software tools. Here we describe one, libSBML, that many projects are using for implementing support for SBML in their software applications.

17.5.1   General Characteristics of libSBML

LibSBML is an application programming interface (API) library for reading, writing, and manipulating files and data streams containing SBML content. Developers can embed the library in their applications, saving themselves the work of implementing their own parsing, manipulation, and validation software. At the API level, the library provides the same interface to data structures independently of whether the model originated in SBML Level 1 or 2. The library currently also offers the ability to translate SBML Level 1 models to SBML Level 2.

LibSBML is written in ISO standard C and C++ and is highly portable. It is currently supported on the Linux, Solaris, MacOS X, and Microsoft Windows operating systems. The library provides language bindings for C, C++, Java, Python, Perl, MATLAB, and Common Lisp, with support for other languages planned for the future. We distribute the package in both source-code form and as precompiled dynamic libraries for the Microsoft Windows, Linux, and Apple MacOS X operating systems; they are available under terms of the LGPL (Free Software Foundation, 1999) from the *sbml* project on SourceForge.net (SourceForge.net, 2002), the world's largest open-source software repository and project hosting service. LibSBML is at release version 2.3.4 as of October 2005.

17.5.2 Advantages of a Dedicated Library for SBML

An often-repeated question is, why not simply use a generic XML parsing library? After all, SBML is usually expressed in XML, and there exist plenty of XML parsers, so why not simply tell people to use one of them, rather than develop a specialized library? The answer is: while it is true that developers *can* use general-purpose XML libraries, there are many reasons why using a system such as libSBML is a vastly better choice.

One of the features of libSBML is its facilities for manipulating mathematical formulas supporting differences in representation between SBML Level 1 and SBML Level 2. As discussed in more detail below, libSBML provides an API that allows working with formulas in both text-string and MathML (Ausbrooks et al., 2001) form, and to interconvert mathematical expressions between these forms. The utility of this facility extends beyond converting between SBML Level 1 and 2. Many software packages provide users with the ability to express formulas for such things as reaction rate expressions, and these packages' interfaces often let users type in the formulas directly as text strings. LibSBML saves application programmers the work of developing formula manipulation and translation functionality. It makes it possible to translate those formula strings directly into Abstract Syntax Trees (ASTs), manipulate them using AST operations, and write them out in the MathML format of SBML Level 2.

As discussed in Section 17.5.5, another feature of libSBML is the validation it performs on SBML inputs at the time of parsing files and data streams. This helps verify the correctness of models in a way that goes beyond simple syntactic validation. Still another invaluable feature of libSBML is the domain-specific operations it provides beyond simple SBML-specific accessor facilities. Examples of such operations include obtaining a count of the number of boundary condition species, determining the modifier species of a reaction (assuming the reaction provides kinetics), and constructing the stoichiometric matrix for all reactions in a model.

Finally, libSBML is solidly written and tested. The entire library has been written by seasoned, professional software engineers using the test-driven approach (Beck, 2002). The libSBML source code currently has 760 unit tests and over 3,400 individual assertions. It represents a robust and well-tested system that others can build upon.

17.5.3 Manipulating Mathematical Formulas

In SBML Level 1, mathematical formulas are represented as text strings using a C-like syntax. We chose this representation because of its simplicity, widespread familiarity, and use in applications such as Gepasi (Mendes, 1997) and Jarnac (Sauro, 2000), whose authors contributed to the initial design of SBML. For SBML Level 2, there was a need to expand the mathematical vocabulary of Level 1 to include additional functions (both built-in and user-defined), mathematical constants, logical operators, relational operators, and a special symbol to represent time. Rather than growing the simple C-like syntax into something more complicated and esoteric in order to support these features, and consequently having to manage two standards in two different formats (XML and text string formulas), we chose to leverage an existing standard for expressing mathematical formulas in Level 2: the content portion of MathML (Ausbrooks et al., 2001).

Using MathML in SBML has at least two advantages. First, instead of reinventing the wheel, we are building upon an existing and well-established W3C standard. Second, since the entirety of a model is expressed in XML, SBML is now more amenable to tools that can process, manipulate, and store XML, such as (for example) XSLT (Clark and DeRose, 1999), XQuery (Fernández et al., 2005), XPath (Fernández et al., 2005), and other XML technologies. That said, there are some disadvantages to using MathML. By introducing MathML part-way through the evolution of SBML, we have created a legacy support problem by having two formula representations with which to contend and interconvert. Also, most simulator packages cannot parse and understand MathML directly (but, we should point out the same would hold true had we chosen to expand the lowest-common-denominator C-like syntax of Level 1). Overcoming both of these disadvantages is easy with libSBML.

Abstract Syntax Trees (ASTs) are well-known in the computer science community; they are simple recursive data structures useful for representing the syntactic structure of sentences in certain kinds of languages (mathematical or otherwise). Much as libSBML allows programmers to manipulate SBML at the level of domain-specific objects, regardless of SBML level or version, it also allows programmers to work with mathematical formula at the level of ASTs regardless of whether the original format was C-like infix notation or MathML. LibSBML goes one step further by allowing programmers to work exclusively with infix formula strings and instantly convert them to the appropriate MathML whenever needed.

LibSBML ASTs provide a canonical, in-memory representation for all mathematical formulas regardless of their original format (that is, C-like infix strings or MathML). In libSBML, an AST is a collection of one or more ASTNodes. ASTNodes represent the most basic, indivisible part of a mathematical formula and come in many types. For instance, there are node types to represent numbers (with subtypes to distinguish integer, real, and rational numbers), names (for example, constants or variables), simple mathematical operators, logical or relational operators, and functions. Each ASTNode node may have none, one, two, or more child ASTNodes depending on its type. For instance, table 17.1 illustrates how the mathematical expression $1 + 2$, is represented as an AST with one *plus* node with two *integer* child nodes for the numbers 1 and 2, and the corresponding MathML representation.

17.5.4   Performance of LibSBML

XML parsers come in two popular varieties: Document Object Model (DOM) based and event-based. DOMs (Le Hors et al., 2000) are very generic in-memory structures that nearly duplicate the tree-like structure of the XML on disk. Using

Infix	AST	MathML
$1 + 2 \iff$	$\iff$	```<math xmlns="http://www.w3.org/1998/Math/MathML"> <apply> <plus/> <cn type="integer"> 1 </cn> <cn type="integer"> 2 </cn> </apply></math>```

Table 17.1 Illustration of a simple mathematical expression represented in both libSBML's AST structure and MathML (Ausbrooks et al., 2001).

a DOM simply moves the parsing bump under the rug. Instead of parsing a file, one now has to parse an in-memory data structure. Moreover, because DOMs are generic, needing to handle any XML that comes their way, one pays a penalty in terms of large memory consumption. Event-based parsers, on the other hand, allow programmers to intercept specific XML events (tags) and act on them. Event-based parsers are memory-efficient, but are often too low-level and fined-grained. They therefore lack the convenience of manipulating XML data in larger logical units.

LibSBML aims to strike a balance between DOM and event-based models of XML parsing. It provides the conveniences of a domain-specific object model while keeping memory usage to a minimum. Below, we compare the performance of libSBML, which uses the Xerces-C++ (Apache Software Foundation, 2004) event-based SAX parser under the hood, to parsing SBML with the Xerces-C++ DOM.

We obtained memory consumption statistics by writing two simple programs to read an SBML model from file into memory. One program used libSBML to read the model into domain-specific SBML objects and the other program used Xerces-C++ 2.6 to read the model into the W3C XML DOM format (Le Hors et al., 2000). Each program recorded its total resident memory consumption immediately before and after reading the model and reported the difference between these two numbers.

Total resident memory gives an estimate not only of the size of the model in memory, but also the size of the library and all supporting code that must be loaded into memory (often of concern to programmers). LibSBML was compiled with Xerces 2.6, so the amount of memory consumed by the Xerces library itself is the same for both programs.

We ran both programs over the 10,000+ models in the SBML Test Suite (SBML Team, 2005a) and models used in the first SBML Hackathon. Individual file sizes varied from 600 bytes to 5.76 MBytes. The runs were performed on computers running SuSE Linux 9.1 (Novell, Inc., 2005) with dual 64-bit AMD Opteron 2.2 GHz processors (Advanced Micro Devices, Inc., 2005).

Figure 17.5 shows a plot of the file size on disk versus the object model size in memory. While the Xerces-C++ 2.6 DOM is more efficient than previous implementations, the DOM consumed nearly five times as much memory for large multi-megabyte files. For small files (under five kilobytes), the DOM is ever so slightly more efficient. This is likely because Xerces uses string pooling and other reference counting techniques to optimize memory usage. For SBML files larger than five kilobytes and especially files larger than one megabyte, libSBML is the clear performance winner.

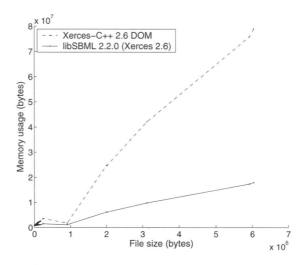

Figure 17.5 A plot of memory consumption by libSBML (solid line) and the Xerces-C++ Document Object Model (dotted line), when each is used to read SBML models into computer memory. Data are based on over 10,000 sample models taken from the SBML Semantic Validation Suite and the first SBML Hackathon of 2003. File sizes (horizontal axis) varied from 600 bytes to 5.76 MBytes.

17.5.5 Helping Ensure Correctness and Consistency

Syntactic validation involves verifying that the SBML input is well-formed, and, for example, that data values are of the correct types. Consistency checking involves verifying the *contents* of an SBML model for self-consistency, referential integrity, and adherence to the SBML specifications. The tests are implemented as individual constraints within libSBML; the library reports back validation failures to the calling application via the libSBML API. The constraint checking system is modular, and the constraint set can be easily extended. We describe the design and intent of the constraint syntax below.

The design of SBML is driven by data models instead of the specifics of XML representation. To that end, the SBML specification is first described using UML static class diagrams. These class descriptions are mapped to XML representations using SCHUCS (Hucka, 2000), a technique we developed, tailored to producing efficient, reasonably succinct, and quasi-human-readable XML. We wanted to parallel our emphasis on data over representation, with a declarative language to express SBML model constraints (declarative languages state the *what* without specifying the *how*). For this, we took inspiration from the UML community and its development of OCL, the Object Constraint Language (Object Management Group, Inc., 2002; Warmer and Kleppe, 2003).

Although libSBML consistency checks are not expressed directly in OCL, we have created an OCL-like language on top of the libSBML C++ API. This language balances the readability of OCL with the efficiency and expressiveness of C++, which is sometimes necessary for more complicated validation procedures. The language allows the manipulation of only `constant` C++ objects, which much like OCL, guarantees operations will be side-effect free. Further, this guarantee is enforced at compile time. Being side-effect free is an important property as we do not want the process of consistency checking to change the state of a model. An example will help make these concepts more clear.

One of the 50 consistency checks currently implemented ensures that if a model author overrides the default definition of the *substance* unit, a special unit name in SBML, the resulting unit definition is consistent with the notion of a substance. The consistency check constraint is written as:

```
START_CONSTRAINT (1202, UnitDefinition, ud)
{
  msg =
    "A 'substance' UnitDefinition must simplify to a single "
    "Unit of kind 'mole' or 'item' with an exponent of '1' "
    "(L2v1 Section 4.4.3).";

  pre( ud.getId() == "substance" );

  inv( ud.getNumUnits() == 1                        );
  inv( ud.getUnit(0).isMole() || ud.getUnit(0).isItem() );
  inv( ud.getUnit(0).getExponent() == 1             );
}
END_CONSTRAINT
```

The **START_CONSTRAINT** macro takes three arguments. The first is a number that uniquely identifies this constraint (that is, 1202). Assigning such identifiers to each constraint facilitates traceability and allows programmers to easily determine which rules have been violated. The next two parameters indicate the type of SBML object to which this rule applies (that is, **UnitDefinition**) and a shorthand name to use for the object being checked (that is, **ud**).

The body of the constraint consists of a message (**msg**) to be logged should the SBML object fail the check. After the message, zero or more preconditions (**pre**) may be listed. In order for the rule to apply to the SBML object in question, all preconditions must hold (in the order listed). If a precondition does not hold, the check is aborted without logging either a passage or failure. Finally, assuming all preconditions hold, the object's state must adhere to a set of one or more invariants (**inv**). Should any invariant fail, the constraint immediately fails and a message is logged.

In the above example, notice that preconditions and invariants are specified on the (lib)SBML object model. Each method invocation (operation) does not change the state of the model and specifies *what* not *how* (with apologies made for the standard names used for getter methods, for example, `getUnit()`, which arguably describes *how* and not *what*; even OCL falls victim to this slight, purely esthetic inconsistency.)

Finally, it's worth describing a case where OCL-like statements are not enough and having the full expressive power of C++ to write rules is advantageous. In SBML, compartments may be nested inside one another, with the limitation that this nesting may not be cyclic (an example of a cycle: compartment A is in B which is in C which is in A). While it is relatively easy to encode this constraint in the OCL-like language demonstrated above, reporting a user-friendly error message is another matter. Upon violation of this constraint, instead of simply stating that a cycle exists, it is better to indicate the chain of compartments that was followed to detect this cycle, thereby enabling the model author to quickly track down the cause of the error. Constructing such an informative error message is awkward in a purely declarative language like OCL. However, in C++, with its built-in Standard Template Library (STL) strings, sets, and the ability to iterate over collections, constructing an informative error message is straightforward.

17.5.6 Open-Source Development

We note with satisfaction that the open-source model of software development has been yielding dividends for libSBML. The user community has contributed not only several bug fixes, but new code as well. These include: support for the Expat parser library (Drake and Clark, 2005), a full Perl API, a full Lisp API, and an extension to support the use of a provisional SBML standard for storing model diagrams (see Section 17.4.3).

The libSBML open-source license allows it to be incorporated freely into other programs in whole or part. Several simulator programs and projects developed in academia already make use of libSBML to support both SBML import and export. Such simulator programs include: Gepasi (Mendes, 1997), COPASI (Mendes, 2003), Jarnac (Sauro, 2000), and the DARPA Bio-SPICE project (Kumar and Feidler, 2003). It is worth mentioning that since libSBML is distributed under the terms of the Lesser GNU Public License (LGPL) it may also be used without restriction in commercial applications (Free Software Foundation, 1999). We currently know of two commercial software applications using libSBML.

17.6 Validating Application Behavior

When we first developed SBML, we expected that most of the difficulties faced by developers in implementing software support would stem from issues of constructing and parsing valid model structures. We knew it would be impossible to write perfectly clear specifications for the language, but we expected that once issues of ambiguities and other problems in SBML's definition were overcome, interchange of models between software tools would naturally follow. And to a surprising extent, this was true for a few early applications such as Jarnac and Gepasi—exactly the same applications that informed the definition of SBML in the first place. It was not until a large number of other software developers began working with SBML that it became clear the community faced more subtle issues of model interpretation and consensual agreement about expected behaviors of simulation tools.

17.6.1   Types of Validation

At the highest level, we can partition the question of validity into two main categories:

1. *Syntactic*: does the software accept well-formed SBML input, and reject all syntactically invalid SBML input? (Note that a software package may reject some valid SBML inputs because it detects the presence of constructs it is not designed to handle. For the purposes of syntactic verification, such behavior is acceptable and presumably can be distinguished from a failure to accept well-formed SBML.)

2. *Semantic*: does the software interpret well-formed SBML correctly? This can be further divided:

   (a) *Model structure*: does the software construct the correct model structure based on the SBML input, independent of what it does with that structure?

   (b) *Model behavior*: does the software correctly interpret or generate the intended model behavior?

The difference between the two types of semantic validation is about structure versus dynamics. Going beyond verification of conformance to SBML syntax, the semantic interpretation of a model involves both creating the intended constructs based on the SBML and analyzing or simulating the model in the intended way. In both cases, correctness is something that has to be carefully specified.

Some models can only be evaluated based on their structure. For example, molecular interaction models may not contain any kinetic information, so it is not clear that there is a definable model behavior per se. In that case, the model may be only evaluable based on the model structure. Other models have dynamics, and software tools can be evaluated based on whether they produce agreed-upon simulation or analysis results.

17.6.2   A Problem Not Addressed by Definitions Alone

The problem of achieving "agreed-upon simulation and analysis results" goes deeper than stipulating the required syntactic and semantic aspects of SBML and providing model structure-based verification of consistency of the sort now available in libSBML (Section 17.5.5). At least two issues must be addressed. One is the problem of reaching a consensus in a community about how to to interpret different classes of models. This is a problem of education and communication, which in the case of SBML is being helped tremendously by the biannual SBML face-to-face events

(SBML Forums and Hackathons). A second problem is providing a way for software developers to verify the behaviors of their software tools vis-a-vis the consensus view of simulator behaviors. This requires testing the behavior of software that interprets and manipulates models encoded in SBML.

To help address this latter problem, we have recently introduced the first version of the SBML Semantic Validation Suite, described in the next section.

17.6.3   The SBML Semantic Validation Suite

The Semantic Validation Suite consists of (1) a set of valid SBML models each with representative, simulated time-course data, and (2) a scripted, automated testing framework for running software tools through the suite. This suite is designed to be used by software developers to check that their simulators produce results that are consistent with the SBML standard and thus with each other.

In the general case, verifying the interpretation of SBML by an arbitrary software package is an extremely challenging problem, since different applications use models in different ways, generate different types of outputs, and provide different user interfaces. The only realistic way to approach this problem systematically is to tackle different application types separately, treating ODE-based simulators as one type, stochastic simulators as another, pathway analysis tools as another, etcetera. We chose to develop tests for ODE-based simulators first because: (a) this kind of simulation software makes up a significant proportion of the applications that support SBML; (b) simulation is one of the more complex types of analysis that can be applied to SBML; and (c) apart from metadata, almost all SBML features impact a model's behavior in simulation.

The set of models in the SBML Semantic Validation Suite is still incomplete, but the current version covers the majority of SBML features. The suite is divided into categories of tests, where each category deals with a set of related features of SBML. The scripts in the suite allow a simulator to be tested systematically against the test set. Each test in the suite comes with: (1) the correct simulation output in a consistent documented format; (2) plots of correct simulation output, and (3) documentation for the test. The beta version of the test suite was announced in October 2004. Several developers have begun using the suite as part of their work and communicating feedback to us about the suite itself; this feedback process is helping us to improve every aspect of it.

Our long-term goal in this effort is to eventually produce a highly automated software evaluation system. We hope to be able to generate an in-depth guide that categorizes different tools along different dimensions related to their purposes and coverage of SBML features. This will be an important aid both to potential users (who will be able to easily compare the functionality of different software packages) and to developers (who will be able to use the evaluation tools to help guide their implementation of SBML support during software development). We also believe the content will be useful for researchers wishing to understand SBML on its own.

17.7   Summary

Computational modeling is becoming crucial for making sense of the vast quantities of complex experimental data that are now being collected. The systems biology community needs agreed-upon information standards if models are to be shared, evaluated, and developed cooperatively. The Systems Biology Markup Language (SBML) is an XML-based format for representing computational models in a way that can be used by different software systems to communicate and exchange those models. It is supported today by over 80 software tools worldwide and a vibrant community of modelers and software authors. A variety of resources are available for working with SBML; there is also an Internet MIME type defined for SBML (Kovitz, 2004) and a new public database of models based around SBML (BioModels Team, 2005).

In support of SBML and its community, we continue to develop and make available software infrastructure, including programming libraries, conversion utilities, interface packages for commonly-used software environments, and easy-to-access online tools. All of our software development follows the open-source tradition to maximize the accessibility and utility of the products.

The success of SBML has led to requests from the community for new features and continued evolution of the language. We view our role as organizers and editors in the development and evolution of SBML; the process is open and crucially dependent on the involvement of others in the computational modeling field. We invite interested individuals and groups to join the SBML Forum, the informal community of SBML users and developers, to participate in the process and help us improve SBML and its capacity for acting as a common exchange format for computational modeling software in systems biology. Information on this and other aspects of the SBML project is available on the project Web site (SBML Team, 2005b).

17.8   Acknowledgments

We thank Herbert Sauro for his fundamental work on SBML Level 1 as well as crucial discussions and software development work. We also thank Hamid Bolouri for organizing and leading the SBML effort during its first two and a half years, and we thank the SBML development community for their continuing enthusiasm, participation, feedback, and support. The SBML community includes the members of the sbml-discuss@caltech.edu international mailing list and the DARPA Bio-SPICE project Model Definition Language task force. Finally, we thank the following agencies and institutions for their generous support. The development of SBML was originally funded by the Japan Science and Technology Corporation (JST) under the ERATO Kitano Symbiotic Systems Project. Support for the continued development of SBML and associated software, meetings, and activities today comes from the following sources: the National Human Genome Research Institute (USA); the National Institute of General Medical Sciences (USA); the International Joint Research Program of NEDO (Japan); the JST ERATO-SORST Program (Japan); the Japanese Ministry of Agriculture; the Japanese Ministry of Education, Culture, Sports, Science, and Technology; the BBSRC e-Science Initiative (UK); the DARPA IPTO Bio-Computation Program (USA); and the Air Force Office of Scientific Research (USA). Additional support is provided by the California Institute of Technology (USA), the University of Hertfordshire (UK), the Molecular Sciences Institute (USA), and the Systems Biology Institute (Japan).

A   Software Tools for Biological Modeling

The software tools favored by the contributors to this book, all active researchers in biological modeling, are listed in this chapter. The selection is eclectic, practical, and is presented here more as a guide for intrepid readers to help them get their feet wet than as a complete list of canonical tools. We apologize for any omissions but note that any tool used frequently in publications will not want for users. Keep in mind that advances in software occur faster than advances in science. It follows that tools will either evolve in sophistication or users will migrate upwards. For users, open standards (see chapter 17) for model interchange are therefore crucial to avoid being in thrall to an out-dated program.

A.1   Genetic Network Analyzer: GNA

- Description: GUI with network visualization, model editor and visualization of simulation results (de Jong et al., 2003b)
- System requirements: Java, runs under Windows, Unix, Solaris, MacOS
- Features: Qualitative analysis: modeling, simulation, and analysis of genetic regulatory networks described by piecewise-linear differential equation models supplemented by parameter inequality constraints
- Website: http://www-helix.inrialpes.fr/gna

A.2   Gene Interaction Network Simulator: GINsim

- Description: GUI with network visualization, model editor, and visualization of simulation results (Chaouiya et al., 2003)
- System requirements: Java, runs under Windows, Unix, Solaris, MacOS
- Features: Qualitative analysis: modeling, simulation, and analysis of genetic regulatory networks described by discrete, logical models
- Website: http://www.esil.univ-mrs.fr/~chaouiya/GINsim

A.3   Discrete Dynamics Lab: DDLab

- Description: GUI with model editor and visualization of network dynamics (Wuensche, 2003)
- System requirements: Written in C, runs under DOS, Unix, Linux, Irix
- Features: Tools for researching cellular automata, random boolean networks, multi-value discrete dynamical networks
- Website: `http://www.ddlab.com`

A.4   Cellerator

- Description: Cell model generation and simulation, from reaction descriptions, within a powerful computer algebra system (Shapiro et al., 2003)
- System requirements: Mathematica package
- Features: Quick, easy model construction with palette; ODEs shown and solved Luxuriously supports the power math user Extensible: Biologists can add new reaction types (e.g. kMech add–on package for enzyme kinetics)
- Website: `http://www.igb.uci.edu/servers/sb.html`

A.5   Sigmoid

- Description: Pathway modeling database and web pathway simulation environment (Cheng et al., 2005)
- System requirements: Java (1.4+), runs under Windows, Unix, Solaris, MacOS
- Features: Web GUI access to Cellerator and pathway model database; scalability in organizing the great variety of biological mechanisms; flexible mapping from "biological reaction type hierarchy" to "mathematical reaction model type hierarchy"; UML specification of reaction types and reactant types
- Website: `http://www.sigmoid.org`

A.6   Metatool

- Description: Structural network analysis for studying metabolic networks (Pfeiffer et al., 1999)
- System requirements: Java, runs under Windows, Unix, Solaris, MacOS
- Features: Conservation relations; null space analysis; calculation of elementary modes

- Website:
  `http://pgrc-03.ipk-gatersleben.de/tools/phpMetatool/index.php`

A.7 FluxAnalyzer

- Description: Structural network analysis completely embedded in a GUI with (optional) network visualisation (interactive flux maps) (Klamt et al., 2003)
- System requirements: Matlab
- Features: Calculation of graph-theoretical path lengths and network diameter; null space analysis; conservation relations; metabolic flux analysis; flux balance analysis; calculation and detailed analysis of elementary modes and extreme pathways
- Website: `http://www.mpi-magdeburg.mpg.de/projects/fluxanalyzer`

A.8 ScrumPy

- Description: Simulator for general biochemical systems (Poolman et al., 2003)
- System requirements: Python, mixture of command-line tools and GUIs
- Features: Conservation relations; null space analysis; calculation of elementary modes
- Website: `http://bms-mudshark.brookes.ac.uk/ScrumPy`

A.9 Jarnac

- Description: Simulator for general biochemical systems (Sauro, 2000).
- System requirements: Windows 95/98, NT, 2000
- Features: Jarnac is a language for describing and manipulating cellular system models and can be used to describe metabolic, signal transduction, and gene networks, or in fact any physical system which can be described in terms of a network and associated flows.
- Website: `http://www.cds.caltech.edu/~hsauro/Jarnac.htm`

A.10 Gepasi

- Description: GUI simulator for general biochemical systems (Mendes, 1997).
- System requirements: Windows 95 and up; Linux under Wine

- Features: Gepasi is a software package for modeling biochemical systems. It simulates the kinetics of systems of biochemical reactions and provides a number of tools to fit models to data, optimize any function of the model, perform metabolic control analysis and linear stability analysis.

- Website: `http://www.gepasi.org`

A.11 MesoRD

- Description: MesoRD is a tool for stochastic and deterministic simulation of reaction-diffusion systems. Reads SBML model descriptions. (Hattne et al., 2005)

- System requirements: Linux, Mac OS X, NetBSD, Solaris and Windows XP

- Features: Implements the next subvolume method; explicit unit handling; constructive solid geometry is used for compartment geometry descriptions; MathML reaction rate expressions are automatically restructured for fast evaluation; evaluated reaction rates are hashed; licensed under the GNU GPL.

- Website: `http://mesord.sourceforge.net`

A.12 Ingeneue

- Description: Genetic network construction software (Meir et al., 2002)

- System requirements: Java, runs under Windows, Unix, Solaris, MacOS

- Features: Ingeneue is a general-purpose program designed to construct and analyze models of genetic networks, designed so that it can be used by a biologist with only a minimal amount of mathematical training.

- Website: `http://ingeneue.org`

A.13 XPPAUT

- Description: Simulation and exploration of models of dynamical system (Ermentrout, 2002)

- System requirements: All platforms

- Features: Xppaut is a program designed specifically for the needs of dynamical systems. It has many options for integrators and numerical algorithms and includes Auto for simple bifurcation continuations. It has a simple file format for the input of models and versatile graphing capabilities.

- Website: `http://www.math.pitt.edu/~bard/xpp/xpp.html`

A.14   BioSens

- Description: GUI for methods to identify cellular architecture and dynamics from experimental data (Taylor et al., 2005)
- System requirements: Windows, partial installation on Linux using XPP
- Features: Dynamical sensitivity analysis; Fisher information matrix; FIM-based measurement selection
- Website: http://www.chemengr.ucsb.edu/~ceweb/faculty/doyle/biosens/BioSens.htm

A.15   JigCell

- Description: Building models, simulation, comparison to experimental data, parameter estimation (Vass et al., 2004)
- System requirements: Java, runs under Windows, Unix, Solaris, MacOS
- Features: SBML input
- Website: http://jigcell.biol.vt.edu

A.16   Oscill8

- Description: Simulation and advanced bifurcation analysis
- System requirements: Windows, Linux, Mac OS X
- Features: Oscill8 is a suite of tools for analyzing large systems of ODEs, particularly with respect to understanding how the high dimensional parameter space controls the dynamics of the system.
- Website: http://oscill8.sourceforge.net/

A.17   Madonna

- Description: Simulation, sensitivity analysis, optimization
- System requirements: Windows, Mac OS X
- Features: Berkeley Madonna is a general purpose differential equation solver for the modeling and analysis of dynamical systems. Developed on the Berkeley campus under the sponsorship of NSF and NIH, it is currently used for constructing mathematical models for research and teaching.
- Website: http://www.berkeleymadonna.com/

A.18   Systems Biology Workbench

- Description: General frameworks for computational modules: Systems Biology Workbench, Matlab, Mathematica, Maple, Scilab, Octave
- System requirements: All operating systems
- Features: The Systems Biology Workbench is software that uses SBML (chapter 17) to allow communications between diverse software modules. A host of software packages are compatible with SBW (http://www.sys-bio.org). Maple, Mathematica, Matlab, Octave, and Scilab are general purpose mathematical analysis software packages, with Maple and Mathematica more adept at algebraic manipulations and Matlab and Scilab more adept at numerical computations. Octave, Scilab, and the Systems Biology Workbench are free for use while the others are commercial.
- Website: `http://sbml.org/index.psp`

References

Achard, F., Vaysseix, G., and Barillot, E. XML, bioinformatics and data integration. *Bioinformatics*, 17:115–125, 2001.

Advanced Micro Devices, Inc. AMD Opteron processor product overview. Available via the World Wide Web at http://www.amd.com/us-en/Processors, 2005.

Aebersold, R. and Mann, M. Mass spectrometry-based proteomics. *Nature*, 422: 198–207, 2003.

Agou, F., Ye, F., and Veron, M. In vivo protein cross-linking. *Methods Mol Biol*, 261:427–442, 2004.

Akutsu, T., Miyano, S., and Kuhara, S. Inferring qualitative relations in genetic networks and metabolic pathways. *Bioinformatics*, 16:727–734, 2000.

Albert, R., Jeong, H., and Barabási, A.-L. Error and attack tolerance of complex networks. *Nature*, 406:378–382, 2000.

Albert, R. and Othmer, H. The topology of the regulatory interactions the expression pattern of the segment polarity genes in *Drosophila melanogaster*. *J. Theor. Biol.*, 223:1–18, 2003.

Alberts, B., Johnson, A., Lewis, J., Raff, M., Roberts, K., and Walter, P. *Molecular Biology of the Cell*. New York: Garland, 4th edition, 2002.

Aldana, M. and Cluzel, P. A natural class of robust networks. *Proc. Natl. Acad. Sci. USA*, 100:8710–8714, 2003.

Allen, J., Davey, H. M., Broadhurst, D., Heald, J. K., Rowland, J. J., Oliver, S. G., and Kell, D. B. High-throughput classification of yeast mutants for functional genomics using metabolic footprinting. *Nat. Biotechnol.*, 21:692–696, 2003.

Alm, E. and Arkin, A. Biological networks. *Curr. Opin. Struct. Biol.*, 13:193–202, 2003.

Alon, U., Surette, M., Barkai, N., and Leibler, S. Robustness in bacterial chemotaxis. *Nature*, 397:168–171, 1999.

Alterovitz, G., Afkhami, E., and Ramoni, M. Robotics, automation, and statistical learning for proteomics. In Columbus, F., editor, *Focus on Robotics and Intelligent Systems Research*. New York: Nova Science Publishers, Inc., 2006. in press.

Alur, R., Henzinger, T., Lafferriere, G., and Pappas, G. Discrete abstractions of hybrid systems. *Proc. IEEE*, 88:971–984, 2000.

Amaral, L. A. N., Díaz-Guilera, A., Moreira, A. A., Goldberger, A. L., and Lipsitz, L. A. Emergence of complex dynamics in a simple model of signaling networks. *Proc. Natl. Acad. Sci. USA*, 101:15551–15555, 2004.

Ander, M., Beltrao, P., Di Ventura, B., Ferkinghoff-Borg, J., Foglierini, M., Kaplan, A., Lemerle, C., Tomas-Oliveira, I., and Serrano, L. Smartcell, a framework to simulate cellular processes that combines stochastic approximation with diffusion and localisation: Analysis of simple gene networks. *Systems Biology*, 1:129–138., 2004.

Andrec, M., Kholodenko, B. N., Levy, R. M., and Sontag, E. Inference of signaling and gene regulatory networks by steady-state perturbation experiments: structure and accuracy. *J. Theor. Biol.*, 232:427–441, 2005.

Andrews, S. and Bray, D. Stochastic simulation of chemical reactions with spatial resolution and single molecule detail. *Phys. Biol.*, 1:137–151, 2004.

Andronov, A., Leontovich, E., Gordon, I., and Maier, A. *Qualitative Theory of Second-Order Dynamic Systems*. Chichester: John Wiley & Sons, 1973.

Angeli, D., Ferrell Jr., J., and Sontag, E. Detection of multistability, bifurcations, and hysteresis in a large class of biological positive-feedback systems. *Proc. Natl. Acad. Sci. USA*, 101:1822–27, 2004.

Apache Software Foundation. Xerces C++ parser. Available via the World Wide Web at `http://xml.apache.org/xerces-c/`, 2004.

Arita, M. The metabolic world of *Escherichia coli* is not small. *Proc. Natl. Acad. Sci. USA*, 101:1543–1547, 2004.

Arkin, A., Ross, J., and McAdams, H. H. Stochastic kinetic analysis of developmental pathway bifurcation in phage lambda-infected Escherichia coli cells. *Genetics*, 149:1633–1648, 1998.

Arnold, L. On the consistency of the mathematical models of chemical reactions. In Haken, H., editor, *Dynamics of Synergetic Systems*. Berlin: Springer, 1980.

Arnold, L. and Theodosopulu, M. Deterministic limit of the stochastic model of chemical reactions with diffusion. *Adv. Applied Prob.*, 12:367–379, 1980.

Arthur, B. On generalized urn schemes of the polya kind. *Cybernetics*, 19:61–71, 1963.

Artzy-Randrup, Y., Fleishman, S., Ben-Tal, N., and Stone, L. Comment on "network motifs: Simple building blocks of complex networks" and "superfamilies of evolved and designed networks". *Science*, 305:1107c, 2004.

Ascher, U. M. and Petzold, L. R. *Computer Methods for Ordinary Differential Equations and Differential-Algebraic Equations*. Philadelphia: SIAM, 1998.

Atkinson, M. R., Savageau, M. A., Myers, J. T., and Ninfa, A. J. Development of genetic circuitry exhibiting toggle switch or oscillatory behavior in Escherichia coli. *Cell*, 113:597–607, 2003.

Augen, J. Information technology to the rescue! *Nat. Biotechnol.*, 19(Supplement): BE39–BE40, 2001.

Aurell, E. and Sneppen, K. Epigenetics as a first exit problem. *Phys. Rev. Lett.*, 88:048101, 2002.

Ausbrooks, R., Buswell, S., Dalmas, S., Devitt, S., Diaz, A., Hunter, R., Smith, B., Soiffer, N., Sutor, R., and Watt, S. Mathematical Markup Language (MathML) Version 2.0. Available via the World Wide Web at http://www.w3.org/TR/MathML2/., 2001.

Bachelier, L. *Théorie de la spéculation*. Paris: Gauthier-Villars, 1900.

Bagowski, C. and Ferrell Jr., J. Bistability in the JNK cascade. *Curr. Biol.*, 11: 1176–1182, 2001.

Bammler, T., Beyer, R. P., Bhattacharya, S., Boorman, G. A., Boyles, A., Bradford, B. U., Bumgarner, R. E., Bushel, P. R., Chaturvedi, K., and Choi, D. Standardizing global gene expression analysis between laboratories and across platforms. *Nat. Methods.*, 2:351–356, 2005.

Banks, R. E., Dunn, M. J., Hochstrasser, D. F., Sanchez, J. C., Blackstock, W., Pappin, D. J., and Selby, P. J. Proteomics: new perspectives, new biomedical opportunities. *Lancet*, 356:1749–1756, 2000.

Bar-Joseph, Z., Gerber, G., Lee, T., Rinaldi, N., Yoo, Y., Robert, F., Gordon, D., Fraenkel, E., Jaakkola, T., Young, R., and Gifford, D. Computational discovery of gene modules and regulatory networks. *Nat. Biotechnol.*, 21:1337–1342, 2003.

Barabási, A.-L. and Albert, R. Emergence of scaling in random networks. *Science*, 286:509–512, 1999.

Barabási, A.-L. and Oltvai, Z. Network biology: Understanding the cell's functional organization. *Nat. Rev. Genetics*, 5:101–113, 2004.

Baras, F. and Mansour, M. M. Microscopic simulation of chemical instabilities. *Adv. Chem.Phys.*, 100:393–475, 1997.

Barkai, N. and Leibler, S. Robustness in simple biochemical networks. *Nature*, 387: 913–917, 1997.

Barkai, N. and Leibler, S. Circadian clocks limited by noise. *Nature*, 403:267–268, 2000.

Bartel, D. P. MicroRNAs: Genomics, biogenesis, mechanism, and function. *Cell*, 116:281–297, 2004.

Bartholomay, A. F. A stochastic approach to statistical kinetics with applications to enzyme kinetics. *Biochemistry*, 1:223–230, 1962.

Basu, S., Mehreja, R., Thiberge, S., Chen, M.-T., and Weiss, R. Spatiotemporal control of gene expression with pulse-generating networks. *Proc. Natl Acad. Sci. USA*, 101:6355–6360, 2004.

Batsilas, L., Berezhkovskii, A., and Shvartsman, S. Stochastic model of autocrine and paracrine signals in cell culture assays. *Biophys. J.*, 85:3659–3665, 2003.

Batt, G., Casey, R., de Jong, H., Geiselmann, J., Gouzé, J.-L., Page, M., Ropers, D., Sari, T., and Schneider, D. Qualitative analysis of the dynamics of genetic regulatory networks using piecewise-linear models. In Pecou, E., Martinez, S., and Maass, A., editors, *Mathematical and Computational Methods in Biology*. Paris: Editions Hermann, 2005a. In press.

Batt, G., Ropers, D., de Jong, H., Geiselmann, J., Mateescu, R., Page, M., and Schneider, D. Validation of qualitative models of genetic regulatory networks by model checking: Analysis of the nutritional stress response in *Escherichia coli*. *Bioinformatics*, 21:i119–i28, 2005b.

Beard, D., Lang, S., and Qian, H. Energy balance for analysis of complex metabolic networks. *Biophys. J.*, 83:79–86, 2002.

Beck, J. V. and Arnold, K. J. *Parameter Estimation in Engineering and Science*. Chichester: John Wiley & Sons, Inc., 1977.

Beck, K. *Test Driven Development: By Example*. Boston: Addison-Wesley Professional, 2002.

Becskei, A. and Serrano, L. Engineering stability in gene networks by autoregulation. *Nature*, 405:590–593, 2000.

Becskei, A., Séraphin, B., and Serrano, L. Positive feedback in eukaryotic gene networks: Cell differentiation by graded to binary response conversion. *EMBO J.*, 20:2528–2535, 2001.

Beeler, G. W. and Reuter, H. Reconstruction of the action potential of ventricular myocardial fibres. *J. Physiol.*, 268:177–210, 1977.

Benner, S. A. and Sismour, A. M. Synthetic biology. *Nat. Rev. Genet.*, 6:533–543, 2005.

Bennett, S. *A history of control engineering: 1800–1930*. Stevenage, UK: Peter Peregrinus, 1979.

Bentele, M., Lavrik, I., Ulrich, M., Stösser, S., Heermann, D., Kalthoff, H., Krammer, P., and Eils, R. Mathematical modeling reveals threshold mechanism in CD95-induced apoptosis. *J. Cell Biol.*, 166:839–851, 2004.

Benzer, S. Induced synthesis of enzymes in bacteria analysed at the cellular level. *Biochim. Biophys. Acta*, 11:383–395, 1953.

Berg, O. G. A model for the statistical fluctuations of protein numbers in a microbial population. *J. Theor. Biol.*, 71:587–603, 1978a.

Berg, O. G. On diffusion-controlled dissociation. *Chem. Phys.*, 31:47–57, 1978b.

Berg, O. G., Paulsson, J., and Ehrenberg, M. Fluctuations and quality of control in biological cells: Zero-order ultrasensitivity reinvestigated. *Biophys. J.*, 79:1228–1236, 2000a.

Berg, O. G., Paulsson, J., and Ehrenberg, M. Fluctuations in repressor control: Thermodynamic constraints on stochastic focusing. *Biophys. J.*, 79:2944–2953, 2000b.

Berg, O. G., Winter, R. B., and Hippel, P. H. V. Diffusion-driven mechanisms of protein translocation on nucleic acids. 1. Models and theory. *Biochemistry*, 20: 6929–6948, 1981.

Berge, C. *The Theory of Graphs*. New York: Dover Publications, unabridged reprint edition, 2001.

Bernot, G., Comet, J.-P., Richard, A., and Guespin, J. Application of formal methods to biological regulatory networks: Extending Thomas' asynchronous logical approach with temporal logic. *J. Theor. Biol.*, 229:339–348, 2004.

Bertsimas, D. and Tsitsiklis, J. *Introduction to Linear Optimization*. Belmont: Athena Scientific, 1997.

Bhalla, U. Signaling in small subcellular volumes. I. Stochastic and diffusion effects on individual pathways. *Biophys. J.*, 87:733–744, 2004.

Bhalla, U., Ram, P., and Iyengar, R. MAP kinase phosphatase as a locus of flexibility in a mitogen-activated protein kinase signaling network. *Science*, 297: 1018–1023, 2002.

Bharucha-Reid, A. T. *Elements of the theory of markov processes and their applications*. New York: McGraw-Hill, 1960.

Bi, E. and Lutkenhaus, J. Cell division inhibitors SulA and MinCD prevent formation of the FtsZ ring. *J. Bacteriol.*, 175:1118–1125, 1993.

BioModels Team. The BioModels database. Available via the World Wide Web at `http://www.ebi.ac.uk/biomodels.`, 2005.

Biron, P. V. and Malhotra, A. XML Schema part 2: Datatypes (W3C candidate recommendation 24 October 2000). Available via the World Wide Web at `http://www.w3.org/TR/xmlschema-2/.`, 2000.

Blake, W. J., Kaern, M., Cantor, C. R., and Collins, J. J. Noise in eukaryotic gene expression. *Nature*, 422:633–637, 2003.

Blanc, O., Virag, N., Vesin, J. M., and Kappenberger, L. A computer model of human atria with reasonable computation load and realistic anatomical properties. *IEEE Trans. Bio-Medical Eng.*, 48:1229–1237, 2001.

Blank, L., Kuepfer, L., and Sauer, U. Large-scale ^{13}C-flux analysis reveals mechanistic principles of metabolic network robustness to null mutations in yeast. *Genome Biol.*, 6:R49, 2005.

Bock, J. R. and Gough, D. A. Whole-proteome interaction mining. *Bioinformatics*, 19:125–134, 2003.

Bode, H. *Network Analysis and Feedback Amplifier Design*. Princeton, NJ: D. Van Nostrand, 1945.

de Boer, P. A., Crossley, R. E., and Rothfield, L. I. A division inhibitor and a topological specificity factor coded for by the minicell locus determine proper placement of the division septum in *E. coli. Cell*, 56:641–649, 1989.

Böhm, H.-J. and Schneider, G., editors. *Virtual screening for bioactive molecules.* Weinheim, D: Wiley-VCH, 2000.

Bollenbach, T., Kruse, K., Pantazis, P., González-Gaitán, M., and Jülicher, F. Robust formation of morphogen gradients. *Phys. Rev. Lett.*, 94:018103, 2005.

Bosak, J. and Bray, T. XML and the second-generation Web. *Scientific Am.*, 280: 89–93, 1999.

Bosl, W. J. and Li, R. Mitotic-exit control as an evolved complex system. *Cell*, 121:325–333, 2005.

Bottino, D., Penl, R. C., Stamps, A., Traebert, M., Dumotier, B., Georgieva, A., Helmlinger, G., and Lett, G. S. Preclinical cardiac safety assessment of pharmaceutical compounds using an integrated systems-based computer model of the heart. *Progr. Biophys. Mol. Biol.*, 2005. In press.

Bower, J. M. and Bolouri, H. *Computational Modeling of Genetic and Biochemical Networks.* Cambridge, MA: MIT Press, 2001.

Boyett, M. R., Dobrzynski, H., Lancaster, M. K., Jones, S. A., Honjo, H., and Kodama, I. Sophisticated architecture is required for the sinoatrial node to perform its normal pacemaker function. *J. Cardiovasc. Electrophysiol.*, 14:104–106, 2003.

Boyett, M. R., Honjo, H., Yamamoto, M., Nikmaram, M. R., Niwa, R., and Kodama, I. Downward gradient in action potential duration along conduction path in and around the sinoatrial node. *Am. J. Physiol.*, 276:H686–H698, 1999.

Bradley, C. P., Pullan, A. J., and Hunter, P. J. Geometric modeling of the human torso using cubic hermite elements. *Ann. Biomed. Eng.*, 25:96–111, 1997.

Bray, D. Molecular networks: The top-down view. *Science*, 301:1864–1865, 2003.

Bray, T., Paoli, J., Sperberg-McQueen, C. M., and Maler, E. Extensible Markup Language (XML) 1.0 (second edition), W3C recommendation 6 October 2000. Available via the World Wide Web at `http://www.w3.org/TR/1998/REC-xml-19980210`, 2000.

Brenner, S. Biological computation. *Novartis Found. Symp.*, 213:106–116, 1998.

Briggs, G. and Haldane, J. A note on the kinetics of enzyme action. *Biochem. J.*, 19:339, 1925.

Brockmann, D. and Geisel, T. Particle dispersion on rapidly folding random heteropolymers. *Phys. Rev. Lett.*, 91:48303, 2003.

Brogan, W. L. *Modern Control Theory.* Upper Saddle River, NJ: Prentice Hall, 1991.

Brown, T. A., editor. *Genomes.* New York: Wiley-Liss, 2nd edition, 2002.

Bruggeman, F. J., Westerhoff, H. V., Hoek, J. B., and Kholodenko, B. N. Modular response analysis of cellular regulatory networks. *J. Theor. Biol.*, 218:507–520, 2002.

Bryson, J. Modular representations of cognitive phenomena in AI, psychology and neuroscience. In Davis, D., editor, *Visions of Mind: Architectures for Cognition and Affect*. Hershey, PA: IDEA, 2005.

Burgard, A., Nikolaev, E., Schilling, C., and Maranas, C. Flux coupling analysis of genome-scale metabolic network reconstructions. *Genome Res.*, 14:301–312, 2004.

Burgard, A., Pharkya, P., and Maranas, C. OptKnock: A bilevel programming framework for identifying gene knockout strategies for microbial strain optimization. *Biotechnol. Bioeng.*, 84:647–657, 2003.

Burgard, A. P. and Maranas, C. D. Optimization-based framework for inferring and testing hypothesized metabolic objective functions. *Biotech. Bioeng.*, 82:670–677, 2003.

Butte, A., Tamayo, P., Slonim, D., Golub, T., and Kohane, I. Discovering functional relationships between RNA expression and chemotherapeutic susceptibility using relevance networks. *Proc. Natl. Acad. Sci. USA*, 97:12182–12186, 2000.

Bynum, W. F., Browne, E., and Porter, R., editors. *Dictionary of the History of Science*. London: MacMillan Press, 1981.

Cakir, T., Kirdar, B., and Ulgen, K. Metabolic pathway analysis of yeast strengthens the bridge between transcriptomics and metabolic networks. *Biotech. Bioeng.*, 86:251–260, 2004.

Cao, Y., Gillespie, D. T., and Petzold, L. R. The slow-scale stochastic simulation algorithm. *J. Chem. Phys.*, 122:014116, 2005.

Cao, Y., Li, H., and Petzold, L. Efficient formulation of the stochastic simulation algorithm for chemically reacting systems. *J. Chem. Phys.*, 121:4059–4067, 2004a.

Cao, Y., Petzold, L. R., Rathinam, M., and Gillespie, D. T. The numerical stability of leaping methods for stochastic simulation of chemically reacting systems. *J. Chem. Phys.*, 121:12169–12178, 2004b.

Carlson, J. and Doyle, J. Highly optimized tolerance: A mechanism for power laws in designed systems. *Phys. Rev. E*, 60:1412–27, 1999.

Carlson, J. and Doyle, J. Highly optimized tolerance: robustness and design in complex systems. *Phys. Rev. Lett.*, 84:2529–32, 2000.

Carlson, J. and Doyle, J. Complexity and robustness. *Proc. Natl. Acad. Sci. USA*, 99:2538–2545, 2002.

Carmeliet, E. E. Chloride ions and the membrane potential of purkinje fibres. *J. Physiol.*, 156:375–388, 1961.

Cascante, M., Boros, L. G., Comin-Anduix, B., de Atauri, P., Centelles, J. J., and Lee, P. W.-N. Metabolic control analysis in drug discovery and disease. *Nat. Biotechnol.*, 20:243–249, 2002.

Chabrier-Rivier, N., Chiaverini, M., Danos, V., Fages, F., and Schächter, V. Modeling and querying biomolecular interaction networks. *Theor. Comput. Sci.*, 325: 25–44, 2004.

Chaikin, P. M. and Lubensky, T. C. *Principles of Condensed Matter physics*. Cambridge: Cambridge University Press, 1995.

Chan, E. Y., Goncalves, N. M., Haeusler, R. A., Hatch, A. J., Larson, J. W., Maletta, A. M., Yantz, G. R., Carstea, E. D., Fuchs, M., and Wong, G. G. DNA mapping using microfluidic stretching and single-molecule detection of fluorescent site-specific tags. *Genome Res.*, 14:1137–1146, 2004.

Chaouiya, C., Remy, E., Mossé, B., and Thieffry, D. Qualitative analysis of regulatory graphs: A computational tool based on a discrete formal framework. In *Positive Systems (POSTA 2003)*, volume 294 of *LNCIS*, pages 119–126. Berlin: Springer-Verlag, 2003.

Ch'en, F. C., Vaughan-Jones, R. D., Clarke, K., and Noble, D. Modelling myocardial ischaemia and reperfusion. *Prog. Biophys. Mol. Biol.*, 69:515–537, 1998.

Cheng, J., Scharenbroich, L., Baldi, P., and Mjolsness, E. Sigmoid: Towards an intelligent, scalable, software infrastructure for pathway bioinformatics and systems biology. *IEEE Intelligent Systems*, 20:68–75, 2005.

Cherry, J. L. and Adler, F. R. How to make a biological switch. *J. Theor. Biol.*, 203:117–133, 2000.

Chutinan, A. and Krogh, B. Verification of infinite-state dynamic systems using approximate quotient transition systems. *IEEE Trans. Aut. Control*, 46:1401–1410, 2001.

Ciliberto, A., Novak, B., and Tyson, J. J. Steady states and oscillations in the p53/Mdm2 network. *Cell Cycle*, 4:488–493, 2005.

Cinquin, O. and Demongeot, J. Positive and negative feedback: Striking a balance between necessary antagonists. *J. Theor. Biol.*, 216:229–241, 2002.

Clancy, C. E. and Rudy, Y. Linking a genetic defect to its cellular phenotype in a cardiac arrhythmia. *Nature*, 400:566–569, 1999.

Clark, J. and DeRose, S. XML Path Language (XPath) version 1.0: W3C recommendation 16 November 1999. Available via the World Wide Web at http://www.w3.org/TR/1999/REC-xpath-19991116, 1999.

Clarke, B. Stoichiometric network analysis. *Cell Biophys.*, 12:237–253, 1988.

Coen, E. *The Art of Genes*. Oxford: Oxford University Press, 1999.

Conant, G. and Wagner, A. Convergent evolution of gene circuits. *Nat. Genet.*, 34: 264–266, 2003.

Coombes, S., Hinch, R., and Timofeeva, Y. Receptors, sparks and waves in a fire-diffuse-fire framework for calcium release. *Prog. Biophys. Mol. Biol.*, 85:197–216, 2004.

Corne, D., Dorigo, M., and Glover, F., editors. *New Ideas in Optimization.* New York: McGraw-Hill, 1999.

Cornish-Bowden, A. *Fundamentals of Enzyme Kinetics.* London: Portland Press, revised edition, 1995.

Cornish-Bowden, A. Metabolic control analysis in biotechnology and medicine. *Nat. Biotechnol.*, 17:641–643, 1999.

Cornish-Bowden, A. and Hofmcyr, J. The role of stoichiometric analysis in studies of metabolism: An example. *J. Theor. Biol.*, 216:179–191, 2002.

Costa, K. D., Hunter, P. J., Wayne, J. S., Waldman, L. K., Guccione, J. M., and Mcculloch, A. D. A three-dimensional finite element method for large elastic deformations of ventricular myocardium. 2. Prolate spheroidal coord. *J. Biomech. Eng.*, 118:464–472, 1996.

Cover, T. M. and Thomas, J. A. *Elements of Information Theory.* New York: John Wiley & Sons, Inc., 1991.

Covert, M., Knight, E., Reed, J., Herrgard, M., and Palsson, B. Integrating high-throughput and computational data elucidates bacterial networks. *Nature*, 429: 92 96, 2004.

Covert, M. and Palsson, B. Constraints-based models: Regulation of gene expression reduces the steady-state solution space. *J. Theor. Biol.*, 221:309–325, 2003.

Covert, M., Schilling, C., and Palsson, B. Regulation of gene expression in flux balance models of metabolism. *J. Theor. Biol.*, 213:73–88, 2001.

Crampin, E. J., Halstead, M., Hunter, P. J., Nielsen, P., Noble, D., Smith, N., and Tawhai, M. Computational physiology and the physiome project. *Exp. Physiol.*, 89:1–26, 2004.

Cross, M. C. and Hohenberg, P. C. Pattern formation outside of equilibrium. *Rev. Mod. Phys.*, 65:851–1112, 1993.

Csete, M. and Doyle, J. Bow ties, metabolism and disease. *Trends Biotechnol.*, 22: 446–450, 2004.

Csete, M. E. and Doyle, J. C. Reverse engineering of biological complexity. *Science*, 295:1664–1669, 2002.

Cuthrell, J. E. and Biegler, L. On the Optimization of Differential Algebraic Process Systems. *AIChE J.*, 33:1257–1270, 1987.

Czechowski, T., Bari, R. P., Stitt, M., Scheible, W. R., and Udvardi, M. K. Real-time rt-pcr profiling of over 1400 arabidopsis transcription factors: Unprecedented sensitivity reveals novel root- and shoot-specific genes. *Plant J.*, 38:366–379, 2004.

Czernik, A. J., Girault, J. A., Nairn, A. C., Chen, J., Snyder, G., Kebabian, J., and Greengard, P. Production of phosphorylation state-specific antibodies. *Methods Enzymol*, 201:264–283, 1991.

D'Agostini, G. *Bayesian Reasoning in Data Analysis—A Critical Introduction.* Singapore: World Scientific Publishing, 2003.

von Dassow, G., Meir, E., Munro, E., and Odell, G. The segment polarity network is a robust developmental module. *Nature*, 406:188–192, 2000.

Davey, H. and Kell, D. Flow cytometry and cell sorting of heterogeneous microbial populations: the importance of single-cell analysis. *Microbiol. Rev.*, 60:641–696, 1996.

Davies, P. Emergent biological principles and the computational properties of the universe. *Complexity*, 11:11–15, 2004.

Deb, K. *Multi-Objective Optimization using Evolutionary Algorithms*. Chichester, UK: John Wiley & Sons, 2001.

Deck, K. A. and Trautwein, W. Ionic currents in cardiac excitation. *Pflügers Arch.*, 280:65–80, 1964.

Delbruck, M. Statistical fluctuation in autocatalytic reactions. *J. Chem. Phys.*, 8: 120–124, 1940.

Denardo, E. *Dynamic Programming: Models and Applications*. New York: Dover Publications, 2003.

Devreotes, P. and Janetopoulos, C. Eukaryotic chemotaxis: Distinctions between directional sensing and polarization. *J. Biol. Chem.*, 278:20445–20448, 2003.

DiFrancesco, D. A new interpretation of the pace-maker current, ik2, in purkinje fibres. *J. Physiol.*, 314:359–376, 1981.

DiFrancesco, D. and Noble, D. A model of cardiac electrical activity incorporating ionic pumps and concentration changes. *Phil. Trans. Royal Soc. B*, 307:353–398, 1985.

Dobrin, R., Beg, Q., Barabási, A., and Oltvai, Z. Aggregation of topological motifs in the *Escherichia coli* transcriptional regulatory network. *BMC Bioinformatics*, 5:10, 2004.

Dobrzynski, H., Zhang, H., Wright, S. E., Holden, A. V., and Boyett, M. R. Structure-function relationships of the sinoatrial node. *Intl. J. Bifurcation Chaos*, 13:3621–3629, 2003.

Doob, J. L. Markoff chains—denumerable case. *Trans. Am. Math. Soc.*, 58:455–473, 1945.

Dover, G. *Dear Mr Darwin. Letters on the Evolution of Life and Human Nature*. London: Weidenfeld and Nicolson, 2000.

Doyle, J., Francis, B., and Tannenbaum, A. *Feedback Control Theory*. New York: Macmillan, 1992.

Draghici, S., Khatri, P., Eklund, A., and Szallasi, Z. Reliability and reproducibility issues in DNA microarray measurements. *Trends Genet.*, in press, 2006.

Drake, F. and Clark, J. The Expat XML parser, 2005. Available via the World Wide Web at `http://www.libexpat.org/`.

Drew, D. A., Osborn, M. J., and Rothfield, L. I. A polymerization-depolymerization model that accurately generates the self-sustained oscillatory system involved in bacterial division site placement. *Proc. Natl. Acad. Sci. USA*, 102:6114–6118, 2005.

Dupont, G., Berridge, M., and Goldbeter, A. Signal-induced Ca^{2+} oscillations: Properties of a model based on Ca^{2+}-induced Ca^{2+} release. *Cell Calcium*, 12: 73–85, 1991.

Duran, A., Diaz-Meco, M. T., and Moscat, J. Essential role of RelA Ser311 phosphorylation by zetaPKC in NF-kappaB transcriptional activation. *EMBO J.*, 22:3910–3918, 2003.

Earl, D. J. and Deem, M. W. Evolvability is a selectable trait. *Proc. Natl. Acad. Sci. USA*, 101:11531–11536, 2004.

Edwards, J., Ibarra, R., and Palsson, B. *In silico* predictions of *Escherichia coli* metabolic capabilities are consistent with experimental data. *Nat. Biotechnol.*, 19:125–130, 2001a.

Edwards, J. and Palsson, B. The *Escherichia coli* MG1655 *in silico* metabolic phenotype: Its definition, characteristics and capabilities. *Proc. Natl. Acad. Sci. USA*, 97:5528–5533, 2000.

Edwards, R., Siegelmann, H., Aziza, K., and Glass, L. Symbolic dynamics and computation in model gene networks. *Chaos*, 11:160–169, 2001b.

Egan, T., Noble, D., Noble, S. J., Powell, T., Spindler, A. J., and Twist, V. W. Sodium-calcium exchange during the action potential in guinea-pig ventricular cells. *J. Physiol.*, 411:639–661, 1989.

Ehrenberg, M. and Blomberg, C. Thermodynamic constraints on kinetic proofreading in biosynthetic pathways. *Biophys. J.*, 31:333–358, 1980.

Einstein, A. On the motion of small particles suspended in liquids at rest required by the molecular-kinetic theory of heat. *Ann. Physik*, 17:549–560, 1905.

Eisner, D. A., Choi, H. S., Diaz, M. E., and O'Neill, S. C. Integrative analysis of calcium cycling in cardiac muscle. *Circulation Res.*, 87:1087–1094, 2000.

Eissing, T., Conzelmann, H., Gilles, E. D., Allgöwer, F., Bullinger, E., and Scheurich, P. Bistability analyses of a caspase activation model for receptor-induced apoptosis. *J. Biol. Chem.*, 279:36892–36897, 2004.

Eker, S., Knapp, M., Laderoute, K., Lincoln, P., Meseguer, J., and Sonmez, K. Pathway logic: Symbolic analysis of biological signaling. In *Pac. Symp. Biocompy* volume 7, pages 400–412. Singapore: World Scientific Publishing, 2002.

El-Samad, H., Goff, J., and Khammash, M. Calcium homeostasis and par hypocalcemia: An integral feedback perspective. *J. Theor. Biol.*, 214:17–

El-Samad, H. and Khammash, M. Regulated degredation is a mec' suppressing stochastic fluctuations in gene regulatory networks. *under review*, 2006.

El-Samad, H., Kurata, H., Doyle, J., Gross, C., and Khammash, M. Surviving heat shock: Control strategies for robustness and performance. *Proc. Natl. Acad. Sci. USA*, 102:2736–2741, 2005.

Eldar, A., Dorfman, R., Weiss, D., Ashe, H., Shilo, B.-Z., and Barkai, N. Robustness of the BMP morphogen gradient in *Drosophila* embryonic patterning. *Nature*, 419:304–308, 2002.

Eldar, A., Rosin, D., Shilo, B.-Z., and Barkai, N. Self-enhanced ligand degradation underlies robustness of morphogen gradients. *Dev. Cell*, 5:635–646, 2003.

Elf, J. *Intracellular Flows and Fluctuations*. Uppsala: Uppsala Univ., 2004.

Elf, J. and Ehrenberg, M. Fast evaluation of fluctuations in biochemical networks with the linear noise approximation. *Genome Res.*, 13:2475–2484, 2003.

Elf, J. and Ehrenberg, M. Spontaneous separation of bi-stable biochemical systems into spatial domains of opposite phases. *Systems Biol.*, 2:230–236, 2004.

Elf, J. and Ehrenberg, M. Near-critical behavior of aminoacyl-tRNA pools in E. coli at rate limiting supply of amino acids. *Biophys. J.*, 88:132–146, 2005a.

Elf, J. and Ehrenberg, M. What makes ribosome-mediated transcriptional attenuation sensitive to amino acid limitation? *PLoS Comp. Biol.*, 1:e2, 2005b.

Elf, J., Paulsson, J., Berg, O. G., and Ehrenberg, M. Near-critical phenomena in intracellular metabolite pools. *Biophys. J.*, 84:154–170, 2003.

Elowitz, M. and Leibler, S. A synthetic oscillatory network of transcriptional regulators. *Nature*, 403:335–338, 2000.

Elowitz, M. B., Levine, A. J., Siggia, E. D., and Swain, P. S. Stochastic gene expression in a single cell. *Science*, 297:1183–1186, 2002.

Emery, A. and Nenarokomov, A. Optimal experiment design. *Meas. Sci. Technol.*, 9:864–876, 1998.

Erdi, P. and Tóth, J. *Mathematical Models of Chemical Reactions*. Princeton: Princeton University Press, 1989.

Eriksson, H.-E. and Penker, M. *UML Toolkit*. New York: John Wiley & Sons, 1998.

Ermentrout, B. *Simulating, Analyzing, and Animating Dynamical Systems*. Philadelphia, PA: SIAM, 2002.

Fabiato, A. Calcium induced release of calcium from the sarcoplasmic reticulum. *Am. J. Physiol.*, 245:C1–C14, 1983.

Falcke, M. Reading the patterns in living cells—the physics of Ca2+ signaling. *Adv. Phys.*, 53:255–440, 2004.

Fallside, D. C. XML Schema part 0: Primer (W3C candidate recommendation 24 October 2000). Available via the World Wide Web at `http://www.w3.org/TR/xmlschema-0/`, 2000.

Featherstone, D. and Broadie, K. Wrestling with pleiotropy: Genomic and topological analysis of the yeast gene expression network. *Bioessays*, 24:267–274, 2002.

Fedoroff, N. and Fontana, W. Small numbers of big molecules. *Science*, 297:1129–1131, 2002.

Felix, M., Labbé, J., Dorée, M., Hunt, T., and Karsenti, E. Triggering of cyclin degradation in interphase extracts of amphibian eggs by cdc2 kinase. *Nature*, 346:379–382, 1990.

Fell, D. Metabolic control analysis: A survey of its theoretical and experimental development. *Biochem. J.*, 286 (Pt 2):313–330, Sep 1992.

Fell, D. *Understanding the control of metabolism*. London: Portland Press, 1996.

Fell, D. Increasing the flux in metabolic pathways: A metabolic control analysis perspective. *Biotechnol. Bioeng.*, 58:121–124, 1998.

Feng, X.-J., Hooshangi, S., Chen, D., Li, G., Weiss, R., and Rabitz, H. Optimizing genetic circuits by global sensitivity analysis. *Biophys. J.*, 87:2195–2202, 2004.

Ferm, L., Lötstedt, P., and Sjöberg, P. Adaptive, conservative solution of the Fokker-Planck equation in molecular biology. Technical Report 2004-054, Dept. of Information Technology, Scientific Computing, Uppsala Univ., Uppsala, Sweden, 2004.

Fernández, M., Malhotra, A., Marsh, J., Nagy, M., and Walsh, N. XQuery 1.0 and XPath 2.0 data model: W3C working draft 4 April 2005. Available via the World Wide Web at `http://www.w3.org/TR/2005/WD-xpath-datamodel-20050404/`, 2005.

Ferrell Jr., J. Tripping the switch fantastic: How a protein kinase cascade can convert graded inputs into switch-like outputs. *Trends Biochem. Sci.*, 21:460–466, 1996.

Ferrell Jr., J. Self-perpetuating states in signal transduction: Positive feedback, double-negative feedback and bistability. *Curr. Opin. Chem. Biol.*, 6:140–148, 2002.

Ferrell Jr., J. and Machleder, E. The biochemical basis of an all-or-none cell fate switch in *Xenopus* oocytes. *Science*, 280:895–898, 1998.

Feytmans, E., Noble, D., and Peitsch, M. Genome size and numbers of biological functions. *Trans. Comp. Systems Biol.*, 1:44–49, 2005.

Fink, M., Noble, D., and Giles, W. Contributions of inwardly-rectifying K+ currents to repolarization assessed using mathematical models of ventricular myocytes. *In press*, 2005.

Finney, A. and Hucka, M. Systems biology markup language: Level 2 and beyond. *Biochem. Soc. Trans.*, 31:1472–1473, 2003.

Finney, A., Hucka, M., and Bolouri, H. Systems Biology Markup Language (SBML) Level 2: Structures and facilities for model definitions. Available via the World Wide Web at `http://sbml.org/documents/`, 2002.

Fischer, E. and Sauer, U. Large-scale in vivo flux analysis reveals rigidity and suboptimal performance of *Bacillus subtilis* metabolism. *Nat. Genet.*, 37:636–640, 2005.

Fonseca, C. and Fleming, P. Nonlinear system identification with multiobjective genetic algorithms. In *Proc. 13th World Congr. Intl. Fed. Automat. Control, San Francisco, CA*, pages 187–192, 1996.

Forger, D. and Peskin, C. A detailed predictive model of the mammalian circadian clock. *Proc. Natl. Acad. Sci. USA*, 100:14806–14811, 2003.

Förster, J., Famili, I., Palsson, B., and Nielsen, J. Large-scale evaluation of *in silico* gene knockouts in *Saccharomyces cerevisiae*. *OMICS*, 7:193–202, 2003.

Förster, J., Gombert, A., and Nielsen, J. A functional genomics approach using metabolomics and *in silico* pathway analysis. *Biotechnol. Bioeng.*, 79:703–712, 2002.

Fox, J. and Hill, C. From topology to dynamics in biochemical networks. *Chaos*, 11:809–815, 2001.

Francis, B. and Wonham, W. The internal model principle of control theory. *Automatica*, 12:457–465, 1976.

Free Software Foundation. The GNU Lesser General Public License (LGPL). Available via the World Wide Web at `http://www.fsf.org/licenses/licenses.html`, 1999.

Freeman, M. Feedback control of intercellular signalling in development. *Nature*, 408:313–319, 2000.

Freter, R. R. and Savageau, M. A. Proofreading systems of multiple stages for improved accuracy of biological discrimination. *J. Theor. Biol.*, 85:99–123, 1980.

Fricke, T. and Wendt, D. The markoff-automaton a new alorithm for simulating the time-evolution of large stochastic dynamic systems. *Intl. J. Modern Phys. C*, 6:277–306, 1992.

Funamoto, S., Meili, R., Lee, S., Parry, L., and Firtel, R. A. Spatial and temporal regulation of 3-phosphoinositides by PI 3-kinase and PTEN mediates chemotaxis. *Cell*, 109:611–623, 2002.

Fussenegger, M., Bailey, J. E., and Varner, J. A mathematical model of caspase function in apoptosis. *Nat. Biotech.*, 18:768–774, 2000.

Gadkar, K. G., Gunawan, R., and Doyle III, F. J. Iterative approach to model identification of biological networks. *BMC Bioinformatics*, 6:155, 2005a.

Gadkar, K. G., Varner, J., and Doyle III, F. J. Model identification of signal transduction networks from data using a state regulator problem. *IEE Systems Biol.*, 2:17–30, 2005b.

Gagneur, J. and Casari, G. From molecular networks to qualitative cell behavior. *FEBS Lett.*, 579:1867–1871, 2005.

Gagneur, J., Jackson, D., and Casari, G. Hierarchical analysis of dependency in metabolic networks. *Bioinformatics*, 19:1027–1034, 2003.

Gagneur, J. and Klamt, S. Computation of elementary modes: A unifying framework and the new binary approach. *BMC Bioinformatics*, 5:175, 2004.

Gardiner, C. *Handbook of Stochastic Methods*. Berlin: Springer-Verlag, 2nd edition, 1985.

Gardiner, C., McNeil, K., Walls, D., and Matheson, I. Correlations in stochastic theories of chemical reactions. *J. Stat. Phys.*, 14:307–331, 1976.

Gardiner, C. and Steyn-Ross, M. Adiabatic elimination in stochastic systems. II. Application to reaction diffusion and hydrodynamic-like systems. *Phys. Rev. A*, 29:2823–2833, 1984.

Gardner, T., Cantor, C., and Collins, J. Construction of a genetic toggle switch in *Escherichia coli*. *Nature*, 403:339–342, 2000.

Gardner, T. S., di Bernardo, D., Lorenz, D., and Collins, J. J. Inferring genetic networks and identifying compound mode of action via expression profiling. *Science*, 301:102–105, 2003.

Garny, A., Kohl, P., Hunter, P. J., Boyett, M. R., and Noble, D. One-dimensional rabbit sinoatrial node models: Benefits and limitations. *J. Cardiovasc. Electrophysiol.*, 14:S121–S132, 2003.

Garny, A., Kohl, P., Noble, D., and Hunter, P. J. 1-D and 2-D models of the origin and propagation of cardiac excitation from the sino-atrial node into the right atrium. *Phil. Trans. Royal Soc. London B*, 2000.

Gavin, A. C., Bosche, M., Krause, R., Grandi, P., Marzioch, M., Bauer, A., Schultz, J., Rick, J. M., Michon, A. M., and and, C. M. C. Functional organization of the yeast proteome by systematic analysis of protein complexes. *Nature*, 415: 141–147, 2002.

Gee, D. A. and Ramirez, W. F. On-line state estimation and parameter identification for batch fermentation. *Biotechnol. Prog.*, 12:132–140, 1996.

Gesbert, F., Sellers, W. R., Signoretti, S., Loda, M., and Griffin, J. D. BCR/ABL regulates expression of the cyclin-dependent kinase inhibitor p27Kip1 through the phosphatidylinositol 3-kinase/AKT pathway. *J. Biol. Chem.*, 275:39223–39230, 2000.

Ghosh, R. and Tomlin, C. Symbolic reachable set computation of piecewise affine hybrid automata and its application to biological modelling: Delta-Notch protein signalling. *Syst. Biol.*, 1:170–183, 2004.

Gibson, M. and Bruck, J. Efficient exact stochastic simulation of chemical systems with many species and channels. *J. Phys. Chem. A*, 104:1876–1889, 2000a.

Gibson, M. A. and Bruck, J. Exact stochastic simulation of chemical systems with many species and many channels. *J. Phys. Chem.*, 105:1876–1889, 2000b.

Gilbert, G. and Mulkay, N. *Opening Pandora's box : A sociological analysis of scientists' discourse*. Cambridge, UK: Cambridge University Press, 1984.

Gillespie, D. and Petzold, L. Improved leap-size selection for accelerated stochastic simulation. *J. Chem. Phys.*, 119:8229–8234, 2003.

Gillespie, D. T. A general method for numerically simulating the stochastic time evolution of coupled chemical reactions. *J. Comp. Phys.*, 22:403–434, 1976.

Gillespie, D. T. Exact stochastic simulation of coupled chemical reactions. *J. Phys. Chem.*, 81:2340–2361, 1977.

Gillespie, D. T. *Markov Processes: An Introduction for Physical Scientists*. Boston, MA: Academic Press, 1992a.

Gillespie, D. T. A rigorous derivation of the chemical master equation. *Physica A*, 188:404–425, 1992b.

Gillespie, D. T. The chemical Langevin equation. *J. Chem. Phys.*, 113:297–306, 2000.

Gillespie, D. T. Approximate accelerated stochastic simulation of chemically reacting systems. *J. Chem. Phys.*, 115:1716–1733, 2001.

Gillespie, D. T. The chemical Langevin and Fokker-Planck equations for the reversible isomerization reaction. *J. Phys. Chem. A*, 106:5063–5071, 2002.

Girvan, M. and Newman, M. Community structure in social and biological networks. *Proc. Natl. Acad. Sci. USA*, 99:7821–7826, 2002.

Glass, L. and Kauffman, S. The logical analysis of continuous non-linear biochemical control networks. *J. Theor. Biol.*, 39:103–129, 1973.

Glassey, J., Ignova, M., Ward, A. C., Montague, G. A., and Morris, A. J. Bioprocess supervision: Neural networks and knowledge based systems. *J. Biotechnol.*, 52: 201–205, 1997.

Glendinning, P. *Stability, Instability and Chaos: An Introduction to the Theory of Nonlinear Differential Equations*. Cambridge, UK: Cambridge University Press, 1994.

Gold, D., Coombes, K., Medhane, D., Ramaswamy, A., Ju, Z., Strong, L., Koo, J. S., and Kapoor, M. A comparative analysis of data generated using two different target preparation methods for hybridization to high-density oligonucleotide microarrays. *BMC Genomics*, 5:2, 2004.

Goldberg, D. *The Design of Innovation: Lessons from and for Competent Genetic Algorithms*. Boston, MA: Kluwer, 2002.

Goldbeter, A. A model for circadian oscillations in the *Drosophila* period protein PER. *Proc. R. Soc. London Ser. B*, 261:319–324, 1995.

Goldbeter, A. Computational approaches to cellular rhythms. *Nature*, 420:238–245, 2002.

Goldbeter, A. and Koshland, D. E. An amplified sensitivity arising from covalent modification in biological systems. *Proc. Natl. Acad. Sci. USA*, 78:6840–6844, 1981.

Goldenfeld, N. and Kadanoff, L. A. Simple lessons from complexity. *Science*, 284: 87–89, 1999.

Golyandina, N., Nekrutkin, V., and Zhigljavsky, A. *Analysis of Time Series Stucture. SSA and Related Techniques*. Boca Raton, FL: Chapman and Hall/CRC, 2001.

Gonze, D., Halloy, J., and Goldbeter, A. Robustness of circadian rhythms with respect to molecular noise. *Proc. Natl. Acad. Sci. USA*, 99:673–678, 2002.

Goodwin, B. C. Oscillatory behavior in enzymatic control processes. *Adv. Enzyme Regul.*, 3:425–438, 1965.

Goto, S., Okuno, Y., Hattori, M., Nishioka, T., and Kanehisha, M. LIGAND: Database of chemical compounds and reactions in biological pathways. *Nucleic Acids Res.*, 30:402–404, 2002.

Gouzé, J.-L. Positive and negative circuits in dynamical systems. *J. Biol. Sys.*, 6: 11–15, 1998.

Gouzé, J.-L. and Sarı, T. A class of piecewise linear differential equations arising in biological models. *Dyn. Syst.*, 17:299–316, 2002.

de Groot, S. R. and Mazur, P. *Non-Equilibrium Thermodynamics*. New York: Dover Publications, 1984.

Gu, Y., Steinmetz, L., Gu, X., Scharfe, C., Davis, R., and Li, W.-H. Role of duplicate genes in genetic robustness against null mutations. *Nature*, 421:63–66, 2003.

Guelzim, N., Bottani, S., Bourgine, P., and Képès, F. Topological and causal structure of the yeast transcriptional regulatory network. *Nat. Genet.*, 31:60–63, 2002.

Guet, C., Elowitz, M., Hsing, W., and Leibler, S. Combinatorial synthesis of genetic networks. *Science*, 296:1466–1470, 2002.

Guimerà, R., Arenas, A., and Díaz-Guilera, A. Communication and optimal hierarchical networks. *Physica A*, 299:247–252, 2001.

Guimera, R., Sales-Pardo, M., and Amaral, L. Modularity from fluctuations in random graphs and complex networks. *Phys. Rev. E*, 70:025101, 2004.

Gunawan, R., Cao, Y., Petzold, L., and Doyle III, F. J. Sensitivity analysis of discrete stochastic systems. *Biophys. J.*, 88:2530–2540, 2005.

Gunawan, R., Jung, M. Y. L., Seebauer, E. G., and Braatz, R. D. Maximum a posteriori estimation of transient enhanced diffusion energetics. *AIChE J.*, 49: 2114–2123, 2003.

Gygi, S., Rist, B., Gerber, S., Turecek, F., Gelb, M., and Aebersold, R. Quantitative analysis of complex protein mixtures using isotope coded affinity tags. *Nat. Biotechnol.*, 17:994–999, 1999.

Haab, B. B., Dunham, M. J., and Brown, P. O. Protein microarrays for highly parallel detection and quantitation of specific proteins and antibodies in complex solutions. *Genome Biol.*, 2:0004, 2001.

D' haeseleer, P., Liang, S., and Somogyi, R. Genetic network inference: From co-expression clustering to reverse engineering. *Bioinformatics*, 16:707–26, 2000.

Hale, C. A., Meinhardt, H., and de Boer, P. A. Dynamic localization cycle of the cell division regulator MinE in *Escherichia coli*. *EMBO J.*, 20:1563–1572, 2001.

Hall, A. E., Hutter, O. F., and Noble, D. Current-voltage relations of Purkinje fibres in sodium-deficient solutions. *J. Physiol.*, 166:225–240, 1963.

Halloy, J., Lauzeral, J., and Goldbeter, A. Modeling oscillations and waves of cAMP in Dictyostelium discoideum cells. *Biophys. Chem.*, 72:9–19, 1998.

Hanusse, P. and Blanché, A. A Monte Carlo method for large reaction-diffusion systems. *J. Chem. Phys.*, 74:6148–6153, 1981.

Hardiman, G. Microarray platforms—comparisons and contrasts. *Pharmacoge-nomics*, 5:487–502, 2004.

Harlow, E. and Lane, D., editors. *Antibodies : A laboratory manual.* Cold Spring Harbor, NY:Cold Spring Harbor Laboratory, 1988.

Harrild, D. and Henriquez, C. A. Computer model of normal conduction in the human atria. *Circulation Res.*, 87:E25–36, 2000.

Harris, E., Sawhwill, B., Wuensche, A., and Kauffman, S. A model of transcriptional regulatory networks based on biases in the observed regulation rules. *Complexity*, 7:23–40, 2002.

Hartemink, A. J., Gifford, D. K., Jaakola, T. S., and Young, R. Combining location and expression data for principled discovery of genetic regulatory network models. *Proc. Pac. Symp. Biocomput.*, 7:437–449, 2002.

Hartman, J., Garvik, B., and Hartwell, L. Principles of the buffering of genetic variation. *Science*, 291:1001–1004, 2001.

Hartwell, L. H., Hopfield, J. J., Leibler, S., and Murray, A. W. From molecular to modular cell biology. *Nature*, 402:C47–52, 1999.

Hastings, M. Circadian clockwork: Two loops are better than one. *Nat. Rev. Neurosci.*, 1:143–146, 2000.

Hasty, J., McMillen, D., and Collins, J. Engineered gene circuits. *Nature*, 420: 224–230, Nov. 14 2002.

Hasty, J., Pradines, J., Dolnik, M., and Collins, J. Noise-based switches and amplifiers for gene expression. *Proc. Natl. Acad. Sci. USA*, 97:2075–2080, 2000.

Hattne, J., Fange, D., and Elf, J. Stochastic reaction-diffusion simulation with MesoRD. *Bioinformatics*, 21:2923–2924, 2005.

Hawley, T. S., Telford, W. G., Ramezani, A., and Hawley, R. G. Four-color flow cytometric detection of retrovirally expressed red, yellow, green, and cyan fluorescent proteins. *Biotechniques*, 30:1028–1034, 2001.

Hedley, W. J., Nelson, M. R., Bullivant, D., Cuellar, A., Ge, Y., Grehlinger, M., Jim, K., Lett, S., Nickerson, D., Nielsen, P., and Yu, H. CellML specification. Available via the World Wide Web at http://www.cellml.org., 2001a.

Hedley, W. J., Nelson, M. R., Bullivant, D. P., and Nielson, P. F. A short introduction to CellML. *Phil. Trans. Royal Soc. London A*, 359:1073–1089, 2001b.

Heidtke, K. and Schulze-Kremer, S. Design and implementation of a qualitative simulation model of λ phage infection. *Bioinformatics*, 14:81–91, 1998.

van der Heijden, R., Heijnen, J., Hellinga, C., Romein, B., and Luyben, K. Linear constraint relations in biochemical reaction systems: 1. classification of the calculability and the balancebility of conversion rates. *Biotechnol. Bioeng.*, 43:3–10, 1994.

Heinrich, R. and Schuster, S. *The Regulation of Cellular Systems*. New York: Chapman & Hall, 1996.

Hekstra, D., Taussig, A. R., Magnasco, M., and Naef, F. Absolute mRNA concentrations from sequence-specific calibration of oligonucleotide arrays. *Nucleic Acids Res.*, 31:1962–1968, 2003.

Hemby, S. E., Ginsberg, S. D., Brunk, B., Arnold, S. E., Trojanowski, J. Q., and Eberwine, J. H. Gene expression profile for schizophrenia. *Arch. Gen. Psychiat.*, 59:631–640, 2002.

Hilgemann, D. W. Extracellular calcium transients and action potential configuration changes related to post-stimulatory potentiation in rabbit atrium. *J. Gen. Physiol.*, 87:675–706, 1986.

Hilgemann, D. W. and Noble, D. Excitation-contraction coupling and extracellular calcium transients in rabbit atrium: Reconstruction of basic cellular mechanisms. *Proc. Royal Soc. B*, 230:163–205, 1987.

Hinch, R. A mathematical analysis of the generation and termination of calcium sparks. *Biophys. J.*, 86:1293–1307, 2004.

Ho, Y., Gruhler, A., Heilbut, A., Bader, G. D., Moore, L., Adams, S. L., Millar, A., Taylor, P., Bennett, K., and and, K. B. Systematic identification of protein complexes in saccharomyces cerevisiae by mass spectrometry. *Nature*, 415:180–183, 2002.

Hoffmann, A., Levchenko, A., Scott, M. L., and Baltimore, D. The IκB-NF-κB signaling module: temporal control and selective gene activation. *Science*, 298:1241–1245, 2002.

Hofmeyr, J., Kacser, H., and van der Merwe, K. Metabolic control analysis of moiety-conserved cycles. *Eur. J. Biochem.*, 155:631–641, 1986.

Hofmeyr, J.-H. S. Metabolic control analysis in a nutshell. In *Proc. 2nd Intl. Conf. Systems Biol.*, pages 291–300, 2000.

Hofstadter, D. *Gödel, Escher, Bach: An Eternal Golden Braid*. New York: Basic Books, 1979.

Holland, J. *Emergence*. Reading, MA: Helix, 1998.

Holland, M. J. Transcript abundance in yeast varies over six orders of magnitude. *J. Biol. Chem.*, 277:14363–14366, 2002.

Holme, P., Huss, M., and Jeong, H. Subnetwork hierarchies of biochemical pathways. *Bioinformatics*, 19:532–538, 2003.

Holzhütter, H.-G. The principle of flux minimization and its application to estimate stationary fluxes in metabolic networks. *Eur. J. Biochem.*, 271:2905–2922, 2004.

Hooshangi, S., Thiberge, S., and Weiss, R. Ultrasensitivity and noise propagation in a synthetic transcriptional cascade. *Proc. Natl. Acad. Sci. USA*, 102:3581–3586, 2005.

Hopfield, J. J. Kinetic proofreading: A new mechanism for reducing errors in biosynthetic processes requiring high specificity. *Proc. Natl. Acad. Sci. USA*, 71:4135–4139, 1974.

Horsthemke, W. and Lefever, R. *Noise-Induced Transitions. Theory and Applications in Physics, Chemistry, and Biology*. Berlin: Springer Verlag, 1984.

Howard, M., Rutenberg, A. D., and de Vet, S. Dynamic compartmentalization of bacteria: Accurate division in *E. coli. Phys. Rev. Lett.*, 87:278102, 2001.

Hu, Z. L., Gogol, E. P., and Lutkenhaus, J. Dynamic assembly of MinD on phospholipid vesicles regulated by ATP and MinE. *Proc. Natl. Acad. Sci. USA*, 99:6761–6766, 2002.

Hu, Z. L. and Lutkenhaus, J. Topological regulation of cell division in *Escherichia coli* involves rapid pole to pole oscillation of the division inhibitor MinC under the control of MinD and MinE. *Mol. Microbiol.*, 34:82–90, 1999.

Huang, K. C., Meir, Y., and Wingreen, N. S. Dynamic structures in *Escherichia coli*: Spontaneous formation of MinE rings and MinD polar zones. *Proc. Natl. Acad. Sci. USA*, 100:12724–12728, 2003.

Hucka, M. SCHUCS: A notation for describing model representations intended for XML encoding. Available via the World Wide Web at `http://www.sbml.org/`, 2000.

Hucka, M., Finney, A., Bornstein, B. J., Keating, S. M., Shapiro, B. E., Matthews, J., Kovitz, B. L., Schilstra, M. J., Funahashi, A., Doyle, J. C., and Kitano, H. Evolving a lingua franca and associated software infrastructure for computational systems biology: The Systems Biology Markup Language (SBML) project. *IEE Systems Biol.*, 1:41–53, 2004.

Hucka, M., Finney, A., Sauro, H., Bolouri, H., Doyle, J., Kitano, H., Arkin, A., et al. The Systems Biology Markup Language (SBML): A medium for representation and exchange of biochemical network models. *Bioinformatics*, 19:524–531, 2003.

Hucka, M., Finney, A., Sauro, H. M., and Bolouri, H. Systems Biology Markup Language (SBML) Level 1: Structures and facilities for basic model definitions. Available via the World Wide Web at `http://www.sbml.org`, 2001.

Hunt, J., Lee, M., and Price, C. Applications of qualitative model-based reasoning. *Control Eng. Pract.*, 1:253–266, 1993.

Hunter, P. J., Robbins, P., and Noble, D. The IUPS human physiome project. *Pflügers Arch.*, 445:1–9, 2002.

Husmeier, D. Sensitivity and specificity of inferring genetic regulatory interactions from microarray experiments with dynamic Bayesian networks. *Bioinformatics*, 19:2271–2282, 2003.

Hutter, O. F. and Noble, D. Rectifying properties of heart muscle. *Nature*, 188: 495, 1960.

Ibarra, R., Edwards, J., and Palsson, B. *Escherichia coli* K-12 undergoes adaptive evolution to achieve in silico predicted optimal growth. *Nature*, 420:186–189, 2002.

Ideker, T. and Lauffenburger, D. Building with a scaffold: Emerging strategies for high- and low-level cellular modeling. *Trends Biotechnol.*, 21:255–262, 2003.

Ideker, T. E., Thorsson, V., and Karp, R. M. Discovery of regulatory interactions through perturbations: Inference and experimental design. *Proc. Pac. Symp. Biocomput.*, 5:305–316, 2000.

Iglesias, P. Feedback control in intracellular signaling pathways: Regulating chemotaxis in Dictyostelium discoideum. *Eur. J. Control*, 9:216–225, 2003.

Iglesias, P. and Levchenko, A. Modeling the cell's guidance system. *Sci. STKE*, 2002:RE12, 2002.

Ihekwaba, A., Broomhead, D., Grimley, R., Benson, N., and Kell, D. Sensitivity analysis of parameters controlling oscillatory signalling in the NF-κB pathway: The roles of IKK and IκBα. *IEE Systems Biol.*, 1:93–103, 2004.

Ihmels, J., Bergmann, S., and Barkai, N. Defining transcription modules using large-scale gene expression data. *Bioinformatics*, 20:1993–2003, 2004a.

Ihmels, J., Friedlander, G., Bergmann, S., Sarig, O., Ziv, Y., and Barkai, N. Revealing modular organization in the yeast transcriptional network. *Nat. Genet.*, 31:370–377, 2002.

Ihmels, J., Levy, R., and Barkai, N. Principles of transcriptional control in the metabolic network of *Saccharomyces cerevisiae*. *Nat. Biotechnol.*, 22:86–92, 2004b.

Iijima, M. and Devreotes, P. Tumor suppressor PTEN mediates sensing of chemoattractant gradients. *Cell*, 109:599–610, 2002.

Ingalls, B. A frequency domain approach to sensitivity analysis of biochemical systems. *J. Phys. Chem. B*, 108:1143–1152, 2004.

International Human Genome Sequencing Consortium. Initial sequencing and analysis of the human genome. *Nature*, 409:860–921, 2001.

International Human Genome Sequencing Consortium. Finishing the euchromatic sequence of the human genome. *Nature*, 431:931–945, 2004.

Irish, J. M., Hovland, R., Krutzik, P. O., Perez, O. D., Bruserud, O., Gjertsen, B. T., and Nolan, G. P. Single cell profiling of potentiated phospho-protein networks in cancer cells. *Cell*, 118:217–228, 2004.

Isaacs, F. J., Hasty, J., Cantor, C. R., and Collins, J. J. Prediction and measurement of an autoregulatory genetic module. *Proc. Natl. Acad. Sci. USA*, 100:7714–7719, 2003.

Isidori, A. and Byrnes, C. Output regulation of nonlinear systems. *IEEE Trans. Automat. Control*, 35:131–140, 1990. ISSN 0018-9286.

Jafri, S., Rice, J. J., and Winslow, R. L. Cardiac Ca2+ dynamics: The roles of ryanodine receptor adaptation and sarcoplasmic reticulum load. *Biophys. J.*, 74: 1149–1168, 1998.

James, S., Nilsson, P., James, G., Kjelleberg, S., and Fagerström, T. Luminescence control in the marine bacterium Vibrio fischeri: An analysis of the dynamics of lux regulation. *J. Mol. Biol.*, 296:1127–1137, 2000.

Jaynes, E. Entropy and search theory. In Smith, C. R. and Grandy, W. T., editors, *Maximum-Entropy and Bayesian Methods in Inverse Problems*. Dordrecht, NL: D. Reidel, 1985.

Jaynes, E. *Probability Theory: The Logic of Science*. Cambridge: Cambridge University Press, 2003.

Jeong, H., Mason, S., Barabasi, A., and Oltvai, Z. Lethality and centrality in protein networks. *Nature*, 411:41–42, 2001.

Jeong, H., Tombor, B., Albert, R., Oltvai, Z., and Barabási, A.-L. The large-scale organization of metabolic networks. *Nature*, 407:651–654, 2000.

Johnson, J. M., Edwards, S., Shoemaker, D., and Schadt, E. E. Dark matter in the genome: Evidence of widespread transcription detected by microarray tiling experiments. *Trends Genet.*, 21:93–102, 2005.

Johnson, N. L. and Kotz, S. *Urn Models and their Application: An Approach to Modern Discrete Probability Theory*. New York: Wiley, 1977.

Johnson, S. *Emergence*. New York: Scribner, 2001.

de Jong, H. Modeling and simulation of genetic regulatory systems: A literature review. *J. Comput. Biol.*, 9:67–103, 2002.

de Jong, H. Qualitative simulation and related approaches for the analysis of dynamical systems. *Knowl. Eng. Rev.*, 19:93–132, 2005.

de Jong, H., Geiselmann, J., Batt, G., Hernandez, C., and Page, M. Qualitative simulation of the initiation of sporulation in B. subtilis. *Bull. Math. Biol.*, 66: 261–299, 2004a.

de Jong, H., Geiselmann, J., Hernandez, C., and Page, M. Genetic Network Analyzer: Qualitative simulation of genetic regulatory networks. *Bioinformatics*, 19:336–344, 2003a.

de Jong, H., Geiselmann, J., Hernandez, C., and Page, M. Genetic Network Analyzer: Qualitative simulation of genetic regulatory networks. *Bioinformatics*, 19:336–344, 2003b.

de Jong, H., Gouzé, J.-L., Hernandez, C., Page, M., Sari, T., and Geiselmann, J. Qualitative simulation of genetic regulatory networks using piecewise-linear models. *Bull. Math. Biol.*, 66:301–340, 2004b.

Jordan, B. R. How consistent are expression chip platforms? *Bioessays*, 26:1236–1242, 2004.

Jordan, M., editor. *Learning in Graphical Models*. Cambridge, MA: MIT Press, 1998.

Joshi, A. and Palsson, B. Metabolic dynamics in the human red cell. Part I: A comprehensive kinetic model. *J. Theor. Biol.*, 141:515–528, 1989.

Jülicher, F., Ajdari, A., and Prost, J. Modeling molecular motors. *Rev. Mod. Phys.*, 69:1269–1281, 1997.

Jurinke, C., Oeth, P., and van den Boom, D. MALDI-TOF mass spectrometry: A versatile tool for high-performance DNA analysis. *Mol. Biotechnol.*, 26:147–164, 2004.

Kacser, H. and Burns, J. A. The control of flux. In *Symp. Soc. Exp. Biol.*, volume 27, pages 65–104, 1973.

Kalagnanam, J., Simon, H., and Iwasaki, Y. The mathematical bases for qualitative reasoning. *IEEE Expert*, 6:11–19, 1991.

van Kampen, N. G. *Stochastic Processes in Physics and Chemistry*. Amsterdam: Elsevier, 2nd edition, 1992.

van Kampen, N. A power series expansion of the master equation. *Can. J. Phys.*, 39:551, 1961.

Kane, M. D., Jatkoe, T. A., Stumpf, C. R., Lu, J., Thomas, J. D., and Madore, S. J. Assessment of the sensitivity and specificity of oligonucleotide (50mer) microarrays. *Nucleic Acids Res.*, 28:4552–4557, 2000.

Kaplan, D. and Glass, L. *Understanding Nonlinear Dynamics*. Berlin: Springer-Verlag, 1995.

Karin, M., Cao, Y., Greten, F. R., and Li, Z. W. NF-kappaB in cancer: From innocent bystander to major cul. *Nat. Rev. Cancer*, 2:301–310, 2002.

Karp, P., Riley, M., Paley, S., and Pellegrini-Toole, A. The MetaCyc database. *Nucleic Acids Res.*, 30:59–61, 2002.

Kauffman, K., Prakash, P., and Edwards, J. Advances in flux balance analysis. *Curr. Opin. Biotechnol.*, 14:491–496, 2003a.

Kauffman, S. Metabolic stability and epigenesis in randomly constructed genetic nets. *J. Theor. Biol.*, 22:437–467, 1969.

Kauffman, S. *The Origins of Order: Self-Organization and Selection in Evolution*. Oxford: Oxford University Press, 1993.

Kauffman, S. *Investigations*. Oxford: Oxford University Press, 2000.

Kauffman, S. A proposal for using the ensemble approach to understand genetic regulatory networks. *J. Theor. Biol.*, 230:581–590, 2004.

Kauffman, S., Lobo, J., and Macready, W. Optimal search on a technology landscape. *J. Econ. Behav. Organ.*, 43:141–166, 2000.

Kauffman, S., Peterson, C., Samuelsson, B., and Troein, C. Random Boolean network models and the yeast transcriptional network. *Proc. Natl. Acad. Sci. USA*, 100:14796–14799, 2003b.

Kauffman, S., Peterson, C., Samuelsson, B., and Troein, C. Genetic networks with canalyzing Boolean rules are always stable. *Proc. Natl. Acad. Sci. USA*, 101: 17102–17107, 2004.

Keating, S. M. SBMLToolbox. Available via the World Wide Web at `http://www.sbml.org/software/sbmltoolbox`, 2005.

Keizer, J. *Statistical Thermodynamics of Nonequlibrium Processes*. Berlin: Springer-Verlag, 1987.

Kell, D. Metabolomics, machine learning and modelling: Towards an understanding of the language of cells. *Biochem. Soc. Trans.*, 33:520–524, 2005.

Kell, D. and King, R. On the optimization of classes for the assignment of unidentified reading frames in functional genomics programmes: The need for machine learning. *Trends Biotechnol.*, 18:93–98, 2000.

Kell, D. and Welch, G. No turning back, reductonism and biological complexity. *Times Higher Educational Supplement*, 9 August:15, 1991.

Kell, D. and Westerhoff, H. Metabolic control theory: Its role in microbiology and biotechnology. *FEMS Microbiol. Rev.*, 39:305–320, 1986.

Kell, D. B. Genotype:phenotype mapping: Genes as computer programs. *Trends Genet.*, 18:555–559, 2002.

Kell, D. B. Metabolomics and systems biology: Making sense of the soup. *Curr. Opin. Microbiol.*, 7:296–307, 2004.

Kell, D. B. and Oliver, S. G. Here is the evidence, now what is the hypothesis? The complementary roles of inductive and hypothesis-driven science in the post-genomic era. *Bioessays*, 26:99–105, 2004.

Kepler, T. B. and Elston, T. C. Stochasticity in transcriptional regulation: Origins, consequences, and mathematical representations. *Biophys. J.*, 81:3116–3136, 2001.

Kessin, R. *Dictyostelium discoideum: Evolution, cell biology, and the development of multicellularity*. Cambridge: Cambridge Univ. Press, 2001.

Khalil, H. *Nonlinear Systems*. Prentice-Hall, Upper Saddle River, NJ, third edition, 2002.

Kholodenko, B. N. Negative feedback and ultrasensitivity can bring about oscillations in the mitogen-activated protein kinase cascades. *Eur. J. Biochem.*, 267: 1583–1588, 2000.

Kholodenko, B. N., Kiyatkin, A., Bruggeman, F., Sontag, E., Westerhoff, H., and Hoek, J. Untangling the wires: A strategy to trace functional interactions in signaling and gene networks. *Proc. Natl. Acad. Sci. USA*, 99:12841–12846, 2002.

Kimura, J., Noma, A., and Irisawa, H. Na-Ca exchange current in mammalian heart cells. *Nature*, 319:596–597, 1986.

Kimura, S., Ide, K., Kashihara, A., Kano, M., Hatakeyama, M., Masui, R., Nakagawa, N., Yokoyama, S., Kuramitsu, S., and Konagaya, A. Inference of S-systems models of genetic networks using a cooperative coevolutionary algorithm. *Bioinformatics*, 21:1154 1163, 2005.

Kirschner, M. and Gerhart, J. Evolvability. *Proc. Natl. Acad. Sci. USA*, 95:8420–8427, 1998.

Kitano, H. Computational systems biology. *Nature*, 420:206–210, 2002a.

Kitano, H. Systems biology: A brief overview. *Science*, 295:1662–1664, 2002b.

Kitano, H. Biological robustness. *Nat. Rev. Genetics*, 5:826 836, 2004a.

Kitano, H. Cancer as a robust system: Implications for anticancer therapy. *Nat. Rev. Cancer*, 4:227–235, 2004b.

Klamt, S. and Gilles, E. Minimal cut sets in biochemical reaction networks. *Bioinformatics*, 20:226–234, 2004.

Klamt, S., Schuster, S., and Gilles, E. Calculability analysis in underdetermined metabolic networks illustrated by a model of the central metabolism in purple nonsulfur bacteria. *Biotechnol. Bioeng.*, 77:734–751, 2002.

Klamt, S. and Stelling, J. Combinatorial complexity of pathway analysis in metabolic networks. *Mol. Biol. Rep.*, 29:233–236, 2002.

Klamt, S. and Stelling, J. Two approaches for pathway analysis in metabolic networks? *Trends Biotechnol.*, 21:64–69, 2003.

Klamt, S., Stelling, J., M.Ginkel, and Gilles, E. FluxAnalyzer: Exploring structure, pathways and flux distributions in metabolic networks on interactive flux maps. *Bioinformatics*, 19:261–269, 2003.

Klebe, G., editor. *Virtual screening: An alternative or complement to high-throughput screening.* Dordrecht, NL: Kluwer Academic Publishers, 2000.

Klipp, E., Herwig, R., Kowald, A., Wierling, C., and Lehrach, H. *Systems Biology in Practice: Concepts, Implementation and Clinical Application.* Berlin: Wiley-VCH, 2005.

Knuth, D. *The Art of Computer Programming.* Reading, MA: Addison-Wesley, 1997.

Kobayashi, H., Kærn, M., Araki, M., Chung, K., Gardner, T. S., Cantor, C. R., and Collins, J. J. Programmable cells: Interfacing natural and engineered gene networks. *Proc. Natl. Acad. Sci. USA*, 101:8414–8419, 2004.

Koch, A. J. and Meinhardt, H. Biological pattern formation: From basic mechanisms to complex structures. *Rev. Mod. Phys.*, 66:1481–1507, 1994.

Koch, I., Junker, B., and Heiner, M. Application of Petri net theory for modelling and validation of the sucrose breakdown pathway in the potato tuber. *Bioinformatics*, 21:1219–1226, 2005.

Kohl, P., Noble, D., Winslow, R., and Hunter, P. J. Computational modelling of biological systems: Tools and visions. *Phil. Trans. Royal Soc. A*, 358:579–610, 2000.

Kohl, P. and Sachs, F. Mechanoelectric feedback in cardiac cells. *Phil. Trans. Royal Soc. A*, 359:1173–1185, 2001.

Korobkova, E., Emonet, T., Vilar, J. M., Shimizu, T. S., and Cluzel, P. From molecular noise to behavioural variability in a single bacterium. *Nature*, 428: 574–578, 2004.

Koshland Jr., D., Goldbeter, A., and Stock, J. Amplification and adaptation in regulatory and sensory systems. *Science*, 217:220–225, 1982.

Kostelich, E. and Schreiber, T. Noise-reduction in chaotic time-series data: A survey of common methods. *Phys. Rev. E*, 48:1752–1763, 1993.

Kovacs, S., McQueen, D. M., and Peskin, C. S. Modelling cardiac fluid dynamics and diastolic function. *Phil. Trans. Royal Soc. A*, 359:1299–1314, 2001.

Kovitz, B. RFC 3823: MIME media type for the Systems Biology Markup Language (SBML), 2004. Available on the Internet at `http://www.faqs.org/rfcs/rfc3823.html`.

Krakauer, D. and Plotkin, J. Redundancy, antiredundancy, and the robustness of genomes. *Proc. Natl. Acad. Sci. USA*, 99:1405–1409, 2002.

Kramers, H. Brownian motion in a field of force and the diffusion model of chemical reactions. *Physica*, 7:284, 1940.

Kremling, A., Jahreis, K., Lengeler, J., and Gilles, E. The organization of metabolic reaction networks: A signal-oriented approach to cellular models. *Metabolic Eng.*, 2:190–200, 2000.

Kruse, K. A dynamic model for determining the middle of *Escherichia coli*. *Biophys. J.*, 82:618–627, 2002.

Kruse, K. and Jülicher, F. Oscillations in cell biology. *Curr. Opin. Cell Biol.*, 17: 20–26, 2005.

Kuepfer, L., Sauer, U., and Blank, L. Metabolic functions of duplicate genes in *Saccharomyces cerevisiae*. *Genome Res.*, 15:1421–1430, 2005.

ter Kuile, B. and Westerhoff, H. Transcriptome meets metabolome: Hierarchical and metabolic regulation of the glycolytic pathway. *FEBS Lett.*, 500:169–171, 2001.

Kuipers, B. *Qualitative Reasoning: Modeling and Simulation with Incomplete Knowledge*. Cambridge, MA: MIT Press, 1994.

Kumagai, A. and Dunphy, W. Control of the Cdc2/cyclin B complex in Xenopus egg extracts arrested at a G2/M checkpoint with DNA synthesis inhibitors. *Mol. Biol. Cell*, 6:199–213, 1995.

Kumar, S. and Feidler, J. C. BioSPICE: A computational infrastructure for integrative biology. *OMICS*, 7:225, 2003.

Kuramoto, Y. Effects of diffusion on the fluctuations in open systems. *Prog. Theor. Phys.*, 52:711–713, 1974.

Kurosawa, G., Mochizuki, A., and Iwasa, Y. Comparative study of circadian clock models, in search of processes promoting oscillation. *J. Theor. Biol.*, 216:193–208, 2002.

Kuznetsov, A., Kærn, M., and Kopell, N. Synchrony in a population of hysteresis-based genetic oscillators. *SIAM J. Appl. Math.*, 65:392–425, 2005.

Kyoda, K., Baba, K., Onami, S., and Kitano, H. DBRF-MEGN method: An algorithm for deducing minimum equivalent gene networks from large-scale gene expression profiles of gene deletion mutants. *Bioinformatics*, 20:2662–2675, 2004.

Laemmli, U. K. Cleavage of structural proteins during the assembly of the head of bacteriophage T4. *Nature*, 227:680–685, 1970.

Lahav, G., Rosenfeld, N., Sigal, A., Geva-Zatorsky, N., Levine, A. J., Elowitz, M. B., and Alon, U. Dynamics of the p53-Mdm2 feedback loop in individual cells. *Nat. Genet.*, 36:147–150, 2004.

Landau, L. D. and Lifshitz, E. M. *Fluid Mechanics*, volume 6 of *Course of Theoretical Physics*. Oxford: Butterworth-Heinemann, 2nd edition, 1995.

Lander, A. A calculus of purpose. *PLoS Biol.*, 2:0712, 2004.

Langer, T. and Hoffmann, R. Virtual screening: An effective tool for lead structure discovery? *Curr. Pharm. Design*, 7:509–527, 2001.

Laub, M. and Loomis, W. A molecular network that produces spontaneous oscillations in excitable cells of Dictyostelium. *Mol. Biol. Cell.*, 9:3521–3532, 1998.

Laubenbacher, R. and Stigler, B. A computational algebra approach to the reverse engineering of gene regulatory networks. *J. Theor. Biol.*, 229:523–537, 2004.

Lauffenburger, D. Cell signaling pathways as control modules: Complexity for simplicity? *Proc. Natl. Acad. Sci. USA*, 97:5031–5033, 2000.

Lax, M. Fluctuations from the nonequlibrium steady state. *Rev. Mod. Phys.*, 32:25–64, 1960.

Le Hors, A., Le Hégaret, P., Wood, L., Nicol, G., Robie, J., Champion, M., and Byrne, S. Document Object Model (DOM) Level 2 core specification. Available via the World Wide Web at http://www.w3.org/TR/DOM-Level-2-Core/, 2000.

Lee, B. and Cardy, J. Renormalization-group study of the A+B $\to \phi$ diffusion-limited reaction. *J. Stat. Phys.*, 80:971–1007, 1995.

Lee, I., Date, S. V., Adai, A. T., and Marcotte, E. M. A probabilistic functional network of yeast genes. *Science*, 306:1555–1558, 2004.

Lee, J. and Lee, J. Approximate dynamic programming strategies and their applicability for process control: A review and future directions. *Int. J. Control Automat. Syst.*, 2:263–278, 2004.

Lee, J., Phalakornkule, C., Domach, M., and Grossmann, I. Recursive MILP model for finding all the alternate optima in LP models for metabolic networks. *Comput. Chem. Eng.*, 24:711–716, 2000.

Lee, T., Rinaldi, N., Odom, D., Bar-Joseph, Z., Gerber, G., Hannett, N., Harbison, C., Thompson, C., Simon, I., Zeitlinger, J., Jennings, E., Murray, H., Gordon, D., Ren, B., Wyrick, J., Tagne, J., Volkert, T., Fraenkel, E., Clifford, D., and Young, R. Transcriptional regulatory networks in *Saccharomyces cerevisiae*. *Science*, 298:799–804, 2002.

LeGrice, I. J., Hunter, P. J., Young, A. A., and Smaill, B. H. The architecture of the heart: A data-based model. *Phil. Trans. Royal Soc. A*, 359:1217–1232, 2001.

LeGuennec, J. Y. and Noble, D. The effects of rapid perturbation of external sodium concentration at different moments of the action potential in guinea-pig ventricular myocytes. *J. Physiol.*, 478:493–504, 1994.

Lehner, B., Williams, G., Campbell, R. D., and Sanderson, C. M. Antisense transcripts in the human genome. *Trends Genet.*, 18:63–65, 2002.

Leloup, J.-C. and Goldbeter, A. A model for circadian rhythms in *Drosophila* incorporating the formation of a complex between the PER and TIM proteins. *J. Biol. Rhythms*, 13:70–87, 1998.

Leloup, J.-C. and Goldbeter, A. Chaos and birhythmicity in a model for circadian oscillations of the PER and TIM proteins in *Drosophila*. *J. Theor. Biol.*, 198: 445–459, 1999.

Leloup, J.-C. and Goldbeter, A. Toward a detailed computational model for the mammalian circadian clock. *Proc. Natl. Acad. Sci. USA*, 100:7051–7056, 2003.

Leontovich, M. The basic equations of the kinetic theory of gases from the viewpoint of stochastic processes. *Zh. Eksp. Teor. Fiz.*, 5:211–231, 1935.

Levchenko, A. and Iglesias, P. Models of eukaryotic gradient sensing: Application to chemotaxis of amoebae and neutrophils. *Biophys. J.*, 82:50–63, 2002.

Liang, R. Q., Li, W., Li, Y., Tan, C. Y., Li, J. X., Jin, Y. X., and Ruan, K. C. An oligonucleotide microarray for microRNA expression analysis based on labeling RNA with quantum dot and nanogold probe. *Nucleic Acids Res.*, 33:e17, 2005.

Liang, S., Fuhrman, S., and Somogyi, R. REVEAL: A general reverse engineering algorithm for inference of genetic network architectures. In *Pac. Symp. Biocomput.*, volume 3, pages 18–29. Singapore: World Scientific Publishing, 1998.

Lipton, P. Testing hypotheses: Prediction and prejudice. *Science*, 307:219–221, 2005.

Ljung, L. Model validation and model error modeling. In Wittenmark, B. and Rantzer, A., editors, *The Åström Symposium on Control*, pages 15–42. Lund, Sweden: Studentlitteratur, 1999a.

Ljung, L. *System Identification : Theory for the User*. Upper Saddle River, NJ: Prentice Hall PTR, 2nd edition, 1999b.

Lloyd, C. M., Halstead, M. D. B., and Nielsen, P. F. CellML: Its future, present and past. *Prog. Biophys. Mol. Biol.*, 85:433–450, 2004.

Luo, C. and Rudy, Y. A dynamic model of the cardiac ventricular action potential: II afterdepolarizations, triggered activity and potentiation. *Circulation Res.*, 74: 1097–1113, 1994.

Luo, C. H. and Rudy, Y.-. A model of the ventricular cardiac action potential: Depolarization, repolarization, and their interaction. *Circulation Res.*, 68:1501–1526, 1991.

Luscombe, N. M., Babu, M. M., Yu, H., Snyder, M., Teichmann, S. A., and Gerstein, M. Genomic analysis of regulatory network dynamics reveals large topological changes. *Nature*, 431:308–312, 2004.

Lusted, L. *Introduction to Medical Decision Making*. Springfield, IL: Charles C. Thomas, 1968.

Lutz, R. and Bujard, H. Independent and tight regulation of transcriptional units in *escherichia coli* via the LacR/O, the TetR/O and AraC/I1-I2 regulatory elements. *Nucleic Acids Res.*, 25:1203–1210, 1997.

Lynn, P. *An introduction to the analysis and processing of signals*. London: Macmillan, 2nd edition, 1982.

Ma, H.-W., Kumar, B., Ditges, U., Gunzer, F., Buer, J., and Zeng, A.-P. An extended transcriptional regulatory network of *Escherichia coli* and analysis of its hierarchical structure and network motifs. *Nucleic Acids Res.*, 32:6643–6649, 2004a.

Ma, L. and Iglesias, P. Quantifying robustness of biochemical network model. *BMC Bioinformatics*, 3:38, 2002.

Ma, L., Janetopoulos, C., Yang, L., Devreotes, P., and Iglesias, P. Two complementary, local excitation, global inhibition mechanisms acting in parallel can explain the chemoattractant-induced regulation of $PI(3,4,5)P_3$ response in Dictyostelium cells. *Biophys. J.*, 87:3764–3774, 2004b.

Ma, L., Wagner, J., Rice, J., Hu, W., Levine, A., and Stolovitzky, G. A plausible model for the digital response of p53 to DNA damage. *Proc. Natl. Acad. Sci. USA*, 102:14266–14271, 2005.

MacBeath, G. Protein microarrays and proteomics. *Nat. Genet-*, 32 Suppl:526–532, 2002.

Mahadevan, R., Edwards, J. S., and Doyle III, F. J. Dynamic flux balance analysis of diauxic growth in *E. coli. Biophys.J.*, 83:1331–1340, 2002.

Mahadevan, R. and Schilling, C. The effects of alternate optimal solutions in constraint-based genome-scale metabolic models. *Metab. Eng.*, 5:264–276, 2003.

Malek-Mansour, M. and Houard, J. A new approximation scheme for the study of fluctuations in nonuniform nonequilibrium systems. *Phys. Lett. A*, 70:366–368, 1979.

Mangan, S. and Alon, U. Structure and function of the feed-forward loop network motif. *Proc. Natl. Acad. Sci. USA*, 100:11980–11985, 2003.

Mangan, S., Zaslaver, A., and Alon, U. The coherent feedforward loop serves as a sign-sensitive delay element in transcription networks. *J. Mol. Biol.*, 334:197–204, 2003.

Maniatis, T., Fritsch, E. F., and Sambrook, J. *Molecular cloning : A laboratory manual.* Cold Spring Harbor, N.Y.: Cold Spring Harbor Laboratory, 1982.

Martiel, J. and Goldbeter, A. A model based on receptor desensitization for cyclic AMP signaling in *Dictyostelium cells. Biophys. J.*, 52:807–828, 1987.

Maslov, S. and Sneppen, K. Specificity and stability in topology of protein networks. *Science*, 296:910–913, 2002.

Matsuno, H., Doi, A., Nagasaki, M., and Miyano, S. Hybrid Petri net representation of gene regulatory network. *Pac. Symp. Biocomput.*, 5:341–352, 2000.

Matsuoka, S., Sarai, N., Jo, H., and Noma, A. Simulation of ATP metabolism in cardiac excitation-contraction coupling. *Prog. Biophys. Mol. Biol.*, 85:279–299, 2004.

Maynard-Smith, J. *Models in Ecology.* Cambridge, UK: Cambridge University Press, 1974.

McAdams, H. and Arkin, A. Stochastic mechanisms in gene expression. *Proc. Natl. Acad. Sci. USA*, 94:814–819, 1997.

McAdams, H. and Arkin, A. It's a noisy business! Genetic regulation at the nanomolar scale. *Trends Genet.*, 15:65–69, 1999.

McAllister, R. E., Noble, D., and Tsien, R. W. Reconstruction of the electrical activity of cardiac purkinje fibres. *J. Physiol.*, 251:1–59, 1975.

McQuarrie, D. Stochastic approach to chemical kinetics. *J. Appl. Probability*, 4: 413–478, 1967.

Meacci, G. and Kruse, K. Min-oscillations in *Escherichia coli* induced by interactions of membrane-bound proteins. *Phys. Biol.*, 2:89–97, 2005.

Mecham, B. H., Klus, G. T., Strovel, J., Augustus, M., Byrne, D., Bozso, P., Wetmore, D. Z., Mariani, T. J., Kohane, I. S., and Szallasi, Z. Sequence-matched probes produce increased cross-platform consistency and more reproducible biological results in microarray-based gene expression measurements. *Nucleic Acids Res.*, 32:e74, 2004.

Medawar, P. *Pluto's Republic.* Oxford: Oxford University Press, 1982.

Meinhardt, H. Orientation of chemotactic cells and growth cones: Models and mechanisms. *J. Cell. Sci.*, 112:2867–2874, 1999.

Meinhardt, H. and de Boer, P. A. Pattern formation in *Escherichia coli*: A model for the pole-to-pole oscillations of Min proteins and the localization of the division site. *Proc. Natl. Acad. Sci. USA*, 98:14202–14207, 2001.

Meir, E., von Dassow, G., Munro, E., and Odell, G. M. Ingeneue: A versatile tool for reconstituting genetic networks in silico, with examples from the segment polarity network. *J. Exp. Zool.*, 294:216–251, 2002.

Mendes, P. Biochemistry by numbers: Simulation of biochemical pathways with Gepasi 3. *Trends Biochem. Sci.*, 22:361–363, 1997.

Mendes, P. COPASI (complex pathway simulator), 2003. URL http://mendes. vbi.vt.edu/tiki-index.php?page=COPASI. Available on the Internet at http: //mendes.vbi.vt.edu/tiki-index.php?page=COPASI.

Mendes, P. and Kell, D. Non-linear optimization of biochemical pathways: Applications to metabolic engineering and parameter estimation. *Bioinformatics*, 14: 869–883, 1998.

Mendes, P. and Kell, D. MEG (Model Extender for Gepasi): A program for the modelling of complex, heterogeneous, cellular systems. *Bioinformatics*, 17:288–289, 2001.

Mendoza, L., Thieffry, D., and Alvarez-Buylla, E. Genetic control of flower morphogenesis in *Arabidopsis thaliana*: A logical analysis. *Bioinformatics*, 15:593–606, 1999.

von Mering, C., Zdobnov, E., Tsoka, S., Ciccarelli, F., Pereira-Leal, J., Ouzonis, C., and Bork, P. Genome evolution reveals biochemical networks and functional modules. *Proc. Natl. Acad. Sci. USA*, 100:15428–15433, 2003.

Mesarovic, M., Macko, D., and Takahara, Y. *Theory of hierarchical, multilevel, systems.* New York and London: Academic Press, 1970.

Mestl, T., Plahte, E., and Omholt, S. A mathematical framework for describing and analysing gene regulatory networks. *J. Theor. Biol.*, 176:291–300, 1995.

Michaelis, L. and Menten, M. Die Kinetik der Invertinwirkung. *Biochem. Z.*, 49: 333–369, 1913.

Mihalcescu, I., Hsing, W., and Leibler, S. Resilient circadian oscillator revealed in individual cyanobacteria. *Nature*, 430:81–85, 2004.

Milo, R., Itzkovitz, S., Kashtan, N., Levitt, R., and Alon, U. Response to comment on "Network motifs: Simple building blocks of complex networks" and "Superfamilies of evolved and designed networks". *Science*, 305:1107d, 2004a.

Milo, R., Itzkovitz, S., Kashtan, N., Levitt, R., Shen-Orr, S., Ayzenshtat, I., Sheffer, M., and Alon, U. Superfamilies of evolved and designed networks. *Science*, 303: 1538–1542, 2004b.

Milo, R., Shen-Orr, S., Itzkovitz, S., Kashtan, N., Chklovskii, D., and Alon, U. Network motifs: Simple building blocks of complex networks. *Science*, 298:824–827, 2002.

Mitchell, M. R., Powell, T., Terrar, D. A., and Twist, V. A. The effects of ryanodine, EGTA and low-sodium on action potentials in rat and guinea-pig ventricular myocytes: Evidence for two inward currents during the plateau. *Br. J. Pharmacol.*, 81:543–550, 1984.

Modrek, B., Resch, A., Grasso, C., and Lee, C. Genome-wide detection of alternative splicing in expressed sequences of human genes. *Nucleic Acids Res.*, 29:2850–2859, 2001.

Moles, C., Mendes, P., and Banga, J. Parameter estimation in biochemical pathways: A comparison of global optimization methods. *Genome Res.*, 13:2467–2474, 2003.

Monod, J. and Jacob, F. General conclusions: Teleonomic mechanisms in intracellular metabolism, growth, and differentiation. *Cold Spring Harbour Symp. Quant. Biol.*, 26:389–401, 1961.

Montroll, E. W. and Shlesinger, M. F. On the wonderful world of random walks. In Lebowitz, J., editor, *Studies in Statistical Mechanics*, volume 11, pages 1–123. Amsterdam: North-Holland Physics Publishing, 1984.

Morohashi, M., Winn, A., Borisuk, M., Bolouri, H., Doyle, J., and Kitano, H. Robustness as a measure of plausibility in models of biochemical networks. *J. Theor. Biol.*, 216:19–30, 2002.

Morowitz, H. J. *The Emergence of Everything*. Oxford University Press, Oxford, 2002.

Morris, K. *Introduction to Feedback Control*. London: Harcourt Academic Press, 2001.

Munchbach, M., Quadroni, M., Miotto, G., and James, P. Quantitation and facilitated de novo sequencing of proteins by isotopic N-terminal labeling of peptides with a fragmentation-directing moiety. *Anal. Chem.*, 72:4047–4057, 2000.

Murray, J. *Mathematical Biology: I: An Introduction*. Berlin/Heidelberg: Springer-Verlag, 3rd edition, 2002a.

Murray, J. *Mathematical Biology: II: Spatial Models and Biomedical Applications*. Berlin/Heidelberg: Springer-Verlag, 3rd edition, 2002b.

Muzikant, A. L. and Penl, R. C. Models for profiling the potential QT prolongation risk of drugs. *Curr. Opin. Drug Disc. Dev.*, 5:127–135, 2000.

Nagano, S. Modeling the model organism Dictyostelium discoideum. *Dev. Growth Differ.*, 42:541–550, 2000.

Nagasaki, M., Doi, A., Matsuno, H., and Miyano, S. A versatile Petri net based architecture for modeling and simulation of complex biological processes. *Genome Informatics*, 15:180–197, 2004.

Nagel, E. and Newman, J. *Gödel's Proof.* New York: New York University Press, 2002.

Narang, A., Subramanian, K., and Lauffenburger, D. A mathematical model for chemoattractant gradient sensing based on receptor-regulated membrane phospholipid signaling dynamics. *Ann. Biomed. Eng.*, 29:677–691, 2001.

Neidhardt, F., Ingraham, J., and Schaechter, M. *Physiology of the Bacterial Cell: A Molecular Approach.* Sunderland, MA: Sinauer Associates, 1990.

Nelson, D., Ihekwaba, A. E. C., Elliott, M., Johnson, J., Gibney, C., Foreman, B., Nelson, G., See, V., Horton, C., Spiller, D., Edwards, S., McDowell, H., Unitt, J., Sullivan, E., Grimley, R., Benson, N., Broomhead, D., Kell, D., and White, M. R. H. Oscillations in NF-κB signaling control the dynamics of gene expression. *Science*, 306:704–708, 2004.

Newman, M. The structure and function of complex networks. *SIAM Rev.*, 45: 167–256, 2003.

Nicholl, D. S. T. *An Introduction to Genetic Engineering.* Cambridge, UK: Cambridge University Press, 1994.

Nicolis, G. and Prigogine, I. *Self-organization in nonequilibrium systems.* New York. John Wiley & Sons, 1977.

Nielsen, J. Metabolic engineering: Techniques for analysis of targets for genetic engineering. *Biotechnol. Bioeng.*, 58:125–132, 1998.

Ninio, J. Kinetic amplification of enzyme discrimination. *Biochimie*, 57:587–595, 1975.

Noble, D. Cardiac action and pacemaker potentials based on the Hodgkin-Huxley equations. *Nature*, 188:495–497, 1960.

Noble, D. A modification of the Hodgkin-Huxley equations applicable to Purkinje fibre action and pacemaker potentials. *J. Physiol.*, 160:317–352, 1962.

Noble, D. The surprising heart: A review of recent progress in cardiac electrophysiology. *J. Physiol.*, 353:1–50, 1984.

Noble, D. Modelling the heart: From genes to cells to the whole organ. *Science*, 295:1678–1682, 2002a.

Noble, D. Modelling the heart: Insights, failures and progress. *BioEssays*, 24: 1155–1163, 2002b.

Noble, D. Simulation of Na-Ca exchange activity during ischaemia. *Ann. NY Acad. Sci.*, 976:431–437, 2002c.

Noble, D. Unravelling the genetics and mechanisms of cardiac arrhythmia. *Proc. Natl. Acad. Sci. USA*, 99:5755–5756, 2002d.

Noble, D. The future: Putting Humpty-Dumpty together again. *Biochem. Soc. Trans.*, 31:156–158, 2003a.

Noble, D. Will genomics revolutionise pharmaceutical research and development? *Trends Biotechnol.*, 21:333–337, 2003b.

Noble, D. and Colatsky, T. J. A return to rational drug discovery: Computer-based models of cells, organs and systems in drug target identification. *Emerg. Therapeut. Targets*, 4:39–49, 2000.

Noble, D., Levin, J., and Scott, W. Biological simulations in drug discovery. *Drug Disc. Today*, 4:10–16, 1999.

Noble, D., Noble, S. J., Bett, G. C. L., Earm, Y. E., Ho, W. K., and So, I. S. The role of sodium-calcium exchange during the cardiac action potential. *Ann. NY Acad. Sci.*, 639:334–353, 1991.

Noble, D. and Rudy, Y. Models of cardiac ventricular action potentials: Iterative interaction between experiment and simulation. *Phil. Trans. Royal Soc. A*, 359: 1127–1142, 2001.

Noble, D. and Tsien, R. W. The kinetics and rectifier properties of the slow potassium current in cardiac Purkinje fibres. *J. Physiol.*, 195:185–214, 1968.

Noble, D. and Tsien, R. W. Outward membrane currents activated in the plateau range of potentials in cardiac Purkinje fibres. *J. Physiol.*, 200:205–231, 1969.

Noble, D., Varghese, A., Kohl, P., and Noble, P. J. Improved guinea-pig ventricular cell model incorporating a diadic space, iKr and IKs, and length- and tension-dependent processes. *Can. J. Cardiol.*, 14:123–134, 1998.

Noble, P. J. and Noble, D. Reconstruction of the cellular mechanisms of cardiac arrhythmias triggered by early after-depolarizations. *Japan. J. Electrocardiol.*, 20 (Suppl 3):15–19, 2000.

Novartis Foundation. *Complexity in Biological Information Processing*. Chichester: Wiley, 2001.

Novartis Foundation. *The hERG Cardiac Potassium Channel: Structure, Function and Long QT Syndrome*. Chichester: Wiley, 2005.

Novell, Inc. SuSE LINUX professional 9.2. Available via the World Wide Web at `http://www.novell.com/linux/suse/index.html`, 2005.

Noyes, R. M. Effects of diffusion rates on chemical kinetics. *Progr. Reaction Kinetics*, 1:129–160, 1961.

Object Management Group, Inc. Response to the UML 2.0 OCL RfP, version 1.6. OMG document ad2003-01-06, 2002.

O'Donovan, C., Apweiler, R., and Bairoch, A. The human proteomics initiative (HPI). *Trends Biotechnol.*, 19:178–181, 2001.

Oestereich, B. *Developing Software with UML: Object-Oriented Analysis and Design in Practice*. Amsterdam: Addison-Wesley, 1999.

Ogata, K. *Modern Control Engineering*. Upper Saddle River, NJ: Prentice Hall, 2001.

Oltvai, Z. and Barabási, A.-L. Life's complexity pyramid. *Science*, 298:763–764, 2002.

Onsager, L. Reciprocal relations in irreversible processes. I. *Phys. Rev.*, 37:405–426, 1931a.

Onsager, L. Reciprocal relations in irreversible processes. II. *Phys. Rev.*, 38:2265–2279, 1931b.

Oosawa, C. and Savageau, M. Effects of alternative connectivity on behavior of randomly-constructed Boolean networks. *Phys. D*, 170:143–161, 2002.

Ovchinnikov, A., Timashev, S., and Belyy, A. *Kinetics of Diffusion Controlled Chemical Processes*. New York: Nova Science Publishers, 1989.

Ozbudak, E. M., Thattai, M., Kurtser, I., Grossman, A. D., and van Oudenaarden, A. Regulation of noise in the expression of a single gene. *Nat. Genet.*, 31:69–73, 2002.

Ozbudak, E. M., Thattai, M., Lim, H. N., Shraiman, B. I., and Oudenaarden, A. V. Multistability in the lactose utilization network of *Escherichia coli*. *Nature*, 427:737–740, 2004.

Panfilov, A. and Kerkhof, P. Quantifying ventricular fibrillation: In silico research and clinical implications. *IEEE Trans. Biomed. Eng.*, 51:195–196, 2004.

Papin, J. and Palsson, B. Topological analysis of mass balanced signaling networks: A framework to obtain network properties including crosstalk. *J. Theor. Biol.*, 227:283–297, 2004.

Papin, J., Price, N., Edwards, J., and Palsson, B. The genome-scale metabolic extreme pathway structure in *Haemophilus influenzae* shows significant network redundancy. *J. Theor. Biol.*, 215:67–82, 2002.

Papin, J., Price, N., Wiback, S., Fell, D., and Palsson, B. Metabolic pathways in the post-genome era. *Trends Biochem. Sci.*, 28:250–258, 2003.

Papin, J., Reed, J., and Palsson, B. Hierarchical thinking in network biology: The unbiased modularization of biochemical networks. *Trends Biochem. Sci.*, 29:641–647, 2004a.

Papin, J., Stelling, J., Price, N., Klamt, S., Schuster, S., and Palsson, B. Comparison of network-based pathway analysis methods. *Trends Biotechnol.*, 22:400–405, 2004b.

Papin, J. A., Hunter, T., Palsson, B. O., and Subramaniam, S. Reconstruction of cellular signalling networks and analysis of their properties. *Nat. Rev. Mol. Cell. Biol.*, 6:99–111, 2005.

Papp, B., Pál, C., and Hurst, L. D. Metabolic network analysis of the causes and evolution of enzyme dispensability in yeast. *Nature*, 429:661–664, 2004.

Paulsson, J. Summing up the noise in gene networks. *Nature*, 427:415–418, 2004.

Paulsson, J. *Models of Stochastic Gene Expression*. 2005. To appear.

Paulsson, J., Berg, O. G., and Ehrenberg, M. Stochastic focusing: Fluctuation-enhanced sensitivity of intracellular regulation. *Proc. Natl. Acad. Sci. USA*, 97:7148–7153, 2000.

Paulsson, J. and Ehrenberg, M. Random signal fluctuations can reduce random fluctuations in regulated components of chemical regulatory networks. *Phys. Rev. Lett.*, 84:5447–5450, 2000.

Paulsson, J. and Ehrenberg, M. Noise in a minimal regulatory network: Plasmid copy number control. *Q. Rev. Biophys.*, 34:1–59, 2001.

Pearl, J. *Causality: Models, Reasoning and Inference.* Cambridge, UK: Cambridge University Press, 2000.

Peccoud, J. and Ycart, B. Markovian modelling of gene-product synthesis. *Theor. Pop. Biol.*, 48:222, 1995.

Pedraza, J. M. and van Oudenaarden, A. Noise propagation in gene networks. *Science*, 307:1965–1969, 2005.

Pe'er, D., Regev, A., Elidan, G., and Friedman, N. Inferring subnetworks from perturbed expression profiles. *Bioinformatics*, 17:S215–S224, 2001.

Peng, J., Schwartz, D., Elias, J. E., Thoreen, C. C., Cheng, D., Marsischky, G., Roelofs, J., Finley, D., and Gygi, S. P. A proteomics approach to understanding protein ubiquitination. *Nat. Biotechnol.*, 21:921–926, 2003.

Peri, S., Navarro, J. D., Amanchy, R., Kristiansen, T. Z., Jonnalagadda, C. K., Surendranath, V., Niranjan, V., Muthusamy, B., Gandhi, T. K., and Gronborg, M. Development of human protein reference database as an initial platform for approaching systems biology in humans. *Genome Res.*, 13:2363–2371, 2003.

Perkins, T., Hallett, M., and Glass, L. Inferring models of gene expression dynamics. *J. Theor. Biol.*, 230:289–299, 2004.

Perrin, B.-E., Ralaivola, L., Mazurie, A., Bottani, S., Mallet, J., and D'Alché-Buc, F. Gene networks inference using dynamic Bayesian networks. *Bioinformatics*, 19:II138–II148, 2003.

Peskin, C. S. and McQueen, D. M. Cardiac fluid dynamics. In *High-Performance Computing in Biomedical Research*, pages 51–59. Boca Raton, FL: CRC Press, 1993.

Pfeiffer, T., Sanchez-Valdenebro, I., Nuno, J., Montero, F., and Schuster, S. META-TOOL: For studying metabolic networks. *Bioinformatics*, 15:251–257, 1999.

Pfeiffer, T., Schuster, S., and Bonhoeffer, S. Cooperation and competition in the evolution of ATP-producing pathways. *Science*, 292:504–507, 2001.

Plahte, E., Mestl, T., and Omholt, S. Feedback loops, stability and multistationarity in dynamical systems. *J. Biol. Sys.*, 3:409–413, 1995.

Pomerening, J., Kim, S., and Ferrell Jr., J. Systems-level dissection of the cell cycle oscillator: Bypassing positive feedback produces damped oscillations. *Cell*, 122: 565–578, 2005.

Pomerening, J., Sontag, E., and Ferrell Jr., J. Building a cell cycle oscillator: Hysteresis and bistability in the activation of Cdc2. *Nat. Cell Biol.*, 5:346–351, 2003.

Poola, K., Kargonekar, P., Tikku, A., Krause, J., and Nagpal, K. A time-domain approach to model validation. *IEEE Trans. Automat. Contr.*, 39:951–959, 1994.

Poolman, M., Fell, D., and Raines, C. Elementary modes analysis of photosynthate metabolsim in the chloroplast stroma. *Eur. J. Biochem.*, 270:430–439, 2003.

Popper, K. *Conjectures and refutations: the growth of scientific knowledge*. London: Routledge & Kegan Paul, 5th edition, 1992.

Postma, M. and Van Haastert, P. A diffusion-translocation model for gradient sensing by chemotactic cells. *Biophys. J.*, 81:1314–1323, 2001.

Press, W., Flannery, B., Teukolsky, S., and Vetterling, W. *Numerical Recipes: The Art of Scientific Computing*. Cambridge, UK: Cambridge University Press, 1986.

Price, N., Papin, J., and Palsson, B. Determination of redundancy and systems properties of the metabolic network of *Helicobacter pylori* using genome-scale extreme pathway analysis. *Genome Res.*, 12:760–769, 2002.

Price, N., Reed, J., and Palsson, B. Genome-scale models of microbial cells: Evaluating the consequences of constraints. *Nat. Rev. Microbiol.*, 2:886–897, 2004.

Price, N., Reed, J., Papin, J, Wiback, S., and Palsson, B. Network-based analysis of metabolic regulation in the human red blood cell. *J. Theor. Biol.*, 225:185–194, 2003.

Pritchard, L. and Kell, D. B. Schemes of flux control in a model of *Saccharomyces cerevisiae* glycolysis. *Eur. J. Biochem.*, 269:3894–3904, 2002.

Ptashne, M. *A Genetic Switch: Phage λ and Higher Organisms*. Cambridge, MA: Blackwell Science, 1992.

Puglisi, J. L., Wang, F., and Bers, D. M. Modeling the isolated cardiac myocyte. *Prog. Biophys. Mol. Biol.*, 85:163–178, 2004.

Qian, H., Saffarian, S., and Elson, E. L. Concentration fluctuations in a mesoscopic oscillating chemical reaction system. *Proc. Natl. Acad. Sci. USA*, 99:10376–10381, 2002.

Rao, C., Kirby, J., and Arkin, A. Design and diversity in bacterial chemotaxis: A comparative study in *Escherichia coli* and *Bacillus subtilis*. *PLoS Biol.*, 2:E49, 2004.

Rao, C., Wolf, D., and Arkin, A. Control, exploitation and tolerance of intracellular noise. *Nature*, 420:231–237, 2002.

Raser, J. M. and O'Shea, E. K. Control of stochasticity in eukaryotic gene expression. *Science*, 304:1811–1814, 2004.

Raskin, D. M. and de Boer, P. A. MinDE-dependent pole-to-pole oscillation of division inhibitor MinC in *Escherichia coli*. *J. Bacteriol.*, 181:6419–6424, 1999a.

Raskin, D. M. and de Boer, P. A. Rapid pole-to-pole oscillation of a protein required for directing division to the middle of *Escherichia coli*. *Proc. Natl. Acad. Sci. USA*, 96:4971–4976, 1999b.

Rathinam, M., Cao, Y., Petzold, L., and Gillespie, D.   Stiffness in stochastic chemically reacting systems: The implicit tau-leaping method. *J. Chem. Phys.*, 119:12784–12794, 2003.

Rathinam, M., Cao, Y., Petzold, L., and Gillespie, D. Consistency and stability of tau-leaping schemes for chemical reaction systems. *SIAM Multiscale Model. & Sim.*, 4:867–895, 2005.

Ravasz, E., Somera, A., Mongru, D., Oltvai, Z., and Barabási, A.-L. Hierarchical organization of modularity in metabolic networks. *Science*, 297:1551–1555, 2002.

Rawlins, G. *Compared to What? An Introduction to the Analysis of Algorithms.* New York: Computer Science Press, 1991.

Reddy, V., Liebman, M., and Mavrovouniotis, M. Qualitative analysis of biochemical reaction systems. *Comp. Biol. Med.*, 26:9–24, 1996.

Reder, C. Metabolic control theory: A structural approach. *J. Theor. Biol.*, 135: 175–201, 1986.

Reed, J. and Palsson, B. Thirteen years of building constraint-based in silico models of *Escherichia coli*. *J. Bacteriol.*, 185:2692–2699, 2003.

Regev, A., Silverman, W., and Shapiro, E.   Representation and simulation of biochemical processes using the π-calculus process algebra.   In *Pac. Symp. Biocomput.*, volume 6, pages 459–470. Singapore: World Scientific Publishing, 2001.

Remy, E., Mossé, B., Chaouiya, C., and Thieffry, D. A description of dynamical graphs associated with elementary regulatory circuits. *Bioinformatics*, 19:ii172–78, 2003.

Ren, B., Robert, F., Wyrick, J. J., Aparicio, O., Jennings, E. G., Simon, I., Zeitlinger, J., Schreiber, J., Hannett, N., and Kanin, E. Genome-wide location and function of DNA binding proteins. *Science*, 290:2306–2309, 2000.

Renyi, A. Treating chemical reaction using the theory of stochastic processes. *MTA Alk. Mat. Int. Közl.*, 2:93–101, 1954.

Reppert, S. and Weaver, D.   Comparing clockworks: Mouse versus fly.   *J. Biol. Rhythms*, 15:357–364, 2000.

Reuter, H.   The dependence of slow inward current in Purkinje fibres on the extracellular calcium concentration. *J. Physiol.*, 192:479–492, 1967.

Rigney, D. Note on the kinetics and stochastics of induced protein synthesis as influenced by various models for messenger RNA degradation. *J. Theor. Biol.*, 79:247–257, 1979a.

Rigney, D. Stochastic models of cellular variability. In Thomas, R., editor, *Kinetic Logic: A Boolean Approach to the Analysis of Complex Regulatory Systems*, volume 29 of *Lecture Notes in Biomathematics*, pages 237–280. Berlin: Springer-Verlag, 1979b.

Rigney, D. R. Stochastic model of constitutive protein levels in growing and dividing bacterial cells. *J. Theor. Biol.*, 76:453–480, 1979c.

Rigney, D. R. and Schieve, W. C. Stochastic model of linear, continuous protein synthesis in bacterial populations. *J. Theor. Biol.*, 69:761–766, 1977.

Risken, H. *The Fokker-Planck Equation.* Berlin: Springer Verlag, 1984.

Rockafellar, R. *Convex Analysis.* Princeton: Princeton University Press, 1970.

Ronen, M., Rosenberg, R., Shraiman, B. I., and Alon, U. Assigning numbers to the arrows: Parametrizing a gene regulation network by using accurate expression kinetics. *Proc. Natl. Acad. Sci. USA*, 99:10555–10560, 2002.

Ropers, D., de Jong, H., Page, M., Schneider, D., and Geiselmann, J. Qualitative simulation of the carbon starvation response in *Escherichia coli*. *Biosystems*, 2005. To appear.

Rosenfeld, N., Young, J. W., Alon, U., Swain, P. S., and Elowitz, M. B. Gene regulation at the single-cell level. *Science*, 307:1962–1965, 2005.

Rouse, R. and Hardiman, G. Microarray technology—An intellectual property retrospective. *Pharmacogenomics*, 4:623–632, 2003.

Rush, J., Moritz, A., Lee, K. A., Guo, A., Goss, V. L., Spek, E. J., Zhang, H., Zha, X. M., Polakiewicz, R. D., and Comb, M. J. Immunoaffinity profiling of tyrosine phosphorylation in cancer cells. *Nat. Biotechnol.*, 23:94–101, 2005.

Russell, S. and Norvig, P. *Artificial Intelligence: A Modern Approach.* Englewood Cliffs, NJ: Prentice Hall, 2003.

Sachs, K., Perez, O., Pe'er, D., Lauffenburger, D. A., and Nolan, G. P. Causal protein-signaling networks derived from multiparameter single-cell data. *Science*, 308:523–529, 2005.

Saha, S., Sparks, A. B., Rago, C., Akmaev, V., Wang, C. J., Vogelstein, B., Kinzler, K. W., and Velculescu, V. E. Using the transcriptome to annotate the genome. *Nat. Biotechnol.*, 20:508–512, 2002.

Saltelli, A., Chan, K., and Scott, E., editors. *Sensitivity Analysis.* Chichester, UK: Wiley, 2000.

Sánchez, L., van Helden, J., and Thieffry, D. Establishment of the dorso-ventral pattern during embryonic development of *Drosophila melanogaster*: A logical analysis. *J. Theor. Biol.*, 189:377–389, 1997.

Sánchez, L. and Thieffry, D. A logical analysis of the *Drosophila* gap-gene system. *J. Theor. Biol.*, 211:115–141, 2001.

Sánchez, L. and Thieffry, D. Segmenting the fly embryo: A logical analysis of the pair-rule cross-regulatory module. *J. Theor. Biol.*, 224:517–537, 2003.

Sanguinetti, M. C. and Jurkiewicz, N. K. Two components of cardiac delayed rectifier K+ current: Differential sensitivity to block by class III antiarrhythmic agents. *J. Gen. Physiol.*, 96:195–215, 1990.

Santillán, M. and Mackey, M. Dynamic regulation of the tryptophan operon: A modeling study and comparison with experimental data. *Proc. Natl. Acad. Sci. USA*, 98:1364–1369, 2001.

Sasai, M. and Wolynes, P. G. Stochastic gene expression as a many-body problem. *Proc. Natl. Acad. Sci. USA*, 100:2374–2379, 2003.

Saucerman, J. J. and McCulloch, A. D. Mechanistic systems models of cell signaling networks: A case study of myocyte adrenergic regulation. *Prog. in Biophys. Mol. Biol.*, 85:261–278, 2004.

Sauer, U. Evolutionary engineering of industrially important microbial phenotypes. *Adv. Biochem. Eng. Biotechnol.*, 73:129–170, 2001.

Sauer, U. High-throughput phenomics: Experimental methods for mapping fluxomes. *Curr. Opin. Biotechnol.*, 15:58–63, 2004.

Sauro, H. M. Jarnac: A system for interactive metabolic analysis. In *Animating the Cellular Map: Proceedings of the 9th International Meeting on BioThermoKinetics*. Stellenbosch, ZA: Stellenbosch University Press, 2000.

Sauro, H. M., Hucka, M., Finney, A., Wellock, C., Bolouri, H., Doyle, J., and Kitano, H. Next generation simulation tools: The Systems Biology Workbench and BioSPICE integration. *OMICS*, 7:355–372, 2003.

Savageau, M. *Biochemical Systems Analysis.* Reading, MA: Addison-Wesley, 1976.

Savageau, M. A. Introduction to S-systems and the underlying power-law formalism. *Math. Comput. Modelling*, 11:546–551, 1988.

SBML Team. The SBML test suite. Available via the World Wide Web at `http://sbml.org/downloads/`, 2005a.

SBML Team. The SBML.org web site. Available via the World Wide Web at `http://sbml.org`, 2005b.

Schena, M., Shalon, D., Davis, R. W., and Brown, P. O. Quantitative monitoring of gene expression patterns with a complementary DNA microarray. *Science*, 270: 467–470, 1995.

Schilling, C., Letscher, D., and Palsson, B. Theory for the systemic definition of metabolic pathways and their use in interpreting metabolic function from a pathway-oriented perspective. *J. Theor. Biol.*, 203:229–248, 2000.

Schilling, C. and Palsson, B. Assessment of the metabolic capacities of *Helicobacter influenzae* Rd through a genome scale pathway analysis. *J. Theor. Biol.*, 203: 249–283, 2000.

Schoeberl, B., Eichler-Jonsson, C., Gilles, E., and Müller, G. Computational modeling of the dynamics of the MAP kinase cascade activated by surface and internalized EGF receptors. *Nat. Biotechnol.*, 20:370–375, 2002.

Schrödinger, E. *What is life?* Cambridge, UK: Cambridge University Press, 1944.

Schuster, S., Dandekar, T., and Fell, D. Detection of elementary flux modes in biochemical networks: A promising tool for pathway analysis and metabolic engineering. *Trends Biotechnol.*, 17:53–60, 1999.

Schuster, S., Fell, D., and Dandekar, T. A general definition of metabolic pathways useful for systematic organization and analysis of complex metabolic networks. *Nat. Biotechnol.*, 18:326–332, 2000.

Schuster, S. and Hilgetag, C. On elementary flux modes in biochemical reaction systems at steady state. *J. Biol. Syst.*, 2:165–182, 1994.

Schuster, S., Hilgetag, C., Woods, J., and Fell, D. Exploring the pathway structure of metabolism: Decomposition into subnetworks and application to *Mycoplasma pneumoniae*. *Bioinformatics*, 18:351–361, 2002a.

Schuster, S., Hilgetag, C., Woods, J., and Fell, D. Reaction routes in biochemical reaction systems: Algebraic properties, validated calculation procedure and example from nucleotide metabolism. *J. Math. Biol.*, 45:153–181, 2002b.

Schuster, S., Klamt, S., Weckwerth, W., Moldenhauer, F., and Pfeiffer, T. Use of network analysis of metabolic systems in bioengineering. *Bioproc. Biosyst. Eng.*, 24:363–372, 2002c.

Sedra, A. and Smith, K. *Microelectronic circuits*. New York: Oxford University Press, 5th edition, 2004.

Segal, E., Shapira, M., Regev, A., Pe'er, D., Botstein, D., Koller, D., and Friedman, N. Module networks: Identifying regulatory modules and their condition-specific regulators from gene expression data. *Nat. Genetics*, 34:166–176, 2003.

Segel, L. On the validity of the steady state assumption of enzyme kinetics. *Bull. Math. Biol.*, 50:579–593, 1988.

Seger, R. and Krebs, E. G. The MAPK signaling cascade. *FASEB J.*, 9:726–735, 1995.

Segre, D., Vitkup, D., and Church, G. Analysis of optimality in natural and perturbed metabolic networks. *Proc. Natl. Acad. Sci. USA*, 99:15112–15117, 2002.

Selinger, D., Wright, M., and Church, G. On the complete determination of biological systems. *Trends Biotechnol.*, 21:251–254, 2003.

Selkov, E. Self-oscillations in glycolosis. 1. A simple kinetic model. *Eur. J. Biochem.*, 4:79–86, 1968.

Sha, W., Moore, J., Chen, K., Lassaletta, A., Yi, C., Tyson, J., and Sible, J. Hysteresis drives cell-cycle transitions in *Xenopus laevis* egg extracts. *Proc. Natl. Acad. Sci. USA*, 100:975–980, 2003.

Shapiro, B., Levchenko, A., Meyerowitz, E., Wold, B., and Mjolsness, E. Cellerator: Extending a computer algebra system to include biochemical arrows for signal transduction simulations. *Bioinformatics*, 19:677–678, 2003.

Shapiro, B. E., Hucka, M., Finney, A., and Doyle, J. MathSBML: A package for manipulating SBML-based biological models. *Bioinformatics*, 20:2829–2831, 2004.

Shen, J., Xu, X., Cheng, F., Liu, H., Luo, X., Chen, K., Zhao, W., Chen, X., and Jiang, H. Virtual screening on natural products for discovering active compounds and target information. *Curr. Med. Chem.*, 10:2327–2342, 2003.

Shen-Orr, S., Milo, R., Mangan, S., and Alon, U. Network motifs in the transcriptional regulation network of *Escherichia coli*. *Nat. Genet.*, 31:64–68, 2002.

Shmulevich, I., Dougherty, E., Kim, S., and Zhang, W. Probabilistic Boolean networks: A rule-based uncertainty model for gene regulatory networks. *Bioinformatics*, 18:261–274, 2002a.

Shmulevich, I., Dougherty, E., and Zhang, W. From Boolean to probabilistic Boolean networks as models of genetic regulatory networks. *Proc. IEEE*, 90: 1788–1792, 2002b.

Simpson, T., Follstad, B., and Stephanopoulos, G. Analysis of the pathway structure of metabolic networks. *J. Biotechnol.*, 71:207–223, 1999.

Simutis, R. and Lübbert, A. Exploratory analysis of bioprocesses using artificial neural network-based methods. *Biotechnol. Prog.*, 13:479–487, 1997.

Singer, K. Application of the theory of stochastic processes to the study of irreproducible chemical reactions and nucleation processes. *J. Royal Statist. Soc. (B)*, 15:92–106, 1953.

Skilling, J. Probabilistic data analysis: An introductory guide. *J. Microscopy*, 190: 28–36, 1998.

Smith, A. E., Slepchenko, B. M., Schaff, J. C., Loew, L. M., and Macara, I. G. Systems analysis of Ran transport. *Science*, 295:488–491, 2002.

Smith, E. and Morowitz, H. J. Universality in intermediary metabolism. *Proc. Natl. Acad. Sci. USA*, 101:13168–13173, 2004.

Smith, N. P. and Crampin, E. J. Development of models of active ion transport of whole-cell modelling: Cardiac sodium-potassium pump as a case study. *Prog. Biophys. Mol. Biol.*, 85:387–405, 2004.

Smith, N. P., Mulquiney, P. J., Nash, M. P., Bradley, C. P., Nickerson, D. P., and Hunter, P. J. Mathematical modelling of the heart: Cell to organ. *Chaos, Solitons Fractals*, 13:1613–1621, 2001.

Smith, N. P., Pullan, A. J., and Hunter, P. J. Generation of an anatomically based geometric coronary model. *Ann. Biomed. Eng.*, 28:14–25, 2000.

Smolen, P., Baxter, D., and Byrne, J. Modeling circadian oscillations with interlocking positive and negative feedback loops. *J. Neurosc.*, 21:6644–6656, 2001.

von Smoluchowski, M. Versuch einer mathematischen theorie der koagulationskinetik kolloider lösungen. *Z. Phys. Chemie*, 92:129, 1917.

Snoussi, E. Necessary conditions for multistationarity and stable periodicity. *J. Biol. Sys.*, 6:3–9, 1998.

Soeller, C. and Cannell, M. B. Analysing cardiac excitation-contraction coupling with mathematical models of local control. *Prog. Biophys. Mol. Biol.*, 85:141–162, 2004.

Solé R. and Goodwin, B. *Signs of Life: How Complexity Pervades Biology.* New York: Basic Books, 2000.

Somogyi, R. and Sniegoski, C. Modeling the complexity of genetic networks: Understanding multigenic and pleiotropic regulation. *Complexity*, 1:45–63, 1996.

Sontag, E. For differential equations with r parameters, $2r + 1$ experiments are enough for identification. *J. Nonlinear Sci.*, 12:553–583, 2002.

Sontag, E. Adaptation and regulation with signal detection implies internal model. *Systems Control Lett.*, 50:119–126, 2003.

Sontag, E., Kiyatkin, A., and Kholodenko, B. N. Inferring dynamic architecture of cellular networks using time series of gene expression, protein and metabolite data. *Bioinformatics*, 20:1877–1886, 2004.

Soule, C. Graphic requirements for multistationarity. *ComPlexUs*, 1:123–133, 2003.

SourceForge.net. The SourceForge open source software development Web site, 2002. Available via the World Wide Web at http://www.sourceforge.net/.

Spirin, V. and Mirny, L. Protein complexes and functional modules in molecular networks. *Proc. Natl. Acad. Sci. USA*, 100:12123–12128, 2003.

Spiro, P. A., Parkinson, J. S., and Othmer, H. G. A model of excitation and adaptation in bacterial chemotaxis. *Proc. Natl. Acad. Sci. USA*, 94:7263–7268, 1997.

Stelling, J. Mathematical models in microbial systems biology. *Curr. Opin. Microbiol.*, 7:513–518, 2004.

Stelling, J., Gilles, E., and III, F. D. Robustness properties of circadian clock architectures. *Proc. Natl. Acad. Sci. USA*, 101:13210–13215, 2004a.

Stelling, J., Klamt, S., Bettenbrock, K., Schuster, S., and Gilles, E. Metabolic network structure determines key aspects of functionality and regulation. *Nature*, 420:190–193, 2002.

Stelling, J., Sauer, U., Szallasi, Z., Doyle III, F., and Doyle, J. Robustness of cellular functions. *Cell*, 118:675–685, 2004b.

Stephanopoulos, G., Aristidou, A., and Nielsen, J. *Metabolic Engineering.* San Diego, CA: Academic Press, 1998.

Stephanopoulos, G. and San, K. Y. Studies on on-line bioreactor identification. I. Theory. *Biotechnol. Bioeng.*, 26:1176–1188, 1984.

Stiles, J., Bartol Jr., M., Salpeter, E., and Salpeter, M. Monte Carlo simulation of neurotransmitter release using MCell, a general simulator of cellular physiological processes. In *Computational Neuroscience*, pages 279–284. New York: Plenum Press, 1998.

Stock, J., Lukat, G., and Stock, A. Bacterial chemotaxis and the molecular logic of intracellular signal transduction networks. *Annu. Rev. Biophys. Biophys. Chem.*, 20:109–136, 1991.

Strang, G. *Linear Algebra and its Applications*. New York: Academic Press, 1980.

Strang, G. *Introduction to Applied Mathematics*. Wellesley, MA: Wellesley-Cambridge Press, 1986.

Strogatz, S. *Nonlinear Dynamics and Chaos: With Applications to Physics, Biology, Chemistry and Engineering*. Reading, MA: Perseus Publishing, 2000.

Strogatz, S. Exploring complex networks. *Nature*, 410:268–276, 2001.

Stumpf, M., Wiuf, C., and May, R. Subnets of scale-free networks are not scale-free: Sampling properties of networks. *Proc. Natl. Acad. Sci. USA*, 102:4221–4224, 2005.

Swain, P. S. Efficient attenuation of stochasticity in gene expression through post-transcriptional control. *J: Mol: Biol:*, 344:965–976, 2004.

Swain, P. S., Elowitz, M. B., and Siggia, E. D. Intrinsic and extrinsic contributions to stochasticity in gene expression. *Proc. Natl. Acad. Sci. USA*, 99:12795–12800, 2002.

Szallasi, Z. Genetic network analysis in the light of massively parallel biological data acquisition. *Pac. Symp. Biocomput.*, 4:5–16, 1999.

Tang, Z., Coleman, T., and Dunphy, W. Two distinct mechanisms for negative regulation of the Wee1 protein kinase. *EMBO J.*, 12:3427–3436, 1993.

Tapaswi, P. K., Roychoudhury, R. K., and Prasad, T. A stochastic model of gene activation and RNA synthesis during embryogenesis. *Sankhya, Ind. J. Stat.*, 49: 51–67, 1987.

Taylor, S., Gunawan, R., and Doyle III, F. Biosens user guide. Technical report, University of Santa Barbara, CA, 2005.

Tegner, J., Yeung, M., Hasty, J., and Collins, J. Reverse engineering gene networks: Integrating genetic perturbations with dynamical modeling. *Proc. Natl. Acad. Sci. USA*, 100:5944–5949, 2003.

TERANODE, Inc. TERANODE Design Suite. Available via the World Wide Web at http://www.teranode.com, 2005.

Thattai, M. and van Oudenaarden, A. Intrinsic noise in gene regulatory networks. *Proc. Natl. Acad. Sci. USA*, 98:8614–8619, 2001.

The Mathworks, Inc. MATLAB. Available via the World Wide Web at http://www.mathworks.com, 2005.

Thomas, R. Boolean formalization of genetic control circuits. *J. Theor. Biol.*, 42: 563–585, 1973.

Thomas, R. and d'Ari, R. *Biological Feedback*. Boca Raton, FL: CRC Press, 1990.

Thompson, H. S., Beech, D., Maloney, M., and Mendelsohn, N. XML Schema part 1: Structures (W3C candidate recommendation 24 October 2000). Available via the World Wide Web at `http://www.w3.org/TR/xmlschema-1/`, 2000.

Tirosh, I. and Barkai, N. Computational verification of protein-protein interactions by orthologous co-expression. *BMC Bioinformatics*, 6:40, 2005.

Tomioka, R., Kimura, H., Kobayashi, J., and Aihara, K. Multivariate analysis of noise in genetic regulatory networks. *J. Theor. Biol.*, 229:501–521, 2004.

Tong, A. II., Lesage, G., Bader, G. D., Ding, H., Xu, H., Xin, X., Young, J., Berriz, G. F., Brost, R. L., and Chang, M. Global mapping of the yeast genetic interaction network. *Science*, 303:808–813, 2004.

Tsien, R. Y. The green fluorescent protein. *Annu. Rev. Biochem.*, 67:509–544, 1998.

Tsypkin, Y. Z. *Relay Control Systems*. Cambridge, UK: Cambridge University Press, Cambridge, 1984.

Tufillaro, N. B., Abbott, T., and Reilly, J. *An Experimental Approach to Nonlinear Dynamics and Chaos*. Cambridge, MA: Perseus Publishing, 1992.

Turing, A. M. The chemical basis of morphogenesis. *Phil. Trans. Roy. Soc. (Lond.)*, 237:37–72, 1952.

Tyson, J. Modeling the cell division cycle: *cdc2* and cyclin interaction. *Proc. Natl. Acad. Sci. USA*, 88:7328–7332, 1991.

Tyson, J., Chen, K., and Novak, B. Sniffers, buzzers, toggles and blinkers: Dynamics of regulatory and signaling pathways in the cell. *Curr. Opin. Cell Biol.*, 15:221–231, 2003.

Tyson, J. J., Chen, K., and Novak, B. Network dynamics and cell physiology. *Nat. Rev. Mol. Cell. Biol.*, 2:908–916, 2001.

Tyson, J. J., Csikasz-Nagy, A., and Novak, B. The dynamics of cell cycle regulation. *Bioessays*, 24:1095–1109, 2002.

Ueda, H., Hagiwara, M., and Kitano, H. Robust oscillations within the interlocked feedback model of *Drosophila* circadian rhythm. *J. Theor. Biol.*, 210:401–406, 2001.

Usseglio Viretta, A. and Fussenegger, M. Modeling the quorum sensing regulatory network of human-pathogenic *Pseudomonas aeruginosa*. *Biotech. Prog.*, 20:670–678, 2004.

Varghese, A. and Winslow, R. L. Dynamics of abnormal pacemaking activity in cardiac Purkinje fibres. *J. Theoret. Biol.*, 168:407–420, 1994.

Varigonda, S. and Georgiou, T. Dynamics of relay relaxation oscillators. *IEEE Trans. Automat. Control*, 46:65–77, 2001.

Varma, A. and Palsson, B. Metabolic flux balancing: Basic concepts, scientific and practical use. *Biotechnol. Bioeng.*, 12:994–998, 1993.

Varma, A. and Palsson, B. O. Stoichiometric flux balance models quantitatively predict growth and metabolic by-product secretion in wild-type *Escherichia coli* W3110. *Appl. Environ. Microbiol.*, 60:3724–3731, 1994.

Varner, J. and Ramkrishna, D. Metabolic engineering from a cybernetic perspective-I. Theoretical preliminaries. *Bitechnol. Prog.*, 15:407–425, 1999.

Vass, M., Allen, N., Shaffer, C., Ramakrishnan, N., Watson, L., and Tyson, J. The JigCell model builder and run manager. *Bioinformatics*, 18:3680–3681, 2004.

Vayttaden, S., Ajay, S., and Bhalla, U. A spectrum of models of signaling pathways. *ChemBioChem*, 5:1365–1374, 2004.

Vilar, J. M., Kueh, H. Y., Barkai, N., and Leibler, S. Mechanisms of noise-resistance in genetic oscillators. *Proc. Natl. Acad. Sci. USA*, 99:5988–5992, 2002.

Voit, E. *Computational Analysis of Biochemical Systems*. Cambridge: Cambridge University Press, Cambridge, 2000.

Wagner, A. How to reconstruct a large genetic network from n gene perturbations in fewer than n^2 easy steps. *Bioinformatics*, 17:1183–1197, 2001.

Wagner, A. Estimating coarse gene network structure from large-scale gene perturbation data. *Genome Res.*, 12:309–315, 2002.

Wagner, A. and Fell, D. The small world inside large metabolic networks. *Proc. R. Soc. Lond. B*, 268:1803–1810, 2001.

Wagner, C. Nullspace approach to determine the elementary modes of chemical reaction systems. *J. Phys. Chem. B*, 108:2425–2431, 2004.

Wagner, G. Homologues, natural kinds and the evolution of modularity. *Am. Zoologist*, 36:36–43, 1996.

Wang, X. and Chen, G. Pinning control of scale-free dynamical networks. *Physica A*, 310:521–531, 2002.

Wang, Y., Liu, C. L., Storey, J. D., Tibshirani, R. J., Herschlag, D., and Brown, P. O. Precision and functional specificity in mRNA decay. *Proc. Natl. Acad. Sci. USA*, 99:5860–5865, 2002.

Warmer, J. and Kleppe, A. *The Object Constraint Language: Getting Your Models Ready for MDA*. New York: Addison-Wesley Professional, 2nd edition, 2003.

Watanabe, Y. and Kimura, J. Inhibitory effect of amiodarone on Na+/Ca2+ exchange current in guinea-pig cardiac myocytes. *Br. J. Pharmacol.*, 131:80–84, 2000.

Watts, D. and Strogatz, S. Collective dynamics of "small-world" networks. *Nature*, 393:440–442, 1998.

Weidmann, S. Effect of current flow on the membrane potential of cardiac muscle. *J. Physiol.*, 115:227–236, 1951.

Weiss, R. and Basu, S. The device physics of cellular logic gates. In *NSC-1: The First Workshop of Non-Silicon Computing*, Boston, Massachusetts, Feb. 2002.

Welsh, D. *Codes and Cryptography*. Oxford, UK: Oxford University Press, 1988.

Weng, G., Bhalla, U., and Iyengar, R. Complexity in biological signaling systems. *Science*, 284:92–96, 1999.

Westerhoff, H. and Kell, D. Matrix method for determining the steps most rate-limiting to metabolic fluxes in biotechnological processes. *Biotechnol. Bioeng.*, 30:101–107, 1987.

White, T. and Kell, D. Comparative genomic assessment of novel broad-spectrum targets for antibacterial drugs. *Comp. Func. Genomics*, 5:304–327, 2004.

Wiback, S., Mahadevan, R., and Palsson, B. Reconstructing metabolic flux vectors from extreme pathways: Defining the alpha-spectrum. *J. Theor. Biol.*, 241:313–324, 2003.

Wiback, S. and Palsson, B. Extreme pathway analysis of human red blood cell metabolism. *Biophys. J.*, 83:808–818, 2002.

Wiechert, W. 13C metabolic flux analysis. *Metab. Eng.*, 3:195–206, 2001.

Winslow, R. L., Greenstein, J. L., Tomaselli, G. F., and O'Rourke, B. Computational models of the failing myocyte: Relating altered gene expression to cellular function. *Phil. Trans. Royal Soc. A*, 359:1187–1200, 2001.

Winslow, R. L., Rice, J., Jafri, M. S., Marban, E., and O'Rourke, B. Mechanisms of altered excitation-contraction coupling in canine tachycardia-induced heart failure, II Model studies. *Circulation Res.*, 84:571–586, 1999.

Wolf, D. and Arkin, A. Motifs, modules and games in bacteria. *Curr. Opin. Microbiol.*, 6:125–134, 2003.

Wolf, J., Passarge, J., Somsen, O., Snoep, J., Heinrich, R., and Westerhoff, H. Transduction of intracellular and intercellular dynamics in yeast glycolytic oscillations. *Biophys. J.*, 78:1145–1153, 2000.

Wolpert, L. *Principles of Development*. Oxford: Oxford University Press, 2nd edition, 2002.

Wong, P., Gladney, S., and Keasling, J. D. Mathematical model of the *lac* operon: Inducer exclusion, catabolite repression, and diauxic growth on glucose and lactose. *Biotechnol. Prog.*, 13:132–143, 1997.

Woo, Y., Affourtit, J., Daigle, S., Viale, A., Johnson, K., Naggert, J., and Churchill, G. A comparison of cDNA, oligonucleotide, and Affymetrix GeneChip gene expression microarray platforms. *J. Biomol. Tech.*, 15:276–284, 2004.

Wuchty, S., Oltvai, Z., and Barabasi, A.-L. Evolutionary conservation of motif constituents in the yeast protein interaction network. *Nat. Genet.*, 35:118–119, 2003.

Wuensche, A. Discrete Dynamic Lab: Tools for investigating cellular automata and discrete dynamical networks. *Kybernetes*, 32:77–104, 2003.

Xie, S. Single-molecule approach to dispersed kinetics and dynamic disorder: Probing conformational fluctuation and enzymatic dynamics. *J. Chem. Phys.*, 117:11024–11032, 2002.

Xiong, M., Zhao, J., and Xiong, H. Network-based regulatory pathway analysis. *Bioinformatics*, 20:2056–2066, 2004.

Xiong, W. and Ferrell Jr., J. A positive-feedback-based bistable "memory module" that governs a cell fate decision. *Nature*, 426:460–465, 2003.

Xiu, Z.-L., Zeng, A.-P., and Deckwer, W.-D. Model analysis concerning the effects of growth rate and intracellular tryptophan level on the stability and dynamics of tryptophan biosynthesis in bacteria. *J. Biotechnol.*, 58:125–140, 1997.

Yang, H., Luo, G., Karnchanaphanurach, P., Louie, T. M., Rech, I., Cova, S., Xun, L., and Xie, X. S. Protein conformational dynamics probed by single-molecule electron transfer. *Science*, 302:262–266, 2003.

Yao, K. Z., Shaw, B. M., Kou, B., McAuley, K. B., and Bacon, D. W. Modeling ethylene/butene copolymerization with multi-site catalysts: Parameter estimability and experimental design. *Polymer Reaction Eng.*, 11:563–588, 2003.

Yauk, C. L., Berndt, M. L., Williams, A., and Douglas, G. R. Comprehensive comparison of six microarray technologies. *Nucleic Acids Res.*, 32:e124, 2004.

Yeger-Lotem, E., Sattath, S., Kashtan, N., Itzkovitz, S., Milo, R., Pinter, R., Alon, U., and Margalit., H. Network motifs in integrated cellular networks of transcription-regulation and protein-protein interaction. *Proc. Natl. Acad. Sci. USA*, 101:5934–5939, 2004.

Yelin, R., Dahary, D., Sorek, R., Levanon, E. Y., Goldstein, O., Shoshan, A., Diber, A., Biton, S., Tamir, Y., and Khosravi, R. Widespread occurrence of antisense transcription in the human genome. *Nat. Biotechnol.*, 21:379–386, 2003.

Yeung, M. K. S., Tegnér, J., and Collins, J. J. Reverse engineering gene networks using singular value decomposition and robust regression. *Proc. Natl. Acad. Sci. USA*, 99:6163–6168, 2002.

Yi, T., Huang, Y., Simon, M., and Doyle, J. Robust perfect adaptation in bacterial chemotaxis through integral feedback control. *Proc. Natl Acad. Sci. USA*, 97: 4649–4653, 2000.

Young, M. and Kay, S. Time zones: A comparative genetics of circadian clocks. *Nat. Rev. Genet.*, 2:702–715, 2001.

Yuen, T., Wurmbach, E., Pfeffer, R. L., Ebersole, B. J., and Sealfon, S. C. Accuracy and calibration of commercial oligonucleotide and custom cDNA microarrays. *Nucleic Acids Res.*, 30:e48, 2002.

Yuh, C.-H., Bolouri, H., and Davidson, E. Genomic *cis*-regulatory logic: Experimental and computational analysis of a sea urchin gene. *Science*, 279:1896–1902, 1998.

Zak, D., Doyle III, F., Vlachos, D., and Schwaber, J. Stochastic kinetic analysis of transcriptional feedback models for circadian rhythms. In *Proc. 40th IEEE Conference on Decision and Control*, pages 849–54, 2001.

Zak, D., Gonye, G., Schwaber, J., and Doyle III, F. Importance of input perturbations and stochastic gene expression in the reverse engineering of genetic regulatory networks: Insights from an identifiability analysis of an in silico network. *Genome Res.*, 13:2396–2405, 2003.

Zak, D., Gonye, G., Schwaber, J. S., and Doyle III, F. J. Continuous-time identification of gene expression models. *Omics*, 7:373–386, 2004.

Zak, D. E., Vadigepalli, R., Gonye, G. E., Doyle III, F. J., Schwaber, J. S., and Ogunnaike, B. A. Unconventional systems analysis problems in molecular biology: A case study in gene regulatory network modeling. *Comp. Chem. Eng.*, 29:547–563, 2005.

Zanders, E., Bailey, D., and Dean, P. Probes for chemical genomics by design. *Drug Disc. Today*, 7:711–718, 2002.

Zaslaver, A., Mayo, A. E., Rosenberg, R., Bashkin, P., Sberro, H., Tsalyuk, M., Surette, M. G., and Alon, U. Just-in-time transcription program in metabolic pathways. *Nat. Genet.*, 36:486–491, 2004.

Zhang, J., Finney, R. P., Clifford, R. J., Derr, L. K., and Buetow, K. H. Detecting false expression signals in high-density oligonucleotide arrays by an in silico approach. *Genomics*, 85:297–308, 2005.

van Zon, J. and ten Wolde, P. Simulating biochemical networks at the particle level and in time and space: Green's function reaction dynamics. *Phys. Rev. Lett.*, 94: 128103, 2005.

Contributors

- Benjamin J. Bornstein, NASA Jet Propulsion Laboratory, 4800 Oak Grove Drive, Pasadena, CA 91109, USA.

- Emery Conrad, Department of Biological Sciences, M.C. 0406, Virginia Polytechnic Institute & State University, Blacksburg, VA 24061, USA.

- Hidde de Jong, Institut National de Recherche en Informatique et en Automatique (INRIA), Research unit Rhône-Alpes, 655 avenue de l'Europe, Montbonnot, 38334 Saint Ismier CEDEX, France.

- Francis J. Doyle III, Institute for Collaborative Biotechnologies and Department of Chemical Engineering & Biomolecular Science and Engineering Program, University of California, Santa Barbara, CA 93106, USA.

- John Doyle, Control and Dynamical Systems, California Institute of Technology, Pasadena, CA 91125, USA.

- Johan Elf, Dept. of Chemistry and Chemical Biology, Harvard University, 02138 Cambridge, MA, USA.

- Andrew Finney, Physiomics PLC, Oxford, OX4 4GA, United Kingdom.

- Akira Funahashi, Kitano Symbiotic Systems Project, JST ERATO-SORST, 6-31-15 Jingumae, M-31 Suite 6A, Shibuyaku, Tokyo 150-0001, Japan.

- Kapil G. Gadkar, Institute for Collaborative Biotechnologies and Department of Chemical Engineering & Biomolecular Science and Engineering Program, University of California, Santa Barbara, CA 93106, USA.

- Daniel T. Gillespie, Dan T Gillespie Consulting, 30504 Cordoba Place, Castaic, CA 91384, USA.

- Rudiyanto Gunawan, Institute for Collaborative Biotechnologies and Department of Chemical Engineering & Biomolecular Science and Engineering Program, University of California, Santa Barbara, CA 93106, USA.

- Michael Hucka, Control and Dynamical Systems, California Institute of Technology, Pasadena, CA 91125, USA.

- Pablo A. Iglesias, Department of Electrical & Computer Engineering, Johns Hopkins University, Baltimore, MD 21218, USA.

- Brian P. Ingalls, Department of Applied Mathematics, University of Waterloo, Waterloo, Ontario N2L 3G1, Canada.

- Mads Kærn, Ottawa Institute of Systems Biology, Faculty of Medicine, University of Ottawa, 451 Smyth Road, Ottawa, ON K1H 8M5, Canada.

- Sarah M. Keating, Science and Technology Research Institute, University of Hertfordshire, Hatfield, AL10 9AB, United Kingdom.

- Douglas B. Kell, School of Chemistry, University of Manchester, Manchester M60 1QD, United Kingdom and Manchester Interdisciplinary Biocentre, University of Manchester, Manchester, United Kingdom.

- Hiroaki Kitano, The Systems Biology Institute, 6-31-15 Jingumae, M-31 Suite 6A, Shibuyaku, Tokyo 150-0001, Japan and Sony Computer Science Laboratories, Inc., 3-14-15 Higashi-Gotanda, Shinagawa, Tokyo 141-0022, Japan

- Steffen Klamt, Max Planck Institute for Dynamics of Complex Technical Systems, Sandtorstr. 1, 39106 Magdeburg, Germany.

- Joshua D. Knowles, School of Chemistry, University of Manchester, Manchester M60 1QD, United Kingdom and Manchester Interdisciplinary Biocentre, University of Manchester, Manchester, United Kingdom.

- Ben L. Kovitz, Control and Dynamical Systems, California Institute of Technology, Pasadena, CA 91125, USA.

- Karsten Kruse, Max Planck Institute for the Physics of Complex Systems, Nöthnitzer Str. 38, 01187 Dresden, Germany.

- Joanne Matthews,Science and Technology Research Institute, University of Hertfordshire, Hatfield, AL10 9AB, United Kingdom.

- Denis Noble, Emeritus Professor and Director of Computational Physiology, University Laboratory of Physiology, University of Oxford, United Kingdom.

- Johan Paulsson, Dept. of Systems Biology, Harvard University, Boston, MA, 02115, USA; Dept. of Applied Mathematics and Theoretical Physics, Wilberforce Road, Cambridge, CB3 OWA, United Kingdom.

- Vipul Periwal, Children's Hospital Informatics Program, Harvard Medical School, Boston, MA 02215, USA.

- Linda R Petzold, Department of Computer Science, University of California Santa Barbara, Santa Barbara, California 93106, USA.

- Delphine Ropers, Institut National de Recherche en Informatique et en Automatique (INRIA), Research unit Rhône-Alpes, 655 avenue de l'Europe, Montbonnot, 38334 Saint Ismier CEDEX, France.

- Uwe Sauer, Institute of Molecular Systems Biology, ETH Zürich, CH-8093 Zürich, Switzerland.

- Maria J. Schilstra, Science and Technology Research Institute, University of Hertfordshire, Hatfield, AL10 9AB, United Kingdom.

- Bruce E. Shapiro, NASA Jet Propulsion Laboratory, 4800 Oak Grove Drive, Pasadena, CA 91109, USA.

- Jörg Stelling, Institute of Computational Science, ETH Zürich, CH-8092 Zürich, Switzerland.

- Zoltan Szallasi, Children's Hospital Informatics Program, Harvard Medical School, Boston, MA 02215, USA.

- John J. Tyson, Department of Biological Sciences, M.C. 0406, Virginia Polytechnic Institute & State University, Blacksburg, VA 24061, USA.

- Ron Weiss, Department of Electrical Engineering, Princeton University, Princeton, NJ 08544, USA.

- Tau-Mu Yi, 3208 Natural Sciences 1, University of California Irvine, Irvine, CA 92697-2300, USA.

Index